COMPUTATIONAL PROSPECTS OF INFINITY

Part I: Tutorials

LECTURE NOTES SERIES
Institute for Mathematical Sciences, National University of Singapore

Series Editors: Louis H. Y. Chen and Ka Hin Leung
Institute for Mathematical Sciences
National University of Singapore

ISSN: 1793-0758

Published

*For the complete list of titles in this series, please go to
http://www.worldscibooks.com/series/lnsimsnus_series.shtml

Lecture Notes Series, Institute for Mathematical Sciences,
National University of Singapore

Vol.
14

COMPUTATIONAL PROSPECTS OF INFINITY

Part I: Tutorials

Editors

Chitat Chong
National University of Singapore, Singapore

Qi Feng
Chinese Academy of Sciences, China & National University of Singapore, Singapore

Theodore A. Slaman, W. Hugh Woodin
University of California at Berkeley, USA

Yue Yang
National University of Singapore, Singapore

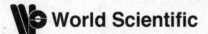
World Scientific

NEW JERSEY · LONDON · SINGAPORE · BEIJING · SHANGHAI · HONG KONG · TAIPEI · CHENNAI

Published by

World Scientific Publishing Co. Pte. Ltd.
5 Toh Tuck Link, Singapore 596224
USA office: 27 Warren Street, Suite 401-402, Hackensack, NJ 07601
UK office: 57 Shelton Street, Covent Garden, London WC2H 9HE

British Library Cataloguing-in-Publication Data
A catalogue record for this book is available from the British Library.

Lecture Notes Series, Institute for Mathematical Sciences,
National University of Singapore – Vol. 14
COMPUTATIONAL PROSPECTS OF INFINITY
Part I: Tutorials

Copyright © 2008 by World Scientific Publishing Co. Pte. Ltd.

ISBN-13 978-981-279-653-0
ISBN-10 981-279-653-3

Printed in Singapore.

CONTENTS

FOREWORD

The Institute for Mathematical Sciences at the National University of Singapore was established on 1 July 2000. Its mission is to foster mathematical research, both fundamental and multidisciplinary, particularly research that links mathematics to other disciplines, to nurture the growth of mathematical expertise among research scientists, to train talent for research in the mathematical sciences, and to serve as a platform for research interaction between the scientific community in Singapore and the wider international community.

The Institute organizes thematic programs which last from one month to six months. The theme or themes of a program will generally be of a multidisciplinary nature, chosen from areas at the forefront of current research in the mathematical sciences and their applications.

Generally, for each program there will be tutorial lectures followed by workshops at research level. Notes on these lectures are usually made available to the participants for their immediate benefit during the program. The main objective of the Institute's Lecture Notes Series is to bring these lectures to a wider audience. Occasionally, the Series may also include the proceedings of workshops and expository lectures organized by the Institute.

The World Scientific Publishing Company has kindly agreed to publish the Lecture Notes Series. This Volume, "Computational Prospects of Infinity, Part I: Tutorials", is the fourteenth of this Series. We hope that through the regular publication of these lecture notes the Institute will achieve, in part, its objective of promoting research in the mathematical sciences and their applications.

January 2008

Louis H.Y. Chen
Ka Hin Leung
Series Editors

PREFACE

The Workshop on *Computational Prospects of Infinity* was held at the Institute for Mathematical Sciences, National University of Singapore, from 20 June to 15 August 2005.

The focus in the first month of the Workshop was on set theory, with two tutorials running in parallel, by John Steel ("Derived models associated to mice") and W. Hugh Woodin ("Suitable extender sequences"). There were also 21 talks given during this period. The second half of the Workshop was devoted to recursion theory, with two tutorials given, respectively, by Rod Downey ("Five lectures on algorithmic randomness") and Ted Slaman ("Definability of the Turing jump"). In addition, there were 42 talks delivered over four weeks.

This volume constitutes Part I of the *proceedings* of the Workshop. It contains written versions of the tutorials given during the set theory and recursion theory sections of the Workshop. Part II of the *proceedings* appears as a separate volume, and comprises refereed articles, both contributed and invited, based on talks given at the Workshop.

During the Workshop, forums on the future of set theory and recursion theory were also organized. Participants discussed technical as well as philosophical and foundational issues related to problems and directions of research in these two fields. The Continuum Hypothesis was very much of central interest in the set theory forum, while in recursion theory the topics covered were more varied and diverse.

The Workshop provided a platform for researchers in the logic community from many parts of the world to meet in Singapore, to discuss mathematics and to experience a city that is a meeting point of East and West. We thank the Department of Mathematics and the Institute for Mathematical

Sciences at the National University of Singapore for their support and hospitality extended to all participants.

November 2007   Chi Tat Chong
                          National University of Singapore, Singapore

                                                                Qi Feng
                        Chinese Academy of Sciences, China and
                        National University of Singapore, Singapore

                                                    Theodore A. Slaman
                        University of California at Berkeley, USA

                                                      W. Hugh Woodin
                        University of California at Berkeley, USA

                                                             Yue Yang
                        National University of Singapore, Singapore
                                                              Editors

RECURSION THEORY TUTORIALS

FIVE LECTURES ON ALGORITHMIC RANDOMNESS

Rod Downey

School of Mathematics, Statistics and Computer Science
Victoria University
PO Box 600, Wellington, New Zealand
E-mail: Rod.downey@vuw.ac.nz

This article is devoted to some of the main topics in algorithmic randomness, at least from my idiosyncratic view.

Contents

1. Introduction

In this article, I plan to give a course around certain highlights in the theory of algorithmic randomness. At least, these will be some highlights as I see them.

I will try not to cover too much of the material already covered in my notes [14]. However, since I plan to make these notes relatively self-contained I will by necessity need to include some introductory material concerning the basics of Martin-Löf randomness and Kolmogorov complexity.

There will also be a certain intersection with the articles Downey, Hirschfeldt, Nies and Terwijn [23] and Downey [13].

I will not be able to cover the background computability theory needed, and here refer the reader to Soare [80] and Odifreddi [70,71]. While I plan to write sketches of proofs within these notes, full details can be found in the forthcoming monograph Downey and Hirschfeldt [18]. Other work on algorithmic randomness can be found in well-known sources such as Calude [4] and Li-Vitanyi [49].

The subject of algorithmic randomness is a vast one, and has been the under intense development in the last few years. With only five lectures I could not hope to cover all that has happened, nor even report on the history. Here I certainly recommend the reader look at the surveys Downey, Hirschfeldt, Nies and Terwijn [23] and Downey [13].

Probably the most brutal omission is the work on triviality and lowness, which has its roots in Solovay's [83], and its first modern incarnation in Kučera-Terwijn [41] and Downey, Hirschfeldt, Nies, and Stephan [22]. This work is of central importance as has been shown especially through the powerful results of (Hirschfeldt and) Nies [66–68], and subsequently Hirschfeldt, Stephan (e.g. [31]), Slaman and others. The reason for this lamentable omission is that I don't think it is possible to give a fair treatment to Martin-Löf lowness and triviality, as well as other important triviality lowness notions for Schnorr and computable randomness, in even one or two lectures. They could have five to themselves! A short account can be found in Downey, Hirschfeldt, Nies, Terwijn [23], and a full account is, or will be soon, found in Downey and Hirschfeldt [18].

Finally these notes will certainly contain more material than I could possibly cover in any set of lectures in the hope that the extra material will help the participants of the meeting *Computational Prospects of Infinity*. This is especially true of the many results I won't have time to prove in the lectures, but whose proofs I have included.

In these notes in Section 2, I will first develop the basic material on Kolmogorov Complexity. Whilst this approach does not follow the historical development of the subject, it does make logical sense. I will include proofs of the fundamental results including Kraft-Chaitin, the Coding Theorem, and Symmetry of Information.

In Section 3, I will discuss the background material on the three basic approaches to randomness, via measure theory, prediction, and compression. Here I will also include the exciting recent results of Miller and Yu classifying 1-randomness in terms of plain complexity.

In Section 4, I will look at some classic theorems concerning randomness for general classes of reals. This material will include the Kučera-Gács Theorem and other results of Kučera. Other central results treated will be van Lambalgen's Theorem, effective 0-1 laws, and results on PA and FPF degrees. We also introduce n-randomness and variations and look at the exciting recent work showing that 2-randomness is the same as Kolmogorov randomness.

In Section 5, I will look at various methods of calibrating randomness using initial segment methods. This will include the Slaman-Kučera Theorem and other work on computably enumerable reals, and the work of Solovay, and Miller-Yu on van Lambalgen reducibility and its relationship with $\leq_K$ and $\leq_C$.

Finally, in Section 6, I plan to give sketches of proofs of the poorly known results of Stuart Kurtz from his thesis [44], and some refinements later by Steven Kautz [33]. None of this work has ever been published aside from the presentations in these theses. The techniques, whose origins go back to work of Paris and of Martin, are powerful and are extremely interesting.

We remark that throughout these notes we will be working over the alphabet $\{0,1\}$, and hence will be looking at $2^{<\omega}$, and 2^{ω} for "reals"meaning members of Cantor Space. This sapce will be equipped with the basis of clopen sets $[\sigma] = \{\alpha : \sigma \prec \alpha\}$, and comes with the usual Lebesgue measure $\mu([\sigma]) = 2^{-|\sigma|}$. The whole development can also be done over other alphabets with little change. Most notation is drawn from Soare [80]. We will use λ to denote the empty string.

2. Lecture 1: Kolmogorov complexity basics

This section is devoted to the analysis of the Kolmogorov complexity (or, rather complexit*ies*) of finite strings. The fundamental idea idea is well-known, and that is a random string should be one that it hard to compress. The compression devices used will generate different kinds of complexities, and we discuss some of these here. We begin with the most basic notion (plain complexity) first articulated by the seminal paper of Kolmogorov [34].

2.1. *Plain complexity*

Thus we can imagine a Turing machine M which acts as a transducer and takes an input string τ and, should it halt, produces an output string σ. Then $M(\tau) = \sigma$. We say that τ is an M-*description* of σ. Since we are interested in the extent that σ can be *compressed*, we can then define the M-complexity as $C_M(\sigma)$ is the *length* of the shortest τ such that $M(\tau) = \sigma$. If no such τ exists, then we regard $C_M(\sigma) = \infty$. We can enumerate all Turing machines $\{M_e : e \in \mathbb{N}\}$, and hence we can define

$$C(\sigma) = \min\{C_{M_e}(\sigma) + e + 1 : e \in \mathbb{N}\},$$

to be *the* Kolmogorov complexity of σ. Notice that we could implement C via machine U where, on input $0^e 1\tau$, U emulates the action of $M_e(\tau)$. Note that we would have that for any machine M, $C(\sigma) \le C_M(\sigma) + O(1)$. Hence we may define a notion of compression *up to a constant*. There will be a unique string z of length $C(\sigma)$ such that

(i) $U(z)[s] = \sigma$,

(ii) s is the first stage where there is a string of length $C(\sigma)$ and (i) holds.

(iii) z is the lexicographically least such string.

Then we will denote z by σ^*. (Strictly speaking, we should use σ_C^*, to indicate that we are using C, since later there will be other measures of complexity. However, we anticipate that things will be clear from context.) The reader will note that x^* contains a lot of information. If I am given x^* then I can generate x (by running U) and $C(x)$ by looking at x^*'s length. For a pair $\langle x, y \rangle$ we denote by

(i) $C(x, y)$ the Kolmogorov complexity of the pair $\langle x, y \rangle$, that is $C(\langle x, y \rangle)$ and

(ii) $C(x|y)$ the Kolmogorov complexity of the string x, *given* the string y. (Here we have machines M which take as input the pair p, q and calculate $C(p)$ given the information q as an oracle. We call $C(x|y)$ the *conditional Kolmogorov complexity of x given y*.) Notice that $C(x) = C(x|\lambda) + O(1)$, where λ denotes the empty string.

Then the observations above show that the following is true.

Proposition 2.1:

(i) $C(x, C(x)) = C(x^*) + O(1)$.

(i) $C(x|x^*) = O(1)$.

(iii) $C(x, C(x)|x^*) = C(x^*|C(x), x) = O(1)$.

(iv) $C(xy) \le C(x, y) + O(1)$ where xy denotes the concatenation of x and y.

The basic theory of plain Kolmogorov complexity is based upon the following straightforward counting result.

Theorem 2.1: (Kolmogorov [34])

(i) $C(x) \le |x| + c$.

(ii) $|\{x : |x| = n \wedge C(x) \le n + c - j\}| = O(2^{n-j})$.

Definition 2.1: (Kolmogorov [34]) We say that a string x is $(C\text{-})$random iff $C(x) \ge |x|$.

We emphasize that Theorem 2.1 is a *combinatorial fact*, which only has an interpretation in algorithmic information theory. Thus we may define the following.

Definition 2.2: (Nies, Stephan and Terwijn [69]) We say that $F : \Sigma^* \to \Sigma^*$ is a *compression function* if for all x $|F(x)| \le C(x)$ and F is 1-1.

Using any compression function, we can define C_F, the "Kolmogorov complexity" relative to F as $C_F(x) = |F(x)|$. Clearly, the Theorem 2.1 holds for any compression function, by the same counting argument. This observation will proves very useful later when we meet the results of Nies, Stephan and Terwijn [69] and of Miller [55] Here is an easy application of the use of compression functions. We begin with a simple observation.

Proposition 2.2: (Folklore)

(i) Consider the "C-overgraph"

$$M_C = \{\langle x, y\rangle : C(x) < y\}.$$

Then C_M is weak truth table complete. Indeed $\overline{R_C}$, the collection of non-C-random strings, is wtt-complete.

(ii) The collection of C-random strings is immune.

Proof: (i) To define a reduction, for each n computably pick a length $g(n)$, so that g is 1-1, and $g(n) > n_0$ is sufficiently large to make the following work. At each stage s, we will have a fixed string $\sigma(n, s)$ of length $g(n)$ and the reduction will be that $n \in \emptyset'$ iff $\sigma(n) = \lim_s \sigma(n, s)$ is not random. Initially we choose $\sigma(n, s)$ to be random at stage s. Whilst $n \notin \emptyset'_t$, should $C_t(\sigma(n, t)) < g(n) + c$ (c being the relevant constant for randomness), we will pick a new string $\sigma(n, t+1)$ to be the next stage t random string of length $g(n)$. Finally, should n enter $\emptyset'$ at some stage u, then we can use the Recursion Theorem to drop the complexity of $\sigma(n, u)$ to below $g(n) + c$. (This is where we would need $g(n)$ to be large enough.)

(ii) Let $A = \{x : C(x) \ge \frac{|x|}{2}\}$. Then A is immune. A is infinite by Theorem 2.1. Suppose that A has an infinite c.e. subset B. Let $h(n)$ be defined as the first element of B to occur in its enumeration of length above n. Then

$$C(h(n)) \ge \frac{|h(n)|}{2} \ge \frac{n}{2}, \text{ but,}$$

$$C(h(n)) \le C(n) + \mathcal{O}(1) \le |n| + \mathcal{O}(1).$$

For large enough n this is a contradiction. $\qquad\qquad\qquad\qquad\qquad\square$

Now note that that the collection of compression functions forms a Π_1^0 class. The function C is easily seen to be Turing complete. (see Theorem 2.2) However, by the Low Basis Theorem, there must be compression functions which have low degrees and contrasts (i) above. Now note that for for such a low F, if a string is C_F-random it is certainly C-random. Thus in spite of the fact that (ii) above says that there is no infinite c.e. set of C-randoms, we do have an amenable class within the set of C-randoms by all of this. To wit:

Proposition 2.3: There is a low infinite collection of C-random strings.

We will see much deeper applications of this idea when we meet the results of Nies, Terwijn and Stephan [69] later.

We remark that there has been significant investigation of the natural c.e. set M_C, particularly by An A. Muchnik. It is relatively easy to see that since the collection of C-randoms is immune, M_C is not m-complete. Kummer however, proved that the adaptive nature of the reduction in Proposition 2.2 if *not* necessary. Kummer's technique is interesting in that the argument, whilst guaranteeing the existence of a tt-reduction, is nonuniform in that we don't know *which* tt-reduction works, and provably so.

Theorem 2.2: (Kummer [43]) $\overline{R}_C$ is truth table complete.

Proof: (Sketch) Kummer's proof is relatively intricate, but the idea is straightforward enough. First, the idea of Proposition 2.2 is replaced by using *blocks* of elements. That is, for each x we will attempt to define a block of elements $S_{i,x}$ so that $x \notin \emptyset'$ iff $S_{i,x}$ only contains C-random reals. Naturally enough, it is within *our* power to be able to drop the complexity of one of the members of any such $S_{i,x}$. But it is also within the power of the opponent to also drop such complexity. The clever part of Kummer's proof is how to be able to select the $S_{i,x}$ so that the opponent can't do this. Roughly speaking, Kummer's idea is to associate (perhaps temporarily) a block of elements from some size n with x, and when they go bad, we use another block, perhaps of a larger length. Different requirements can choose the same length, things being sorted out by priorities. However, the size of blocks are arranged so that $S_{i,x}$ will only be reset at most i times for some $i \le 2^{2d+2}$ where d is a parameter from the Recursion Theorem. (This argument uses the pigeonhole principle.) Then in the end using the parameter i_0, the largest i which is reset infinitely often, we will be able to argue that the *tt*-reduction works for that i_0.

Full details can be found in Kummer [42], and Downey-Hirschfeldt [18].

□

Similar (and easier) methods can be used to show the following. We give its proof as an illustrative example.

Theorem 2.3: (An. A. Muchnik [64]) The collection

$$M = \{(x, y, n) : C(x|y) < n\}$$

is creative[a].

Proof: The construction will have a parameter d which can be worked out in advance, and known by the recursion theorem. For our purposes think of d in the following big enough to make everything work. We will construct a series of (possible) m-reductions g_x for $x \in [1, 2^d]$. Then for each z either we will know that z enters $\emptyset'$ computably, or there will be a unique y such that $g_x(z) = (x, y, d)$ and $x \in \emptyset'$ iff $g_x(z) \in M$. For some maximal x for which we enumerate elements $g_x(v)$ into M infinitely often, this will then give the m-reduction since on those elements which don't computably enter $\emptyset'$, g_x is (computable) and defined.

Construction For each active $y \le s$, find the least $q \in [1, 2^p]$ with

$$(q, y, d) \notin M_s.$$

(Notice that such an x needs to exist since $\{q : (q, y, d) \in M\} < 2^d$.)

If this q is new at stage $s + 1$, (that is, $(q', y, d) \notin M_s$ for some $q' < q$) find the least z with $z \notin \emptyset'[s + 1]$ and define

$$g_q(z) = (q, y, d).$$

Now for any v, if v enters $\emptyset'[s + 1]$, find the largest r, if any, with $g_r(z)$ defined. If one exists, enumerate $g_r(z)$ into M. Find $\hat{y}$ with $g_r = (r, \hat{y}, d)$. Declare that $\hat{y}$ is no longer active. (Therefore, we will do this exactly once for any fixed $\hat{y}$ so the cost is modest, and known by the Recursion Theorem.)

End of Construction

Note that there must a largest $x \le 2^d$ such that $\exists^\infty v(g_x(v) \in M)$. Call this x. We claim that g_x is the required m-reduction. Work in stages after which g_{x+1} enumerates nothing into M.

---

[a]This proof can be made to work for any of the usual complexity measures, such as K which we meet in the next section, and for monotone complexity.

Given z, since g_x is defined on infinitely many arguments and they are assigned in order, we can go to a stage s where either z has entered $\emptyset'[s]$, or $g_x(z)$ becomes defined, and $g_x(z) = (x, y, d)$ for some active y. $g_x(z)$ will be put into M should z enter $\emptyset'$ after s. The result follows. $\qquad \square$

The reductions in these theorems are exponential (or worse) in the number of queries they use, and there has been a lot of fascinating work by Allender and others looking at what can be *efficiently* reducible to sets of random strings. This may seem strange indeed, but these questions seem to have a lot to say about complexity classes. For instance, one open hypothesis is that $PSPACE$ is precisely the collection of computable sets in $\cap_V P^{R_C^V}$, where R_C^V denotes the collection of random strings with universal machine V. (see Allender et. al. [1], Allender, Buhrmann, and Koucký [2].)

2.2. *Symmetry of Information*

We would like that $C(xy) \leq C(x) + C(y) + O(1)$. This, however, is *not* true in general. The problem is how to combine x^* and y^*. In fact, for sufficiently long strings, we always have compressible initial segments.

Lemma 2.1: (Martin-Löf [53]) Let k be given. Suppose that z is sufficiently long. The there is an initial segment x of z whose complexity is below $|x| - k$. Hence if z is C-random and we write $z = xy$,

$$C(z) > C(x) + C(y).$$

Proof: Take some initial segment q of z. This string is the r-th string of $2^{<\omega}$ under the standard length/lex ordering. Let x denote the initial segment of z of length $|q| + r$. Note that $C(x) \leq r + O(1)$, since to figure out x we need only input the final segment m of x following q, read its length which is r, use that to resurrect q and then output $qm = x$. It is easy to arrange q so that $r < |x| - k$. $\qquad \square$

The following powerful theorem gives the precise relationships between $C(xy)$ and $C(x)$ and $C(y)$. The *information content* of a string y in a string x is defined as

$$I(x : y) = C(y) - C(y|x).$$

Theorem 2.4: (Symmetry of Information, Levin and Kolmogorov [101])

$$I(x:y) = I(y:x) \pm O(\log n)$$
$$= I(y:x) \pm O(\log C(x,y))$$

where $n = \max\{|y|, |x|\}$.

Levin and Kolmogorov's Symmetry of Information Theorem will follow from the reformulation below.

Theorem 2.5: (Symmetry of Information-Restated)

$$C(y|x) + C(x) = C(x,y) \pm O(\log C(x,y))$$

Theorem 2.4 will follow from Theorem 2.5 by the following calculation:

$$C(\langle x,y\rangle) = C(y) + C(x|y) \pm O(\log C(x,y))$$
$$= C(x) + C(y|x) \pm O(\log C(x,y))$$
$$\text{then } C(x) - C(x|y) = C(y) - C(y|x) \pm O(\log C(x,y)).$$

Proof: (of Theorem 2.5) The following neat proof follows that of Li-Vitanyi [49]. First it is easy to see that $C(x,y) \leq C(x) + C(y|x) \pm O(\log C(x,y))$, since we can describe $\langle x,y\rangle$ via a description of x, of y given x and an indication of where to delimit the two descriptions.

For the hard direction, that

$$C(\langle x,y\rangle) \geq C(y) + C(x|y) \pm O(\log C(x,y)),$$

define two sets

$$A = \{\langle u,v\rangle : C(\langle u,v\rangle) \leq C(\langle x,y\rangle), \text{ and}$$

$$A_u = \{v : \langle u,v\rangle \in A\}.$$

A is finite and uniformly computably enumerable given $\langle x,y\rangle$, as is A_u for each u. Hence, one can describe y given x and its place in the enumeration order of A_x and given $|C(x,y)|$.

We take $e \in \mathbb{N}$ such that $2^{e+1} > \text{card}(A_x) \geq 2^e$. Then

$$C(y|x) \leq \log \text{card}(A_x) + 2|C(\langle x,y\rangle)| + O(1)$$
$$\leq e + O(\log C(x,y)).$$

Now consider the set $B = \{u : |A_u| \geq 2^e\}$. It is clear that $x \in B$. We see

$$\text{card}(B) \leq \frac{\text{card}(A)}{2^e} \leq \frac{2^{C(\langle x,y \rangle)}}{2^e}.$$

This is independent of the pairing function used, provided the function is 1-1. Note that $\text{card}(\cup_u A_u) \leq \text{card}(A)$.

Then, to specify x, we take $C(x, y)$ and e we can computably enumerate the strings which are possibilities u for x by satisfying

$$A_u = \{z : C(u, z) \leq C(x, y)\} \text{ and,}$$

$$2^e < \text{card}(A_u).$$

Therefore,

$$C(x) \leq |e| + \log \frac{2^{C(\langle x,y \rangle)}}{2^e} + O(\log C(x, y))$$
$$\leq C(\langle x, y \rangle) - e + O(\log C(x, y)),$$

and thus

$$C(x) + C(y|x) \leq C(\langle x, y \rangle) + O(\log C(x, y))$$

as required. □

2.3. *Prefix-free complexity*

We would like that the information contained in the *bits* of x^* encapsulates the complexity of x. Now, Theorem 2.1, says that for C this is not so. x^* gives $|x^*|$ *plus* $\log |x^*|$ many bits of information. It was Levin [50], [101] who finally figured out how to define complexity so that this formal relationship held. He did do using monotone machines (which ask that if $\sigma \prec \tau$, then $M(\sigma) \preceq M(\tau)$, should they both halt), and later both he and Chaitin [8] used the notion of a prefix-free machine. Recall that a set of strings S is called *prefix-free* if whenever $\sigma \in S$ and $\sigma \prec \tau$, then $\tau \notin S$. Then we may define *prefix-free Kolmogorov complexity* as the same as C except that only prefix-free machines, that is ones with prefix-free domains, are allowed. Note that the same proof shows that there are universal prefix-free machines. We let K denote prefix-free complexity.

Prefix-free machines and prefix-free Kolmogorov complexity will play a central role in our story. Note that if a machine M has prefix-free domain, then its domain has (Lebesgue) measure. That is, for $\sigma \in 2^{<\omega}$, we recall that $\mu([\sigma]) = 2^{-|\sigma|}$, we note that by prefix-freeness a prefix-free machine's domain has measure, called its *halting probability*.

$$\mu(M) = \sum_{M(\sigma)\downarrow} 2^{-|\sigma|}.$$

The following result gives an *implicit* way of constructing prefix-free machines and will be used extensively in the rest of these notes. The result says, roughly, if the lengths needed for some prefix-free machine work out, then there is such a machine.

Theorem 2.6: (Kraft [35], Kraft-Chaitin [8, 10]) Let $d_1, d_2, \ldots$ be a collection of lengths, possibly with repetitions, Then $\Sigma 2^{-d_i} \leq 1$ iff there is a prefix-free set A with members σ_i and σ_i has length d_i. Furthermore from the sequence d_i we can effectively compute the set A.

The result as stated here first appears in Chaitin [8], where the result is attributed to Pippinger.

Proof: Consider the correspondence $\Delta : [\sigma] \mapsto [0.\sigma, 0.\sigma + 2^{-|\sigma|})$ taking the string σ to an interval of size $2^{-|\sigma|}$, gives a correspondence between a set of disjoint intervals in $[0, 1)$ and a prefix-free set.

Consider first the following non-effective proof of the result. We are given the lengths $\{d_i : i \in \mathbb{N}\}$ in some random order. But suppose that we re-arrange these lengths in increasing order, say $l_1 \leq l_2 \leq \ldots$. Then we can easily choose disjoint intervals I_j, with the right end-point of I_n as the left endpoint of I_{n+1} and the length of I_{n+1} being $2^{-l_{n+1}}$. Then we can again use the correspondence by setting $[\sigma_n] = \Delta^{-1}(I_n)$.

Now in the case that the intervals are *not* presented in increasing order, how to effectivize this process? The following organizational device for the Chaitin [8] proof was suggested by Joe Miller. We will be given the strings $d_0, d_1, \ldots, d_n$ with d_n given at stage n. The idea is that, at each stage n, we have a mapping $d_i \mapsto [\sigma_i]$, $|\sigma_i| = d_i$, together with a binary string $x[n] = .x_1 x_2 \ldots x_m$ representing the length $1 - \sum_{j \leq n} 2^{-d_j}$, and ask that for each 1 in the string $x[n]$ that there is a string of precisely that length in $2^{<\omega} - \{\sigma_j : j \leq n\}$.

To continue the induction, at stage $n+1$, when a new length d_{n+1} enters, if position $x_{d_{n+1}}$ is a 1, then we can find the corresponding string $\tau_{d_{n+1}}$ in $2^{<\omega} - \{\sigma_j : j \leq n\}$ and set $\sigma_{n+1} = \tau_{d_{n+1}}$. Then of course we make $x_{d_{n+1}} = 0$ in $x[n+1]$.

If position $x_{d_{n+1}}$ is a 0, find the largest $j < d_{n+1}$ with $x_j = 1$, find the lexicographically least string τ extending τ_j of length d_{n+1}, let $\upsilon_{n+1} = \tau$, and let $x[n+1] = x[n] - .\nu$ where ν is the string which is zero except for 1 in position d_{n+1}.

Notice that nothing changes in $x[n+1]$ from $x[n]$ except in positions j to d_{n+1}, and these all change to 1, with the exception of x_j which changes to 0. Since τ was chosen as the lexicographically least string in the cone $[\tau_j]$, there will be corresponding strings in $[\tau_j]$ of lengths $j - 1, \ldots, d_{n+1}$, as required to complete the induction. □

In the proofs in the literature, we often refer to the requests "I would like a string of length m mapping to a string σ" as KC-*axioms*, written as $\langle m, \sigma \rangle$ or sometimes $\langle 2^{-m}, \sigma \rangle$. For a c.e. set of such requests, the result above says that they can be met provided that their measure request is possible; that is provided the sum of the first coordinates is ≤ 1.

What about K-randomness? The counting for C shows that $C(x) \leq |x| + c$ for all x. However, for K we have the following.

Lemma 2.2: (Chaitin [8, 10], Levin [45]) Let $f : \Sigma^* \to \mathbb{N}$. Suppose that $\sum_{\sigma} 2^{-f(\sigma)}$ diverges. (The prototype here is $f(x) = \log x$.) Then $K(\sigma) > |\sigma| + f(\sigma)$ infinitely often.

Proof: Suppose that for all x, $K(x) \leq |x| + f(x)$. Then $\sum_{\sigma \in 2^{<\omega}} 2^{-K(\sigma)} \geq \sum_{\sigma} 2^{-(|\sigma| + f(\sigma))} \geq \sum_n \sum_{|\sigma| = n} 2^{-(n + f(n))} \geq \sum_n 2^n (2^{-(n + f(n))}) \geq \sum_n 2^{-f(n)} = \infty$, a contradiction, since $\sum_{\sigma} 2^{-K(\sigma)} \leq 1$, as the machine is prefix free. □

The analog of $C(x) \leq |x| + O(1)$ is given by the following basic result.

Theorem 2.7: (Counting Theorem, Chaitin [8])

(i) $K(x) \leq |x| + K(|x|) + \mathcal{O}(1)$.

(ii) For all n,

$$\max\{K(x) : |x| = n\} = n + K(n) + \mathcal{O}(1).$$

(iii) For any k,

$$|\{\sigma : |\sigma| = n \wedge K(\sigma) \leq n + K(n) - k\}| \leq 2^{n-k+\mathcal{O}(1)}.$$

Proof: (i) Let U be the universal prefix-free machine. Consider the special prefix-free machine M, which will halt only on strings of the form $z\sigma$, provided that on input $U(z) = |\sigma|$. For such a string $z\sigma$, M outputs σ. Then $K_M(\sigma) = |(|\sigma|^*)| + |\sigma| + \mathcal{O}(1)$. Hence $K(\sigma) \leq |\sigma| + K(|\sigma|) + \mathcal{O}(1)$.□

For the proof of (ii), we will use the "semimeasure" method which originates with the work of Gács and Levin, and is cleverly exploited by Chaitin [10]. Chaitin defined an *information content measure* as a partial function $\widehat{K} : 2^{<\omega} \to \mathbb{N}$ such that

(i) $\sum_{\sigma \in 2^{<\omega}} 2^{-\widehat{K}(\sigma)} \leq 1$, and,
(ii) $\{\langle \sigma, k \rangle : \widehat{K}(\sigma) \leq k\}$ is c.e..

Chaitin's information content measures are more or less the same as the computably enumerable *discrete semimeasures* introduced by Gács [26] and Levin [46].

Definition 2.3: (Discrete semimeasure) A discrete semimeasure is a function $m : 2^{<\omega} \to \mathbb{R}^+ \cup \{0\}$ such that

$$\sum_{\sigma \in 2^{<\omega}} m(\sigma) \leq 1.$$

Here a function g is computably enumerable iff there is a computable function $h(\cdot, \cdot)$ such that h is nondecreasing in both variables, and $g(n) = \lim_s h(n, s)$ for all n, so that it is a special kind of Δ_2^0 function. For instance, the binary expansion of a halting probability of a prefix-free Turing machine gives a c.e. function $f(n)$ which is the value of the first n bits of this expansion.

An equivalent way to think of a discrete semimeasure is to identify $2^{<\omega}$ with $\mathbb{N}$ and think of $\mathbb{N}$ as being our measure space. Notice that under this identification, all strings are incompatible. Clearly, the standard Lebesgue measure we have looked at so far is *not* a discrete semimeasure. The standard Lebesgue measure is a *continuous* measure. (Continuous semi-measures give rise to yet another concept of randomness and relate to semimartingales, as we later see.) The discrete Lebesgue measure is $\lambda(\sigma) = 2^{-2|\sigma|-1}$. There is a computable enumeration $\{\widehat{K}_k : k \in \mathbb{N}\}$ of all information content measures (and similarly one of all computably enumerable discrete semimeasures). Thus, there is a universal minimal one:

$$\widehat{K}(x) = \min_{k \geq 0}\{\widehat{K}_k(x) + k + 1\}.$$

Information content measures and prefix-free machines, are essentially the same. Using Kraft-Chaitin, build a prefix-free machine M which emulates precisely the information content measure $\widehat{K}$; namely for all σ, there is a $\tau \in \text{dom}M$, such that $M(\tau) = \sigma$ and $K(\sigma) = |\tau| - O(1)$, and conversely. (Namely, at stage s, if we see $K_s(\sigma) = k$ and $K_{s+1}(\sigma) = k' < k$ enumerate a Kraft-Chaitin axiom $\langle 2^{-(k'+1)}, \sigma \rangle$ to describe M, and hence generate $\widehat{K} = K_M$. Thus we see that $\widehat{K}$ is within a constant of K, and is thus prefix-free Kolmogorov complexity. Henceforth we will identify K with $\widehat{K}$, without further comment.

Another restatement of all of this is obtained by letting m denote the minimal universal discrete semimeasure. We have the following.

Lemma 2.3: $K(\sigma) = -\log m(\sigma) + O(1)$.

Proof: (ii) and (iii) Now for the proof of (iii), from which (ii) follows. We note that

$$\sum_{n \in \Sigma^*} 2^{-K(n)} = \sum_{n \in \Sigma^*} \sum_{|\sigma| = n} 2^{-K(\sigma)}.$$

(Recall, that we are identifying a string with a unique number.) Thus, as $-\log(\sum_{|\sigma|=n} 2^{-K(\sigma)})$ is an information content measure (using $-\log(\sum_{|\sigma|=n} 2^{-|\sigma|})$ as a dyadic real and letting $\widehat{K}(n)$ as the first nonzero entry in this expansion.) Then as K is minimal, we have

$$2^{-K(n)+O(1)} \geq \sum_{|\sigma|=n} 2^{-K(\sigma)}.$$

Now for the sake of a contradiction suppose that there are more than 2^{n-k+c} strings of length n with $K(\sigma) < n + K(n) - k$. Let $F = \{\sigma : |\sigma| = n \wedge K(\sigma) < n + K(n) - k\}$. Suppose that $|F| = (1 + \epsilon)2^{n-k+c}$. Then

$$2^{-K(n)+c} \geq \sum_{|\sigma|=n} 2^{-K(\sigma)} \geq$$

$$\sum_{\sigma \notin F} 2^{-K(\sigma)} + \sum_{\sigma \in F} 2^{-K(\sigma)} > (1+\epsilon)2^{n-k+c}2^{n-K(n)-k} > 2^{-K(n)+c},$$

a contradiction. $\qquad\square$

2.4. *The Coding Theorem*

We have already seen that $K(\sigma) = -\log m(\sigma)$ up to a constant. There are of course many measures we can put on strings. A particularly useful one is the following.

Definition 2.4: Given a prefix-free machine D, let $Q_D(\sigma) = \mu(D^{-1}(\sigma))$.

$Q_D(\sigma)$ is the probability that D outputs σ. The following is an important and useful basic theorem.

Theorem 2.8: (Coding Theorem) $-\log m(\sigma) = -\log Q(\sigma) + O(1) = K(\sigma) + O(1)$.

Proof: We note that $Q(\sigma) \geq 2^{-K(\sigma)} = 2^{-|\sigma^*|}$, since $D(\sigma^*) = \sigma$. Therefore $-\log Q(\sigma) \leq K(\sigma)$. But, $\sum 2^{-\log Q(\sigma)} \leq \sum_\sigma Q(\sigma) \leq 1$. Hence, by minimality of K, $Q(\sigma) \leq K(\sigma) + O(1)$ and hence $Q(\sigma) = K(\sigma) + O(1)$, as required. $\qquad\square$

From this proof, $-\log Q(\sigma)$ is a measure of complexity, and hence, by the minimality of K among measures of complexity, we know that $2^{-K(\sigma)} \leq Q(\sigma)$. By Theorem 2.8, we know that for some constant d,

$$2^{-K(\sigma)} \leq Q(\sigma) \leq d2^{-K(\sigma)}.$$

Thus we can often replace usage of K by Q.

2.5. *Prefix-free symmetry of information*

Because of its close approximation to information content, prefix-free Kolmogorov complexity can be more pliable than its plain cousin. For instance, we can attach one prefix-free machine M_1 to another M_2 and make a (prefix-free) machine M whose action is $M(\sigma\tau) = M_1(\sigma)M_2(\tau)$. This means that

$$K(xy) \leq K(x) + K(y) + O(1).$$

Additionally Symmetry of Information is more aligned to our intuition in the prefix-free case. We define the K-*information content* as

$$I(x : y) = K(y) - K(y|x).$$

(Here, we are using $K(y|x)$ in the same way as for C, meaning the conditional prefix-free complexity of x given y. Similar comments hold for the use of $x^* = x_K^*$, for instance, in this context.)

Theorem 2.9: (Symmetry of Information, Levin and Gács [26], Chaitin [8]) $I(\langle x, K(x)\rangle : y) = I(\langle y, K(y)\rangle : x) + O(1)$.

Note that, given the relationship $K(z, K(z)) = K(z*) + O(1)$, the Symmetry of Information Theorem for prefix-free complexity may be neatly rewritten as

$$I(x^* : y) = I(y^* : x) + O(1).$$

As with the C case, Levin's Symmetry of Information Theorem follows from a reformulation:

Theorem 2.10: (Symmetry of Information, Levin and Gács [26], Chaitin [8]) $K(x, y) = K(x) + K(y|x, K(x)) + O(1) = K(x) + K(y|x^*) + O(1)$.

Proof: To prove Theorem 2.9 from Theorem 2.10, by Theorem 2.10, we have $K(x, y) = K(x) + K(y|x, K(x)) + O(1) = K(y) + K(x|y, K(y)) + O(1)$, and hence,

$$K(y) - K(y|x, K(x)) = K(x) - K(x|y, K(y)) + O(1),$$

and Theorem 2.9 follows.

Now we turn to the proof of Theorem 2.10.

First we prove that

$$K(x, y) \leq K(x) + K(y|x, K(x)) + O(1).$$

Given x^* and $z = K^*(y|x, K(x))$, we can construct a prefix-free machine M which, upon input x^*z, will compute x and $K(x) = |x^*|$. It will then compute y from x and $K(x)$ and z

To finish we need to prove that

$$K(x, y) \geq K(x) + K(y|x, K(x)) + O(1).$$

To achieve this, we prove that

$$K(y|x^*) \leq K(x, y) - K(x) + O(1).$$

We run the computation of U assuming that exactly one string halts at each stage. Call this p_s at stage s. Then, at each stage s, compute $\langle x_s, y_s \rangle$ with

$$U(p_s) = \langle x_s, y_s \rangle.$$

By the Coding Theorem, Theorem 2.8, there is a constant c such that

$$2^{K(x)-c}(\sum_y Q(\langle x, y \rangle)) \leq 1,$$

for all x. (To see this, imagine we are building a machine V which, each time we see $U(p) \downarrow = \langle x, y \rangle$, declares a Kraft-Chaitin axiom $\langle |p|, x \rangle$. Then, relative to V, $Q_V(x) = \sum_y Q_U(\langle x, y \rangle)$, meaning that $\sum_y Q(\langle x, y \rangle) \leq Q(x) + O(1)$.)

We will now define a new conditional machine M using Kraft-Chaitin. With z on the oracle tape, M tries to compute z' with $U(z) = z'$, and hence with x^* on the tape, computes x. M then simulates the machine M_x with the Kraft-Chaitin set

$$\langle |p_t| - |x^*| + c, y_t \rangle,$$

for each p_t of the form $\langle x, y_t \rangle$. Let W denote the computably enumerable collection of such requirements.

Notice that $\sum_{t \in W} 2^{-(|p_t| - |x^*| + c)} \le 2^{K(x) - c} (\sum_y^* Q(\langle x, y \rangle)) \le 1$, and hence Kraft-Chaitin can be applied to M_x.

For each p with $U(p) = \langle x, y \rangle$, there is a $\widehat{p}$ with $U(\widehat{p}|x^*) = M_x(\widehat{p}) = y$, and with $|\widehat{p}| = |p| - K(x) + c$. This shows that

$$K(y|x^*) \le K(x, y) - K(x) + O(1). \qquad \square$$

Another way to express Theorem 2.10 is

Corollary 2.1: $K(x, y) = K(x) + K(y|x^*) + O(1)$.

2.6. *Prefix-free randomness*

Note that now there are really two possibilities for defining prefix-free randomness for strings. First we might think that a string should be random iff its shortest description is at least as long as it is. This gives what is described, perhaps nonstandardly, in [18] as being *weakly Chaitin random*. That is, σ is weakly Chaitin random iff $K(\sigma) > |\sigma|$. On the other hand, a string could be thought of as being random if it has maximum possible K-complexity, meaning that to describe the string you need $K(|x|)$ for the prefix-free-ness and $|x|$ many bits for the string: to wit, we will say that x is *strongly* Chaitin random iff $K(x) > |x| + K(|x|)$.

The following implications hold between the concepts.

Theorem 2.11: (Solovay [83])

   (i) x is strongly Chaitin random[b] implies x is Kolmogorov random, but not conversely.

   (ii) x is Kolmogorov random implies x is weakly Chaitin random, but not conversely.

---

[b]Strictly speaking this theorem is happening up to fixed constants. In Downey and Hirschfeldt [18], we refer to the "up to constant" behaviour as "essentially" behaviour. That is, for a fixed c, having $C(x) > |x| - c$ is to be "essentially Kolmogorov random," for instance.

The first part of (ii) is immediate since each prefix-free machine is also a plain machine. We will not prove the "but not conversely" statements here since they involve relatively intricate constructions, and simply refer the reader to Downey and Hirschfeldt [18]. We will give a proof of (i)'s statement that every strongly Chaitin random string is Kolmogorov random, as it give the reader some insight into the methods used.

For the proof, the following concepts are useful. We define the *randomness deficiencies* as follows. Let c_C and c_K denote the relevant coding constants.

$$m_C(\sigma) = |\sigma| + c_c - C(\sigma), \text{ and,}$$

$$m_K(\sigma) = |\sigma| + K(|\sigma|) + c_K - K(\sigma).$$

Lemma 2.4: (Solovay [83]) For a string x, we have $m_K(x) \geq m_C(x) - O(\log m_C(x) + 2)$.

Assuming the Lemma, we can get (i) by observing that if x is strongly Chaitin random then for some constant c, $m_K(x) \leq c$. By the Lemma, we see that $m_C(x) - O(\log m_C(x) + 2) \leq c'$ for some fixed constant c' and hence $m_C(x) \leq c''$ for some fixed c''.

Proof: (of Lemma 2.4) We know $C(x) = |x| + c_C - m_C(x)$. Thus, $K(C(x)) = K(|x| + c_C - m_C(x)) \leq K(|x|) + K(c_C - m_C(x)) \leq K(|x|) + O(\log m_C(x) + 2)$. Next we prove the following claim.

$$K(x) \leq C(x) + K(C(x)) + O(1).$$

To see this Let U be a universal prefix-free machine and V a universal machine. We will define a prefix-free machine D via the following.

On input z, D first attempts to simulate U. Hence if $z = z_1 z_2$, then C will first simulate $U(z_1)$. It will then read exactly $U(z_1)$ further bits of input, if possible. These further bits of input will be some word z_3. D will then compute $V(z_3)$, and gives this as its output.

Notice that D is prefix-free because firstly U is, and if C halts on z, then $z = z_1 z_2$ with $U(z_1) \downarrow$, and $|z| = |z_1| + |U(z_1)|$. Thus all extensions of z_1 upon which D halts have the same length, and hence cannot be prefixes of other such strings. Let π_D be the coding constant of D in U.

Let y_3 be a minimal Kolmogorov program for x, and y_1 a minimal prefix-free program for $|y_3|$. Then $U(\pi_D y_1 y_3) = C(y_1 y_3) = V(y_3) = x$. hence $K(x) \leq C(x) + K(C(x)) + |\pi_D|$. This establishes the claim.

To finish the proof, using the claim we get the following calculation.

$$K(x) \leq |x| + K(|x|) + O(1) + O(\log m_C(x) + 2) - m_C(x).$$

Thus $0 \leq m_K(x) + O(\log m_C(x) + 2) - m_C(x)$. Hence,

$$m_K(x) \geq m_C(x) - O(\log m_C(x) + 2). \qquad \square$$

Actually the relationships between C and K are very complex. The following definitive results were obtained by Solovay [83].

Theorem 2.12: (Solovay [83])

$$K(x) = C(x) + C^{(2)}(x) + O(C^{(3)}(x)) \qquad (1)$$

and

$$C(x) = K(x) - K^{(2)}(x) + O(K^{(3)}(x)). \qquad (2)$$

Solovay [83] showed that (1) and (2) above are *sharp!* That is, for instance,

$$K(x) = C(x) + C^{(2)}(x) + C^{(3)}(x) + O(C^{(4)}(x))$$

is *not* true in general. The proof can be found in Downey and Hirschfeldt [18] and is, as you would expect, highly combinatorial.

2.7. *The overgraph functions*

Notice that, again, we can look at the "overgraph" functions. Now we consider

$$M_K = \{\langle x, y \rangle : K(x) < y\},$$

and the conditional case

$$M_K' = \{\langle x, y, z \rangle : K(x|y) < z\},$$

and finally

$$\overline{R}_K = \{x : K(x) \geq |x| + K(|x|) - c\}.$$

Using exactly the same proof, Muchnik proved that M_K' is m-complete. He also considered M_K. We remark that $\overline{R}_K$ is a tricky object to deal with. The point is that both sides of the definition vary in time. Ever since the manuscript of Solovay it has been open whether $\{x : K(x) < |x| + K(|x|) - c\}$ is computably enumerable. This was finally solved by Joe Miller in early 2005.

Theorem 2.13: (Miller [59, 60])

  (i) Fix $c \geq 0$ and let $B = \{v : K(v) < |v| + K(|v|) - c\}$. If A contains B and has property (*) below, then A is not a c.e. set.

  (*) For all n,

$$|A \cap 2^n| < 2^n.$$

  (ii) Hence for all sufficiently large c, $B = \{v : K(v) < |v| + K(|v|) - c\}$ is not Σ_1^0.

We remark that it is not known whether $\overline{R}_K$ *can be tt*-complete. Here I say *can be* since Muchnik prove the following remarkable result which demonstrates that for prefix-complexity things can become machine dependent. Let M_K^Q denote set of strings that are not weakly Chaitin random relative to the universal machine Q, and similarly $\overline{R}_K^Q$.

Theorem 2.14: (Muchnik, An. A. [64]) There exist universal prefix-free machines V and U such that

  (i) M_K^V is *tt*-complete.
  (ii) M_K^U (and hence $\overline{R}_K^U$) is not *tt*-complete.

The proof of (ii) is fairly remarkable. In the construction we will be building a universal machine U and a set B diagonalizing against tt-reductions Γ_e making sure that $\Gamma_e^{M_K^U} \neq B$. To do this, we would pick some follower n and put n into B if there was some way to make $\Gamma_e^{M_K^U}(n) = 0$ and otherwise try to force $\Gamma_e^{M_K^U}(n) = 1$ and keep n out of B. The problem is that at any stage s, U and hence M_K^U are only in a state of formation. It is within the opponent's power to be able to drop the complexity of some string x and hence add some $\langle x, y \rangle$ into M_K^U. This might change the value of $\Gamma_e^{M_K^U}(n)$. We are in control of some part of U (about half in the actual construction) and it would be then in our power to possibly change some complexity to perhaps restore $\Gamma_e^{M_K^U}(n)$ to its previous value.

Muchnik's idea is to view this as a game played on a finite directed graph, with $(\langle x, y \rangle, \langle x, y' \rangle)$ an edge representing the dropping of the complexity of x from y to y'. Then, the whole thing can be viewed as a game played on a finite graph determining the value of $\Gamma_e^{M_K^U}(n)$, by moves alternatively by the opponent then us. There is a computable strategy for such finite games, and this strategy will determine what value to set $B(n)$ to be.

The details are a bit messy, but this is the fundamental idea. We refer the reader to either Muchnik and Positelsky [64] or Downey and Hirschfeldt [18].

3. Lecture 2: Randomness for reals

3.1. *Martin-Löf randomness*

It is a fascinating problem to give mathematical content to our intuition that the real $1111\ldots$ and any other real α are equally likely in terms of measure theory yet our intuition would be that the real of all 1's is not random. The first real attempt to address this question occurs in a remarkable paper by von Mises [93]. Von Mises was a probabilist, and suggested that a stochastic approach to "defining" randomness. To wit, he suggested that given a real $\alpha = a_1 a_2 \ldots$, if we were to "select" some subsequence assuming "acceptable" selection rules, say we choose positions $f(1), f(2) \ldots$, then we should have that the limit as n went to infinity of the number of $a_{f(i)} = 1$ divided by those with $a_{f(i)} = 0$ for $i \leq n$ should be 1. That is, in the limit, we should select equal numbers of 0's and 1's. This approach can be viewed as a generalization of the law of large numbers. Von Mises had no canonical way to formulate the notion of acceptable rule, but did observe that countable collections of selection rules could be dealt with. It remained until the clarification of the notion of a computable function for reasonable classes of f's to be suggested. For instance, we might suggest that something is stochastically random iff it defeats any computable selection rule. With a little care this gives rise to notions like Church randomness. As we will see later, von Mises approach has a lot to say about current research. However, in even the computable formulation of von Mises basic notion, the approach had several drawbacks. These are thoroughly discussed in van Lambalgen's thesis [91]. The first widely acceptable definition of randomness came from the fundamental work of Martin-Löf in 1966 in [53]. Martin-Löf's fundamental observation was that we can think of *effective* statistical tests as being *effective* null sets of reals. Thus a real should be random if it avoids all effective null sets. A computably enumerable open set would be a collection $W = \{[\sigma] : \sigma \in W_e\}$ for some e.

Definition 3.1: (Martin-Löf, [53]) We say that a real is *Martin-Löf random* or 1- *random* iff for all computable collections of c.e. open sets $\{U_n : n \in \omega\}$, with $\mu(U_n) \leq 2^{-n}$, $x \notin \cap_n U_n$.

We call a computable collection of c.e. open sets a *test* since it corresponds to a statistical test, and ones with $\mu(U_n) \leq 2^{-n}$ for all n, a

Martin-Löf test. The usual terminology is to say that a real is Martin-Löf random if it passes all Martin-Löf tests meaning that it is not in the intersection. Since there are only countably many such tests, almost all reals are Martin-Löf random.

Using the enumeration of all c.e. sets, we can enumerate all c.e. tests, $\{W_{e,j,s} : e, j, s \in \mathbb{N}\}$ and stops the enumeration of one if the measure $\mu(W_{e,j,s})$ threatens to exceed $2^{-(j+1)}$ at any stage s of the simultaneous enumeration. Then we can let

$$U_n = \cup_{e \in \mathbb{N}} W_{e,n+e+1}.$$

Then we note that U_n is a Martin-Löf test, and moreover, a real A passes all Martin-Löf tests iff $A \notin \cap_{n \in \mathbb{N}} U_n$. We have established the following.

Theorem 3.1: (Martin-Löf [53]) There exist *universal* Martin-Löf tests: That is there is a Martin-Löf test $\{U_n : n \in \mathbb{N}\}$ such that, for any Martin-Löf test $\{V_n : n \in \mathbb{N}\}$, $x \in \cap_{n \in \mathbb{N}} V_n$ implies $x \in \cap_{n \in \mathbb{N}} U_n$.

The reader should note the following alternative version of Definition 3.1.

A real is Solovay random iff for all computably enumerable collections of intervals $I_n = [\sigma_n] : n \in \omega$, if $\sum_n |I_n| < \infty$, then $x \in I_n$ for at most finitely many n.

The following is an easy exercise.

Theorem 3.2: (Solovay [83]) A real x is Martin-Löf random iff x is Solovay random.

3.2. *Schnorr's Theorem and the computational paradigm*

When we looked at strings, our approach to randomness centered around the computational/incompressibility paradigm. This approach for reals was pioneered by Levin [45, 101] using monotone and prefix-free complexity, the latter also used by Chaitin [8]. We have seen that the plain complexity of sufficiently long strings will always drop, and this can be formalized for reals into the following.

Theorem 3.3: (Li-Vitanyi [49], also Staiger [84]) Let $f : \mathbb{N} \to \mathbb{N}$ be any total computable function. Suppose that $\sum_{n=1}^{\infty} 2^{-f(n)} = \infty$. Then for any real α, $C(\alpha \upharpoonright n|n) \leq n - f(n)$ infinitely often.

Corollary 3.1: (Li-Vitanyi [49], after Martin-Löf [54])   Let $f : \mathbb{N} \to \mathbb{N}$ be any total computable function, such that $\sum_{n=1}^{\infty} 2^{-f(n)} = \infty$, and such that, for all n, $C(n|n - f(n)) = \mathcal{O}(1)$. Then $C(\alpha \upharpoonright n) \leq n - f(n)$ infinitely often.

The prototypical application of Corollary 3.1 is $f(n) = \log n$ allowing us to conclude that $C(\alpha \upharpoonright n) \leq n - \log n$ infinitely often. Again we note that if we use a complexity whose interpretation is equivalent to bit complexity, then we remove the oscillations below n. This allows for the following definition.

Definition 3.2: (Levin(-Gács-Chaitin)) A real α is Levin-Gács-Chaitin random if for all n

$$K(A \upharpoonright n) \geq n - O(1).$$

The following fundamental theorem shows that the two notions of randomness coincide.

Theorem 3.4: (Schnorr) A real x is Levin-Gács-Chaitin random iff it is Martin-Löf random.

Proof: ($\to$) Suppose that x is Martin-Löf random. Let

$$U_k = \{y : \exists n K(y \upharpoonright n) \leq n - k\}.$$

Since the universal machine is prefix-free, we can estimate the size of U_k.

$$\mu(U_k) = \sum \{2^{-|\sigma|} : K(\sigma) \leq n - k\}$$

$$\leq \sum_{n \in \mathbb{N}} 2^{-(n+k)} = 2^{-k}.$$

Hence the sets $\{U_k : k \in \mathbb{N}\}$ form a Martin-Löf test, and if x is Martin-Löf random $x \notin \cap_n U_n$. Thus there is a k such that, for all n, $K(x \upharpoonright n) > n - k$.

For the converse direction, recall from Lecture 1 that K is a minimal information content measure so that for all σ, $K(\sigma) \leq K_k(\sigma) + \mathcal{O}(1)$. Now suppose that x is not Martin-Löf random, and hence $x \in \cap U_n$ with $\{U_n : n \in \mathbb{N}\}$ the universal Martin-Löf test (so that $\mu(U_n) \leq 2^{-n}$). We note that $\sum_{n \geq 3} 2^{-n^2+n} \leq 1$. We use Kraft-Chaitin to build a machine M. Whenever we see some $[\sigma]$ occur in U_{n^2} for $n \geq 3$, we enumerate an axiom for M of the form $|\sigma| - n$. The total cost of the M-axioms is found by the calculation:

$$\sum_{n \geq 3} \sum_{\sigma \in U_{n^2}} 2^{-(|\sigma|-n)} \leq \sum_{n \geq 3} 2^n \mu(U_{n^2}) \leq \sum_{n \geq 3} 2^{-n^2+n} \leq 1.$$

Thus, if c is the coding constant for M in U, we have for all $\sigma \in U_{n^2}$ and $n \geq 3$,

$$K(\sigma) \leq |\sigma| - n + c.$$

Therefore, as $x \in \cap U_{n^2}$ for all $n \geq 3$ we see that $K(x \upharpoonright m) \leq m - n + c$, and hence it drops arbitrarily away from k. Hence, x is not Levin-Gács-Chaitin random. $\square$

We can use Schnorr's Theorem to prove that there are many strings that are (weakly) Chaitin random yet are not Kolmogorov random.

Corollary 3.2: There are infinitely many n and strings x of length n such that

(i) $K(x) \geq n$ and
(ii) $C(x) \leq n - \log n$.

Proof: Let α be Martin-Löf random. Then by Schnorr's Theorem, Theorem 3.4, for all n, $K(\alpha \upharpoonright n) \geq n - O(1)$. But by Corollary 3.1, for infinitely many n, $C(\alpha \upharpoonright n) \leq n - \log n$. $\square$

Schnorr's Theorem also allows us to define some *specific* kinds of random reals. For instance, the class

$$R = \cup_c R_c \text{ where } R_c = \{A : \forall n (K(A \upharpoonright n) \geq n - c)\},$$

is the Σ_2^0 class of all Martin-Löf random reals. From some c onwards, the classes R_c are nonempty Π_1^0 classes of reals. By the Low Basis Theorem and Hyperimmunne-free Basis Theorem we have the following.

Theorem 3.5: (Kučera and others)

(i) There are low Martin-Löf random reals.
(ii) There are Martin-Löf random reals of hyperimmune-free degrees.

Actually, we can come up with a specific Martin-Löf random real. This is the famous example of Chaitin.

Theorem 3.6: (Chaitin [9], Chaitin's Ω) Let U be a universal prefix-free machine. Then Ω, the halting probability below, is Martin-Löf random.

$$\Omega = \sum_{U(\sigma)\downarrow} 2^{-|\sigma|}.$$

Proof: We build a machine M and it has coding constant e given by the recursion theorem. (This means that if we put σ in $\mathrm{dom}(M)$, U *later* puts something of length $|\sigma| + e$ into $\mathrm{dom}(U)$.) Let $\Omega_s = \sum_{M(\sigma)\downarrow[s] \wedge |\sigma| \leq s} 2^{-|\sigma|}$. For $n \leq s$, if we see $K_s(\Omega_s \upharpoonright n) < n - e$, find some σ of $K_s(\Omega_s \upharpoonright n$ with $U_s(\sigma) \mapsto \Omega_s \upharpoonright n$. Declare $M_s(\sigma) \downarrow$, which causes $\Omega_s \upharpoonright n \neq \Omega \upharpoonright n$. Note we cannot put more into the domain of M than U has in its domain and hence we may apply Kraft-Chaitin to build M. □

Solovay was the first to look at other computability-theoretical aspects of Ω. For instance, consider $D_n = \{x : |x| \leq n \wedge U(x) \downarrow\}$. Solovay proved that $K(D_n) = n + O(1)$, where $K(D_n)$ is the K-complexity for an index for D_n. Solovay also proved the following basic relationships between D_n and $\Omega \upharpoonright n$.

Theorem 3.7: (Solovay [83])

   (i) $K(D_n | \Omega \upharpoonright n) = O(1)$. (Indeed $D_n \leq_{wtt} \Omega \upharpoonright n$ via a weak truth table reduction with identity use.)

   (ii) $K(\Omega \upharpoonright n | D_{n+K(n)}) = O(1)$.

Proof: (i) is easy. We simply wait till we have a stage s where $\Omega_s =_{\mathrm{def}} \sum_{U(\sigma)\downarrow[s]} 2^{-|\sigma|}$ is correct on its first n bits. Then we can compute D_n.

The proof of (ii) is more involved. We follow Solovay [83]. Let $\hat{D} = D_{n+K(n)}$. Note that $K(n + K(n) | \hat{D}) = O(1)$. We can simply compute from $\hat{D}$, $K(j)$ for all $j \leq n + K(n)$, by looking for the length of the least $x \in D_n$ with $U(x) = j$. Hence we can find the least j such that $j + K(j) = n + K(n)$. Then we claim that $j - n = O(1)$. To see this note that

$$K(j) - K(n)| \leq K(|j - n|) + O(1).$$

Hence, $|j - n| \leq K(|j-n|) + O(1) \leq 2\log|j - n| + O(1)$. Therefore $|j - n| = O(1)$. Also this means $K(j|\hat{D}) = O(1)$, and hence $K(n|\hat{D}) = O(1)$.

We prove that there is a q such that $K(\Omega \upharpoonright n - q | \hat{D})O(1)$. We construct a machine M that does the following. $M(xy)$ is defined if

   (i) $U(x) = n$.

   (ii) $|y| = n$.

   (iii) $\Omega \geq \frac{y}{2^n}$.

Here of course we are interpreting $y \in \{0, \ldots, 2^n - 1\}$. Now we can find q such that, for all n,

$$|\Pi_M| + K(n - q) + n - q \leq n + K(n).$$

But then,

$$\Omega \geq \frac{y}{2^{n-q}} \text{ iff } \Pi_M(n-q)^* y \in D_{n+K(n)}.$$

Therefore $K(\Omega \restriction n - q | \hat{D}) = O(1)$, since $K(n | \hat{D}) = O(1)$. Clearly, $K(\Omega \restriction n | \Omega \restriction n - q) = O(1)$. Thus $K(\Omega \restriction n | \hat{D}) = O(1)$, as claimed. $\square$

In the light of Schnorr's Theorem, Solovay had asked whether $\liminf_s K(\Omega \restriction n) - n \to \infty$. This was solved affirmatively by Chaitin. However, there is a very attractive generalization of this due to Miller and Yu who show that the complexity of a random real must be above n eventually by "quite a bit."

Theorem 3.8: (Ample Excess Lemma, Miller and Yu [62]) A real α is random iff

$$\sum_{n \in \mathbb{N}} 2^{n - K(\alpha \restriction n)} < \infty.$$

Proof: One direction is easy. Suppose that α is not 1-random. Then we know that for all c, for infinitely many n, $K(\alpha \restriction n) < n - c$. That means that $\sum_{n \in \mathbb{N}} 2^{n - K(\alpha \restriction n)} = \infty$.

Now for the nontrivial direction. For the other direction, note that, for any $m \in \mathbb{N}$,

$$\sum_{\sigma \in 2^m} \sum_{n \leq m} 2^{n - K(\sigma \restriction n)} = \sum_{\sigma \in 2^m} \sum_{\tau \prec \sigma} 2^{|\tau| - K(\tau)}$$

$$= \sum_{\tau \in 2^{\leq m}} 2^{m - |\tau|} 2^{|\tau| - K(\tau)} = 2^m \sum_{\tau \in 2^{\leq m}} 2^{-K(\tau)} \leq 2^m,$$

by Kraft's inequality. Therefore, for any $p \in \mathbb{N}$, there are at most $2^m / p$ strings $\sigma \in 2^m$ for which $\sum_{n \leq m} 2^{n - K(\sigma \restriction n)} \geq p$. This implies that $\mu(\{\alpha \in 2^\omega : \sum_{n \leq m} 2^{n - K(\alpha \restriction n)} \geq p\}) \leq 1/p$. Define $\mathcal{I}_p = \{\alpha \in 2^\omega : \sum_{n \in \mathbb{N}} 2^{n - K(\alpha \restriction n)} \geq p\}$. We can express $\mathcal{I}_p$ as a nested union $\bigcup_{m \in \mathbb{N}} \{\alpha \in 2^\omega \mid \sum_{n \leq m} 2^{n - K(\alpha \restriction n)} \geq p\}$. Each member of the nested union has measure at most $1/p$, so $\mu(\mathcal{I}_p) \leq 1/p$. Also note that $\mathcal{I}_p$ is a Σ_1^0 class. Therefore, $\mathcal{I} = \bigcap_{k \in \mathbb{N}} \mathcal{I}_{2^k}$ is a Martin-Löf test. Finally, note that $\alpha \in \mathcal{I}$ iff $\sum_{n \in \mathbb{N}} 2^{n - K(\alpha \restriction n)} = \infty$. Now assume that $\alpha \in 2^\omega$ is 1-random. Then $\alpha \notin \mathcal{I}$, because it misses all Martin-Löf tests, so $\sum_{n \in \mathbb{N}} 2^{n - K(\alpha \restriction n)}$ is finite. $\square$

The following corollary was proven for f computable by Solovay.

Corollary 3.3: (Miller and Yu [63]) Suppose that f is an arbitrary function with $\sum_{m\in\mathbb{N}} 2^{-f(m)} = \infty$. Suppose that α is 1-random. Then there are infinitely many m with $K(\alpha \restriction m) > m + f(m)$.

To finish this section we remark that it was a longstanding question whether there was a *plain complexity* characterization of 1-randomness. This was also solved by Miller and Yu, having been open for 40 years.

Definition 3.3: (Miller and Yu [62]) Define a computable function $G : \omega \to \omega$ by

$$G(n) = \begin{cases} K_{s+1}(t), & \text{if } n = 2^{\langle s,t \rangle} \text{ and } K_{s+1}(t) \neq K_s(t) \\ n, & \text{otherwise.} \end{cases}$$

Theorem 3.9: (Miller and Yu [62]) For $x \in 2^{\omega}$, the following are equivalent:

    (i) x is 1-random.
    (ii) $(\forall n)\, C(x \restriction n) \geq n - K(n) \pm O(1)$.
    (iii) $(\forall n)\, C(x \restriction n) \geq n - g(n) \pm O(1)$, for every computable $g : \omega \to \omega$ such that $\sum_{n\in\omega} 2^{-g(n)}$ is finite.
    (iv) $(\forall n)\, C(x \restriction n) \geq n - G(n) \pm O(1)$.

3.3. *Martingales and the prediction paradigm*

The very earliest work on randomness was by von Mises [93] and involved *selection*, as we have already mentioned. This encapsulates the general view that random reals should be "unpredictable." Using computable functions, attempts were made to give a definition of randomness in terms of computable or partial computable selection procedures. This gives rise to notions of computable or partial computable stochasticity. It turned out that the correct way to formalize this notion of effective prediction was in terms of betting strategies.

Definition 3.4: (Levy [48]) A *martingale* is a function $f : 2^{<\omega} \to \mathbb{R}^+ \cup \{0\}$ such that for all σ,

$$f(\sigma) = \frac{f(\sigma 0) + f(\sigma 1)}{2}.$$

We say that the martingale *succeeds* on a real α if $\limsup_n F(\alpha \restriction n) = \infty$.

Definition 3.5:

(i) A *supermartingale* is a function $f : 2^{<\omega} \to \mathbb{R}^+ \cup \{0\}$ such that for all σ,

$$f(\sigma) \geq \frac{f(\sigma 0) + f(\sigma 1)}{2}.$$

We say that the supermartingale *succeeds* on a real α if $\limsup_n F(\alpha \restriction n) = \infty$.

(ii) Similarly we can define a *submartingale* and its success if we ask that

$$f(\sigma) \leq \frac{f(\sigma 0) + f(\sigma 1)}{2}.$$

Ville [94] proved that null sets correspond to success sets for martingales. They were used extensively by Doob in the study of stochastic processes.

The principal tool for using martingales is the classical result below. It says that the distribution of capital must be fair level by level.

Theorem 3.10: (Kolmogorov's inequality, see Ville [94])

(i) Let f be a (super-) martingale. For any string σ and prefix-free set $X \subseteq \{x : \nu \preceq x\}$,

$$2^{-|\nu|} f(\nu) \geq \sum_{x \in X} 2^{-|x|} f(x).$$

(ii) Let $S^k(f) = \{\sigma : f(\sigma) \geq k\}$, then

$$\mu(S^k(f)) \leq f(\lambda) \frac{1}{k}.$$

Schnorr showed that Martin-Löf randomness corresponded to effective (super-)martingales failing to succeed.

Definition 3.6: (Schnorr [77]) We will define a (super-, sub-)martingale f as being *effective* or *computably enumerable* if $f(\sigma)$ is a c.e. real, and at every stage we have effective approximations to f in the sense that $f(\sigma) = \lim_s f_s(\sigma)$, with $f_s(\sigma)$ a computable increasing sequence of rationals.

Theorem 3.11: (Schnorr [77]) A real α is Martin-Löf random iff no effective (super-)martingale succeeds on α.

Proof: We show that test sets and martingales are essentially the same. This effectivizes Ville's work. Firstly suppose that f is an effective (super-) martingale. Define open sets

$$V_n = \cup\{\beta : f(\beta) \geq 2^n\}.$$

Then V_n is clearly a c.e. open set. Furthermore, $\mu(V_n) \leq 2^{-n}$ by Kolmogorov's inequality. Thus $\{V_n : n \in \mathbb{N}\}$ is a Martin-Löf test. Moreover, $\alpha \in \cap_n V_n$ iff $\limsup_n f(\alpha \restriction n) = \infty$, by construction. Hence f succeeds on α iff it fails the derived Martin-Löf test.

For the other direction, we show how to build a martingale from a Martin-Löf test. Let $\{U_n : n \in \mathbb{N}\}$ be a Martin-Löf test. We represent U_n by extensions of a prefix-free set of strings σ, and whenever such a σ is enumerated into $\cup_{n,s} U_n^s$, increase $F(\sigma)[s]$ by one. To maintain the martingale nature of F, we also increase F by 1 on all extensions of σ, and by 2^{-t} on the substring of σ of length $(|\sigma| - t)$. □

Corollary 3.4: (Levin [45,101], Schnorr [77]) There is a universal effective martingale. That is there is an effective martingale f, such that for all martingales g, and reals α, f succeeds on α implies g succeeds on α.

Proof: Apply the proof above to the universal Martin-Löf test. □

Notice that any constant multiple of a martingale will succeed on a set exactly if the martingale does. The following strengthens Corollary 3.4 and is an analog to saying that K is a minimal information content measure.

Theorem 3.12: (Schnorr [77]) There is a *multiplicatively optimal* supermartingale. That is there is an effective supermartingale f such that for all effective supermartingales g, there is a constant c such that, for all σ,

$$cf(\sigma) \geq g(\sigma).$$

Proof: It is easy to construct a computable enumeration of all effective supermartingales, g_i for $i \in \mathbb{N}$. (Stop the enumeration when it threatens to fail the supermartingale condition.) Then we can define

$$f(\sigma) = \sum_{i \in \mathbb{N}} 2^{-i} g_i(\sigma).$$

□

3.4. *Supermartingales and continuous semimeasures*

In [47], Levin constructed a universal continuous semi-measure. This can be interpreted as Schnorr's result, as we now see.

Definition 3.7: A *continuous semimeasure* is a function $\delta : [2^{<\omega}] \to \mathbb{R}^+ \cup \{0\}$ satisfying

(i) $\delta([\lambda]) \leq 1$, and
(ii) $\delta([\sigma]) \geq \delta([\sigma 0]) + \delta([\sigma 1])$.

This would seem an appropriate effective analog of normal Lebesgue measure treating the space as 2^ω rather than $2^{<\omega}$.

Levin [47] directly constructed an optimal minimal semimeasure. However, for any supermartingale F, we can define

$$\delta([\sigma]) = 2^{-|\sigma|} F(\sigma),$$

and conversely. Then Schnorr's optimal supermartingale is equivalent to Levin's optimal semimeasure. We can also associate a version of Kolmogorov complexity to a semimeasure.

$$KM(\sigma) = -\log \delta([\sigma]),$$

where δ is the optimal semimeasure. Notice that

$$KM(\sigma) = -\log F(\sigma) + |\sigma| + O(1),$$

where F is an optimal supermartingale.

A cornerstone of algorithmic information theory is the Coding Theorem as we have already seen. This directly shows that the probability a string is output is essentially the same as K. In the case of continuous semimeasures, it is natural to ask for a similar result.

It turns out that the relevant complexity measure is *monotone complexity*. The idea is that attempt to give a real itself a complexity. Thus since we are thinking of our space as 2^ω we are covering the segments of a real by rather than strings. Thus we can imagine a computable real as being output from a machine V in segments. (Here the machines can now have infinite output.) We ask that for all σ, τ if $\sigma \preceq \tau$ and $M(\sigma) \downarrow, M(\tau) \downarrow$, then $M(\sigma) \preceq M(\tau)$. Now we can again develop Kolmogorov complexity using monotone machine and the resulting measure is called Km. Clearly all prefix-free machines are monotone machines. Levin showed that a real α is random iff for all n,

$$Km(\alpha \restriction n) = n + O(1).$$

Levin [47] conjectured that the analogous Coding Theorem held for KM vs Km. That is, that $Km(\sigma) = KM(\sigma) + O(1)$ for all σ.

This attractive conjecture fails.

Theorem 3.13: (Gács [27])

    (i) There exists a function f with $\lim_s f(s) = \infty$, such that for infinitely many σ,

$$Km(\sigma) - KM(\sigma) \geq f(|\sigma|).$$

    (ii) Indeed, we may choose f to be the inverse of Ackermann's function.

Gács proof is very difficult, and it would be nice to have an accessible proof.

3.5. *Schnorr and computable randomness*

Schnorr argued that Theorem 3.11 showed a flaw in the definition of Martin-Löf randomness. He argued that randomness should be concerned with defeating *computable* strategies rather than computably enumerable strategies, since the latter are fundamentally asymmetric, in the same way that a computably enumerable set is semi-decidable rather than decidable. He proposed two variants of the notion of Martin-Löf randomness.

Definition 3.8: (Schnorr randomness, Schnorr [77])

    (i) We say that a Martin-Löf test $\{V_n : n \in \mathbb{N}\}$ is a *Schnorr test* iff for all n,

$$\mu(V_n) = 2^{-n}.$$

    (ii) We say that a real α is *Schnorr random* iff for all Schnorr tests, $\alpha \notin \cap_n V_n$.

Note here 2^{-n} can easily be replaced by any uniformly computable sequence of computable reals effectively converging to 0.

Definition 3.9: (Schnorr [77])

    (i) A martingale f is called computable iff $f : 2^{<\omega} \to \mathbb{R}^+ \cup \{0\}$ is a computable function with $f(\sigma)$ (the index of functions representing the effective convergence of) a computable real. (That is, we will be given indices for a computable sequence of rationals $\{q_i : i \in \mathbb{N}\}$ so that $f(\sigma) = \lim_s q_s$ and $|f(\sigma) - q_s| < 2^{-s}$.)

    (ii) A real α is called *computably random* iff for no computable martingale succeeds on α.

It is possible to give machine characterizations of both of the notions above. The one for computable randomness is a little untidy, but Downey and Griffiths [16] gave a nice characterization of Schnorr randomness in terms of *computable* machines. Here we say that a machine M is computable if the measure of its domain is a computable real. Recall here that a real is called computable iff its dyadic expansion is computable. The domains of prefix-free machines are, in general, only *computably enumerable* or *left computable* in the sense that they are limits of computable nondecreasing sequences of rations. (For example $\Omega = \lim_s \Omega_s = \sum_{U(\sigma)\downarrow[s]} 2^{-|\sigma|}$.) Computably enumerable reals play the same role in this theory as computably enumerable set do in classical computability theory, and will be deal with in more detail later.)

Theorem 3.14: (Downey and Griffiths [16]) A real α is Schnorr random iff for all computable machines M, there is a constant c such that, for all n, $K_M(\alpha \upharpoonright n) \geq n - c$.

Proof: (sketch) Begin by showing that Kraft-Chaitin works to give a computable machine if the measure of the requirements happens to be a computable real. Then run through the translations of tests to machines and conversely to make sure that things work. □

Incidentally, it is possible to look at time bounded variations here and to look at, say, polynomial time computable prefix-free complexity. (Randomness in polynomial time tends to be studied by martingales.) Kraft-Chaitin in this setting would seem to be related to things like $P, NP, PSPACE$, but this remains unexplored.

There is one other related notion of randomness we should mention. We define a *Kurtz test* to be a a Σ_1^0 class of measure 1. Then a real A is called *weakly random* or *Kurtz random*[c] iff it passes all Kurtz tests. That is $A \in U$ for all such U. There is an easy equivalent notion in terms of null tests, implicit in Kurtz's Thesis [44], and explicit in Wang's.

Definition 3.10: (Wang [95]) A *Kurtz null test* is a collection $\{V_n : n \in \mathbb{N}\}$ of c.e. open sets, such that

(i) $\mu(V_n) \leq 2^{-n}$, and

---

[c]Now it could be argued that weak randomness is not really a randomness notion at all, but rather a genericity notion. As we will see, the higher level version is highly relevant to our story.

(ii) There is a computable function $f : \mathbb{N} \to (\Sigma^*)^{<\omega}$ such that $f(n)$ is a canonical index for a finite set of σ's, say, $\sigma_1, \ldots, \sigma_n$ and $V_n = \{[\sigma_1], \ldots, [\sigma_n]\}$.

Theorem 3.15: (Wang [95], after Kurtz [44]) A real α is Kurtz random iff it passes all Kurtz null tests.

Proof: We show how measure 1 open sets correspond to Kurtz null tests. Let U be a c.e. open set with $\mu(U) = 1$. We define V_n in stages. To define V_1, enumerate U until a stage s is found with $\mu(U_s) > 2^{-1}$. The let $V_1 = \overline{U_s}$. Note that V_1 is of the correct form, to be able to define f. Of course for V_n we enumerate enough of U to have $\mu(U_{s_n}) > 2^{-n}$. For the converse reverse the reasoning.   □

There are nice martingale characterizations of both Kurtz and Schnorr randomness.

Theorem 3.16: (Wang [95]) A real α is Kurtz random iff there is no computable martingale F and nondecreasing computable function h, such that for *almost all* n,

$$F(\alpha \restriction n) > h(n).$$

Definition 3.11: We say that a computable martingale *strongly* succeeds on a real x iff there is a computable unbounded nondecreasing function $h : \mathbb{N} \to \mathbb{N}$ such that $F(x \restriction n) \geq h(n)$ infinitely often.

Theorem 3.17: (Schnorr [77]) A real x is Schnorr random iff no computable martingale strongly succeeds on x.

Finally, Downey, Griffiths and Reid [17] gave a machine characterization of Kurtz randomness in terms of "computably layered" machines. It is not hard to see that amongst randomness notions, Martin-Löf implies computable implies Schnorr implies Kurtz. Building on earlier work of Schnorr, Wang, Downey, LaForte, Reid etc, a complete determination of when degree had reals of the various kinds of which were not of others. Nies, Stephan and Terwijn established the following definitive result.

Theorem 3.18: (Nies, Stephan and Terwijn [69]) For every set A, the following are equivalent.

(I) A is high (i.e. $A'' \geq_T \emptyset''$).

(II) $\exists B =_T A$, B is computably random but not Martin-Löf random.

(III) $\exists C \equiv_T A$, C is Schnorr random but not computably random.

Outside the high degrees, things collapse.

Theorem 3.19: (Nies, Stephan and Terwijn [69]) Suppose that a set A is Schnorr random and does not have high degree. Then A is Martin-Löf random.

Proof: Suppose that A is not of high degree and covered by the Martin-Löf test $A \subset \cap_i U_i$. Let f be the function that computes on argument n the stage by which U_n has enumerated a $[\sigma] \in U_{n,s}$ with $A \in [\sigma]$. Note that f is A-computable, and hence computable relative to an oracle which is not high. It follows that there is a computable function g such that $g(n) > f(n)$ for infinitely many n. Then consider the test $\{V_i : i \in \mathbb{N}\}$, found by setting $V_i = U_{i,g(i)}$. The $\cup_i V_i$ is a Schnorr-Solovay test (that is, a Solovay test whose measure is a computable real), and hence A is not Schnorr random.

$\square$

Finally we remark that for some degrees *all* the randomness notions coincide.

Theorem 3.20: (Nies, Stephan, Terwijn [69]) Suppose that A is of hyperimmune-free degree. Then A is Kurtz random iff A is Martin-Löf random[d].

Proof: Suppose that A has hyperimmune free degree, and A is Kurtz random. Suppose that A is not Martin-Löf random. Then there is a Martin-Löf test $\{V_n : n \in \mathbb{N}\}$, such that $A \in \cap_n V_n$. Using A we can compute A-computably compute a stage $g(n)$ such that $A \in V_{g(n)}$, and without loss of generality we can suppose that $V_{g(n+1)} \supseteq V_{g(n)}$. But as A has hyperimmune free degree, we can choose a computable function f so that $f(n) > g(n)$ for all n. Then if we define $W_n = V_{f(n)}$, being a Kurtz null test such that $A \in \cap_n W_n$, a contradiction.

$\square$

We remark that there has been a lot of work trying to see if there is a a possible refutation of Schnorr's criticism by looking at computable supermartingales, where now, we are allowed to bet on the bits of the

---

[d]Actually, Yu Liang observed that the *same* proof shows that A is Kurtz random iff A is weakly 2-random, a notion we meet in the next lecture.

real, but *nonmonotonically*. We refer the reader to the paper of Merkle et. al. [57] and [65], the latter being where Muchnik, An. A., A. Semenov, and V. Uspensky introduced the notion of nonmonotonic betting.

4. Lecture 3: Randomness in general

4.1. *The de Leeuw, Moore, Shannon, Shapiro Theorem, and Sacks' Theorem*

The first relationship of measure theory and computability theory was discovered by de Leeuw, Moore, Shannon, and Shapiro [12] in 1956. Its proof uses an important method called the "majority vote" technique. Recall that an index e (such as the standard construction of $\emptyset'$) is universal if for all indices f and all sets S, there is a finite string σ_f such that

$$W_f^S = W_e^{\sigma_f \frown S}.$$

Definition 4.1: Define the *enumeration probability* of A as

$$P(A) = \mu(\{X \in 2^\omega : W_e^X = A\}).$$

Theorem 4.1: (de Leeuw, Moore, Shannon, and Shapiro, [12]) If $P(A) > 0$ then A is computably enumerable.

Proof: If $P(A) > 0$ then for some e, $D_e = \{X : A = W_e^X\}$ has positive measure. By the Lebesgue Density Theorem, there is a string σ such that the relative measure of D_e above σ is greater than $\frac{1}{2}$. If we let the oracles extending σ vote on membership in D_e, then we get the right answer. That is, enumerate n into A whenever more than half (by measure) of the extensions X of σ put n into W_e^X. $\qquad\square$

Corollary 4.1: (Sacks) If A is noncomputable, then

$$A^{\leq T} =_{\text{def}} \{B : A \leq_T B\},$$

has measure 0.

Solovay [83] examined the relationship between $P(A) > 0$ and the least index for $W_i = A$. Let

$$H(A) = \lceil -\log P(A) \rceil \text{ and,}$$
$$I(A) = \min\{K(i) : W_i = A\}.$$

Theorem 4.2: (Solovay [83])

$$I(A) \leq 3H(A) + K(H(A)) + O(1).$$

The proof is combinatorial, definitely nontrivial, and uses a clever lemma of Martin. It is unknown if the constant 3 can be improved. The point here is that the original proof using majority vote relies on the Lebesgue Density Theorem, and hence is highly noneffective in obtaining the index for A.

There have been a lot of extensions of the results above. For instance, the same technique can be used to prove the following.

Theorem 4.3: (Stillwell [88]) Suppose that $\mu(\{C : D \leq_T A \oplus C\}) > 0$. Then $C \leq_T A$.

At this stage a good exercise to test your understanding is to prove this result. An easy corollary to this result is the following.

Corollary 4.2: (Stillwell [88]) For any $\mathbf{a}, \mathbf{b}$, $(\mathbf{a} \cup \mathbf{b}) \cap (\mathbf{a} \cup \mathbf{c}) = \mathbf{a}$, for almost all $\mathbf{c}$.

Proof: Take $D \leq_T A \oplus B$. Then by Theorem 4.3, if $D \leq_T A \oplus C$ for more than a measure 0 set of C, $D \leq_T A$. Hence for almost all C,

$$D \leq_T A \oplus B \wedge D \leq_T A \oplus C \rightarrow D \leq_T A. \qquad \square$$

Similarly it can be shown that almost all degrees obey

$$\mathbf{a}' = \mathbf{a} \cup \mathbf{0}',$$

That is, almost all degrees are GL_1. Indeed (Stillwell), for almost all $\mathbf{b}$, $(\mathbf{a} \cup \mathbf{b})' = \mathbf{a}' \cup \mathbf{b}$, and $\mathbf{a}^{(n)} = \mathbf{a} \cup \mathbf{0}^{(n)}$. Similarly, for almost all $\mathbf{a}$ and $\mathbf{b}$, $\mathbf{a} \cap \mathbf{b} = \mathbf{0}$. These kind of arguments can be assembled into a nice result of John Stillwell.

Theorem 4.4: The "almost all" theory of degrees is decidable.

Here, variables $\mathbf{a}, \mathbf{b}, \mathbf{c}, \ldots$ vary over arbitrary degrees. Terms are built from $'$ (jump), $\cup, \cap$. An atomic formula is one of the form $t_1 \leq t_2$ for terms t_1, t_2, and formulae in general are built from atomic ones and $\wedge, ^-$ and the quantifier $\forall$ interpreted to mean "for almost all.'

The proof is not difficult. For instance, the Corollary above allows us to compute the meet of two terms of the form $\mathbf{a}_1 \cup \mathbf{a}_2 \cup \cdots \cup \mathbf{0}^{(f)}$ and $\mathbf{a}_1 \cup \mathbf{b}_2 \cup \cdots \cup \mathbf{0}^{(k)}$ as $\mathbf{c}_1 \cup \mathbf{c}2 \cup \cdots \cup \mathbf{0}^{(\min\{f,k\})}$, where $\mathbf{c}_i$ are variables common to both terms. For example $(\mathbf{a}_1 \cup \mathbf{a}_3 \cup \mathbf{0}^{(4)}) \cap (\mathbf{a}_1 \cup \mathbf{a}_5 \cup \mathbf{a}_7 \cup \mathbf{0}^{(6)}) = \mathbf{a}_1 \cup \mathbf{0}^{(4)}$.

These allow for giving normal forms for formulae, with Fubini's Theorem handling nested quantifiers. We refer to Stillwell [88] of Downey-Hirschfeldt [18] for full details.

4.2. *Coding into randoms*

The results above might lead us to suspect that it is not possible to code into random reals, in general. As we will see, this is not the case.

We begin with a famous result often attributed only to Gács [28], but whose first proof was by Kučera [36].

Theorem 4.5: (Kučera [36], Gács [28]) Every set is *wtt* reducible to a Martin-Löf random set.

It is by no means clear that this should be true. The problem is that a random real should not have information easily decodable, or else it would not be random. For instance, we could not expect that the reduction would be, say, an m-reduction. The most attractive proof of this result known to the author is due to Merkle and Mihailovic [56]. The following lemma is the key. It says that for any martingale d and any interval of length k, there are at least k paths extending v of length $\ell(\delta, k)$ where d cannot increase its capital more than a factor of δ while betting on I, no matter how d behaves.

Lemma 4.1: (Folklore, see Merkle and Mihailovic [56]) Given a rational $\delta > 1$ and $k \in \mathbb{Z}^+$, we can compute a length $\ell(\delta, k)$, such that for any martingale d, and any word w,

$$|\{w \in 2^{\ell(\delta,k)} : d(vw) \leq \delta d(v)\}| \geq k.$$

Proof: For any martingale d word w, and k we have

$$d(v) = 2^{-k} \sum_{|u|=k} d(vu).$$

By Kolmogorov's inequality, for any given ℓ and v the average of $d(vw)$ over words of length ℓ is $d(v)$. Thus we have

$$\frac{|\{|w| = \ell : d(vw) > \delta d(v)\}|}{2^\ell} < \frac{1}{\delta}.$$

Since $\delta > 1$, $1 - \delta^{-1} > 0$ and hence it will suffice to have $\ell(\delta, k) \geq \log \frac{k}{1-\delta^{-1}} = \log k + \log \delta - \log(\delta - 1)$. $\qquad\square$

Now we prove the Kučera-Gács Theorem. This proof is due to Merkle and Mihailovic [56]. Let $r_0 > r_1 > \ldots$ be a collection of positive rationals where the sequence β_i $i \in \mathbb{N}$ converges, where

$$\beta_i = \Pi_{j \leq i} r_j.$$

Let $\ell_s = \ell(r_s, 2)$. Partition $\mathbb{N}$ into intervals $\{I_s : s \in \mathbb{N}\}$ with I_s of size ℓ_s. The by Lemma 4.1, there for any word v, and any martingale d, there are at least two words w of length ℓ_s with $d(vw) \leq r_s d(v)$.

We will construct a language R to which a given set X is *wtt* reducible. At step s we will specify R on I_s. Let d be a universal c.e. martingale. We say that w of length I_s is *admissible* if $s = 0$ and $d(w) \leq \beta_0$, and for $s > 0$, if

$$d(vw) \leq \beta_s \text{ where } v = R \restriction (I_0 \cup \cdots \cup I_{s-1})$$

We can argue that at every step there are at least 2 admissible extensions. This is easily seen by induction and the choice of the β_i as the product of the r_j for $j < i$. Now to specify R, armed with $R \restriction (I_0 \cup \cdots \cup I_{s-1})$, we will choose the lexicographically least admissible extension if $s \notin X$ and the lexicographically greatest one if $s \in X$. That ends the proof.

4.3. *Kučera Coding*

Using similar coding with blocks, Kučera was able to prove the following.

Theorem 4.6: (Kučera [36]) Suppose that $\mathbf{a} \geq \mathbf{0}'$. Then $\mathbf{a}$ is Martin-Löf random.

There are several proofs of this result. The first is to derive it as a direct corollary of the Kučera-Gács Theorem.

Proof: (First proof) To derive this result from the proof of the Kučera-Gács Theorem. The point is that in that proof, if $\emptyset' \leq_T X$, then X can figure out R, by calculating the leftmost and rightmost paths. Hence $X \equiv_T R$. □

We will also give Kučera's original proof which is interesting and useful in its own right.

Kučera's proof uses an auxiliary construction of a universal Martin-Löf test. This test has found further applications with particularly ingenious ones being found by Kučera.

Kučera's Construction of a universal Martin-Löf test: Given $n \in \mathbb{N}$, consider all indices $e > n$. For each such e, enumerate all elements of $W_{\{e\}(e)}$ into U_n (where we understand that $W_{\{e\}(e)}$ is empty if $\{e\}(e)$ is undefined) as long as the condition

$$\sum_{w \in W_{\{e\}(e)}} 2^{|w|} < 2^{-e}$$

is satisfied. Then

$$\sum_{w \in U_n} 2^{|w|} \leq \sum_{e>n} 2^{-e} = 2^{-n}.$$

It is not difficult to show that this is indeed a universal Martin-Löf test. The key ingredient in the proof of Theorem 4.6 is a fact about intersections of fat classes which resembles that used in the proof of the Kučera-Gács Theorem.

Lemma 4.2: Let P_n denote the complement of the n-th set of the universal Martin-Löf test constructed above. If A is Π_1^0, then there exists a computable function $\gamma : \Sigma^* \times \mathbb{N} \to \mathbb{Q}^{>0}$ such that for any $w \in \Sigma^*$ and any $n \in \mathbb{N}$ then

$$P_n \cap A \cap [w] \neq \emptyset \quad \to \quad \mu(A \cap [w]) \geq \gamma(w,n).$$

The critical thing, again, is that the function γ is *computable*. Hence, bound the measure of $A \cap [w]$ effectively from below.

The proof of Theorem 4.6 resembles the proof of the Friedberg Cupping Theorem. We construct a perfect tree $T \leq_T \emptyset'$ such that $[T] \subseteq \overline{U_0}$ and every path codes a set $B \subseteq \mathbb{N}$. This coding will be effective due to the preceding lemma, which allows us to compute an effective lower bound for the measure of $\overline{U_0}$.

We B into an infinite path of $\overline{U_0}$. Set $T(\epsilon) = \epsilon$. Assume now for $n \in \mathbb{N}$, $T(\sigma)$ where $\sigma = B \upharpoonright n$ has been constructed, such that $T(\sigma) \prec \overline{U_0}$. To define $T(B \upharpoonright n + 1)$, compute (computably from $\emptyset'$) the smallest number n_σ such that the leftmost and the rightmost path of $[T(\sigma)] \cap \overline{U_0}$ differ (such an n_σ has to exist since a path in $\overline{U_0}$ cannot be isolated). Let these be L_σ and R_σ, respectively. Choose $T(B \upharpoonright n + 1) = L_\sigma \upharpoonright n_\sigma$ if $B(n) = 0$, $T(B \upharpoonright n + 1) = R_\sigma \upharpoonright n_\sigma$, otherwise.

Suppose $B \geq_T \emptyset'$. Claim $B \equiv_T T(B)$. $B \geq_T T(B)$ follows immediately from the construction, which is computable from $\emptyset'$. To show $B \leq_T T(B)$, we employ Lemma 4.2. For instance, to compute $B(0)$, Lemma 4.2 gives us a lower bound on $\mu(\overline{U_0})$, say 2^{-b_0}, $b_0 \in \mathbb{N}$. We know then that the leftmost and the rightmost path of $\overline{U_0} \upharpoonright b_0$ must differ (the tree must branch because its measure is too large). Given $T(B) \upharpoonright b_0$ we compute $\overline{U_0}$ till it turns out to be the left- or rightmost path. Obviously, using Lemma 4.2, this decision procedure can be continued inductively to decide $B(n)$ for any $n \in \mathbb{N}$. That concludes the proof of Theorem 4.6.

The reader will immediately notice the similar use of left and right blocks as in the proof of Kučera-Gács. By the generalized low basis theorem, it can be seen that since the class of random is perfect, there are random reals of all jumps. Below $\mathbf{0}'$ this is still true but requires more elaborate methods.

Theorem 4.7: (Kučera [38], Downey and Miller [24][e]) If $\mathcal{P}$ is a Π_1^0 class such that $\mu(\mathcal{P}) > 0$, then $S \geq_T \emptyset'$ is Σ_2^0 implies that there is a Δ_2^0 real $A \in \mathcal{P}$ such that $A' \equiv_T S$.

Frank Stephan has shown that these random reals above $\mathbf{0}'$ are in essence the only computationally powerful reals.

Theorem 4.8: (Stephan [87]) Suppose that $\mathbf{a}$ is PA and 1-random. Then $\mathbf{0}' \leq_T \mathbf{a}$.

4.4. *n-randomness*

In the same way that the arithmetical hierarchy provides a calibration of computational power, we can analogously calibrate randomness.

Definition 4.2:

(i) A Σ_n^0 test is a computable collection $\{V_n : n \in \mathbb{N}\}$ of Σ_n^0 classes such that $\mu(V_k) \leq 2^{-k}$.

(ii) A real α is Σ_n^0-random or *n*-random iff it passes all Σ_n^0 tests.

(iii) One can similarly define Π_n^0, Δ_n^0 etc tests and randomness.

(iv) A real α is called *arithmetically random* iff for any n, α is *n*-random.

We warn the reader that whilst we can use open sets to define Martin-Löf randomness, we apparently need to be careful to use classes for higher levels. For example, take the Σ_2^0 class consisting of reals that are always zero from some point onwards. This Σ_2^0 class is *not* equivalent to one of the form $\{[\sigma] : \sigma \in W\}$ for some Σ_2^0 set W. However for *n*-randomness, this apparent difference is illusory.

Open sets can be resurrected in the $n > 1$ cases also, as we now see.

Lemma 4.3: (Kurtz [44]) Let $q \in \mathbb{Q}$. The predicate

$$\text{``}\mu(S) > q\text{''}$$

is uniformly Σ_n^0 where S is a Σ_n^0 class. The predicate

$$\text{``}\mu(S) < q\text{''}$$

is uniformly Σ_n^0 where S is a Π_n^0 class.

---

[e]This results was stated without proof in Kučera [38], where he had constructed a high incomplete 1-random real.

This Lemma allows for the following, which says that we are able to use open sets and $\Sigma_1^{\emptyset^{(n)}}$ open classes in place of Σ_{n+1}^0 classes.

Theorem 4.9: (Kurtz [44], Kautz [33]) Let $q \in \mathbb{Q}$.

(i) For S a Σ_n^0 class, we can uniformly (i.e. uniformly in q and a Σ_n^0 index for S) computably compute the index of a $\Sigma_1^{\emptyset^{(n-1)}}$ class which is also an open Σ_n^0 class $U \supseteq S$ and $\mu(U) - \mu(S) < q$.

(ii) For T a Π_n^0 class T, we can uniformly computably compute the index of a $\Pi_1^{\emptyset^{(n-1)}}$ class which is also a closed Π_n^0 class $V \subseteq T$ and $\mu(T) - \mu(V) < q$.

(iii) For each Σ_n^0 class S we can uniformly in $\emptyset^{(n)}$ compute a closed Π_{n-1}^0 class $V \subseteq S$ such that $\mu(S) - \mu(V) < q$. Moreover, if $\mu(S)$ is a real computable from $\emptyset^{(n-1)}$ then the index for V can be found computably from $\emptyset^{(n-1)}$.

(iv) For a Π_n^0 class T we can computably from $\emptyset^{(n)}$ obtain an open Σ_{n-1}^0 class $U \supseteq T$ such that $\mu(U) - \mu(T) < q$. Moreover, if $\mu(S)$ is a real computable from $\emptyset^{(n-1)}$ then the index for U can be found computably from $\emptyset^{(n-1)}$.

The upshot is that a real is $n + 1$-random iff it is 1-random relative to $\emptyset^{(n)}$. We remark that at least for 2-randomness, this characterization is implicit in Solovay's notes [83], where he passes from randomness relativized to $\emptyset'$, and 2-randomness without comment.

With other randomness notions care is needed. Similar relativization will work with Schnorr and computable randomness, but if we define a set to be weakly n-random iff it is in each Σ_n^0 class of measure 1, this is *not* the same as being Kurtz random relative to $\emptyset^{(n-1)}$, which is a genericity notion. The best we can do is the following.

Lemma 4.4: (Kurtz [44], Kautz [33]) Let $n \geq 2$.

(i) Then for any Σ_n^0 class C we can uniformly and computably obtain the index of a $\Sigma_2^{\emptyset^{(n-2)}}$-class $\widehat{C} \subseteq C$ with $\mu(\widehat{C}) = \mu(C)$.

(ii) For any Π_n^0 class V we can uniformly and computably obtain the index of a $\Pi_2^{\emptyset^{(n-2)}}$-class $\widehat{V} \supseteq V$ with $\mu(\widehat{V}) = \mu(V)$.

(iii) Thus, for $n \geq 2$, α is Kurtz n-random iff α is in every $\Sigma_2^{\emptyset^{(n-2)}}$-class of measure 1.

Note that Kurtz 2-randomness is very natural. A real is weakly 2-random iff it avoids all Π_2^0 nullsets. This means that it is not in $\cap_n U_n$

for any computable collection of c.e. open sets with $U_n \to 0$. These are just Martin-Löf tests without the radius of convergence being effective. Similarly, α is Kurtz n-random iff for every $\emptyset^{(n)}$ computable sequence of open $\Sigma_1^{\emptyset^{(n-2)}}$ classes $\{S_i : i \in \mathbb{N}\}$, with $\mu(S_i) \leq 2^{-i}$, $\alpha \notin \cap_i S_i$. Weak 2-randomness was first studied by Gaifman and Snir [29].

The following is more or less immediate.

Theorem 4.10: (Kurtz [44])

   (i) Every n-random real is Kurtz n-random.
   (ii) Every Kurtz $n+1$-random real is n-random.

None of these implications can be reversed even for degrees. (Kurtz [44], Kautz [33]) The most difficult non-containment can be shown by constructing for each $\mathrm{CEA}(\emptyset^{(n)})$ degree $\mathbf{a} > \mathbf{0}^{(n)}$ a weakly $n+1$-random real X computably enumerable in $(\emptyset^{(n)}$, with $X \oplus \emptyset^{(n)}$ of degree $\mathbf{a}$. But then any such $n+1$-random real Y must have $Y \oplus \emptyset^{(n)}$ of degree $\mathbf{0}^{(n+1)}$. This method is due to Downey and Hirschfeldt [18], and is *not* a relativization of the fact that one can have Kurtz random c.e. reals in every nonzero c.e. degree, but rather a finite injury argument over $\mathbf{0}^{(n)}$ argument.

4.5. *Notes on 2-randoms*

Before we turn to the general theory, and 0-1 laws I would like to mention some beautiful results of Nies-Stephan-Terwijn and Miller relating "natural" randomness notions to 2-randomness. I should remark that Veronica Becher and Santiago Figueira have examples of somewhat natural n-random sets, along the lines of Post's Theorem for classical computability. But our concern in the present section are for some completely unexpected results relating plain complexity to 2-randomness.

We have seen that Martin-Löf showed that no real can have $C(\alpha \upharpoonright n) \geq n - d$ for all n, due to complexity oscillations. But it might be possible that some other natural C-condition might guarantee randomness. We have already met one in Theorem 3.9. There was, however, an earlier natural condition which had been analysed in this context. We need the following definitions.

Definition 4.3: We say that a real α is *strongly Chaitin random* iff

$$\exists^\infty n[(K(\alpha \upharpoonright n) > n + K(n) - O(1)].$$

Definition 4.4: We say that α is *Kolmogorov random*[f] iff $\exists^\infty n[(C(\alpha \upharpoonright n) > n - O(1)]$.

We have seen that, should they exist, every strongly Chaitin random real is Kolmogorov random. This follows by Solovay's Theorem 2.11. Strongly Chaitin random reals do exist.

Lemma 4.5: (Yu, Ding, Downey [99], after Solovay [83])

(i) Suppose that α is 3-random. Then

$$\exists^\infty n(K(\alpha \upharpoonright n) = n + K(n) + O(1).$$

(ii) Suppose that α is 3-random. Then

$$\exists^\infty n(C(\alpha \upharpoonright n) = n + O(1).$$

Proof: Consider the test $V_c = \{\alpha : \exists m \forall n(n > m \to K(\alpha \upharpoonright n) \leq n + K(n) - c)\}$. Now $K \leq_T \emptyset'$, and hence V_c is $\Sigma_1^{\emptyset''}$, and hence Σ_3^0. Now we estimate the size of V_c. We show in fact $\mu(V_c) \leq O(2^{-c})$. Let $U_{c,n} = \{x| \ (\forall m \geq n)K(y \upharpoonright m) \leq m + K(n) - c\}$. It suffices to get an estimate $\mu(U_{c,n}) = O(2^{-c})$ uniform in n since $V_c \subseteq \cup_{n \in \omega} U_{c,n}$. But $\mu(U_{c,n}) \leq 2^{-m}|\{\sigma : |\sigma| = m \ \&K(\sigma) \leq m + K(n) - c\}$ for any $m > n$ and by the Counting Theorem this last expression is $O(2^{-c})$. Thus this is a 3-Martin-Löf test. Hence (i) and thus (ii) follow. $\square$

---

[f]There are some problems with terminology here. Kolmogorov did not actually construct or even name such reals, but he was the first more or less to define randomness for *strings* via initial segment plain complexity. The first person to actually construct what we are calling Kolmogorov random strings was Martin-Löf, whose name is already associated with 1-randomness. Schnorr was the first person to show that the notions of Kolmogorov randomness and Martin-Löf randomness were distinct. Again we can't use Schnorr randomness since Schnorr's name is associated with a randomness notion using tests of computable measure. Similar problems occur with our definition of strongly Chaitin random reals. These were never defined by Chaitin, nor constructed by him. They were first constructed by Solovay who has yet another well-known notion of randomness associated with him which is equivalent to 1-randomness. Indeed, Chaitin did have a notion of strongly Chaitin random reals, which is the same as 1-randomness, and is not widely quoted. However, again Chaitin *did* look at the associated notion for *finite* strings, where he proved the fundamental lemma that $K(\sigma) \leq n + K(|\sigma|) + O(1)$ which allows for the definition of the reals. It is also known that Loveland in his 1969 ACM paper proposed equivalent notions via uniform Kolmogorov complexity. Again, Loveland's name is commonly associated with yet another notion of complexity: Kolmogorov-Loveland stochasticity.

Now contrary to claims in the literature, no Δ_2^0 real is Kolmogorov random. (Yu, Ding, Downey [99]) The following clever argument generalized the technique of [99], and proved something much stronger.

Theorem 4.11: (Nies, Stephan and Terwijn [69]) Suppose that α is Kolmogorov random. Then α is 2-random.

Proof: Recall that our universal prefix-free machines are prefix-free relative to *all* oracles. Suppose that α is *not* 2-random. Let K' denote $K^{\emptyset'}$. Then for all c, $\exists^{\infty} n (K'(\alpha \upharpoonright n) < n - c$. Let σ denote the string witnessing this, so that $U^{\emptyset'}(\sigma) = \alpha \upharpoonright n$. Let s be sufficiently large that $U^{\emptyset'}(\sigma) \downarrow [s]$, with $\emptyset'$ correct use. Then consider the plain machine M which runs as follows. M looks at the input ν and attempts to parse it as $\sigma'\tau$, where it runs U with oracle $\emptyset'[t]$ for t steps, where $t = |\tau|$ steps, and for all such simulations, and σ if it gets a result it outputs $U^{\emptyset'}(\sigma')[t]\tau$. Then for inputs with $t > s$, we have $M(\sigma\tau) = \alpha \upharpoonright n + t$, and hence α is not Kolmogorov random. $\square$

Notice that the machine we construct above actually runs (or at least) can run in time polynomial in the size of the input. This then gives to following useful and somewhat remarkable result.

Corollary 4.3: (Nies, Stephan and Terwijn [69]) A real α is 2-random iff for all computable g, with $g(n) > n^2$, $\exists^{\infty} n (C^g(\alpha \upharpoonright n) \geq n - O(1))$, where C^g denotes the plain complexity with time bound $g(n)$ for computations of length n.

Proof: Use the proof above, noting that the plain machine runs in linear time. For the other direction, we lose a quantifier because of the time bound g. $\square$

Miller, then somewhat later, Nies, Stephan and Terwijn were able to prove the following equally remarkable result.

Theorem 4.12: (Miller [55], Nies, Stephan and Terwijn [69]) A real α is 2-random iff α is Kolmogorov random.

The point here is that there is no *a priori* reason that Kolmogorov randomness (defined totally in terms of plain complexity) should coincide with anything involving K, especially given the complex relationships between these two complexities. We recall in Definition 2.2, we defined a compression function which was a function emulating U^{-1}, was $1 - 1$ and for all x

$|F(x)| \leq C(x)$. This notion was invented by Nies, Stephan, and Terwijn for the theorem we are now considering. (The Miller proof was different, and more complex.) For a compression function F we can define F-Kolmogorov complexity C_F: α is F-Kolmogorov random iff $\exists^{\infty} n (C_F(\alpha \upharpoonright n) > n - O(1))$.

The critical lemma was the following.

Lomma 4.0: (Nies, Stephan, and Terwijn [69]) If Z is 2-random relative a compression function F, then Z is Kolmogorov F-random.

The final result if then obtained by using a low compression function to same a quantifier.

4.6. *Kučera strikes again*

We have seen that most random reals are not below $0'$ and hence are not PA. Thus they are computationally feeble. However, Kučera showed that randoms do have *some* computational power. Kučera [38] showed that they can compute FPF functions. Recall that this means that they can compute a function g with $g(e) \neq \varphi_e(e)$ for all e. The point, however, is that in a construction, the ability to compute a DNC $\{0,1\}$-valued function (i.e. being PA) is a *positive* thing in that by saying what something is *not* we will know what it *is*. In the general case, being FPF only allows us to say what something is not. So this computational power is *negative*.

Kučera was able to prove a significant generalization of the fact that random reals are FPF. We need the following definition.

Definition 4.5: (Jockusch, Lerman, Soare, and Solovay [32]) (a) We define a relation $A \sim_n B$ as follows.

   (i) $A = B$ if $n = 0$.
   (ii) $A =^* B$ if $n = 1$.
   (iii) $A^{(n-2)} \equiv_T B^{(n-2)}$, if $n \geq 2$.

(b) A total function f is called *n-fixed point free* (*n*-FPF) iff for all x,

$$W_{f(x)} \not\sim_n W_x.$$

Kučera was able to extend his earlier result that random reals were FPF with the following.

Theorem 4.13: (Kučera [39]) Suppose that A is $n + 1$ random. Then A computes an n-FPF function.

The proof is not really difficult, but is rather clever, and involves extensive use of Kučera's construction of a universal Martin-Löf test. We refer the reader to [39] or Downey and Hirschfeldt [18]. Strictly speaking the proof only uses weak $n + 1$-randomness.

4.7. *van Lambalgen's Theorem*

In this section, we will study independence results for random reals. We begin with a *central* result concerning randomness, which has turned out to play a very important role in the theory.

We begin with a simple Lemma.

Theorem 4.14: (van Lambalgen, Kurtz, Kautz)

  (i) If $A \oplus B$ is n-random so is A.
  (ii) If A is n-random so is $A^{[n]}$, the n-th column of A.
  (iii) If $A \oplus B$ is n-random, then A is $n - B$-random.

Proof: (i) (e.g. $n = 1$.) If A is not random, then $A \in [\sigma]$ for infinitely many $[\sigma]$ in some Solovay test V. Then $A \oplus B$ would be in $\widehat{V}$, where $[\sigma \oplus \tau] \in \widehat{V}$ for all τ with $|\tau| = |\sigma|$ and $\sigma \in V$. The measure of $\widehat{V}$ is the same as V. (iii) and (ii) are similar.   □

Similar methods also who that if $A \oplus B$ is random, then $A \not\leq_T B$ and hence there are no minimal random degrees. The most important fact is that the *converse* of Theorem 4.14 is also true.

Theorem 4.15: (van Lambalgen's Theorem [92])

  (i) If A n-random and B is $n - A$-random, then $A \oplus B$ is n-random.
  (ii) Hence, $A \oplus B$ is n-random iff A n-random and B is $n - A$-random.

Proof: Suppose $A \oplus B$ is not random. We have $A \oplus B \in \bigcap_n W_n$ where W_n is uniformly Σ_1^0 with $\mu(W_n) \leq 1/2^n$. By passing to a subsequence we may assume that $\mu(W_n) \leq 1/2^{2n}$.

Put

$$U_n = \{X \mid \mu(\{Y \mid X \oplus Y \in W_n\}) > 1/2^n\}.$$

Note that U_n is uniformly Σ_1^0. Moreover $\mu(U_n) \leq 1/2^n$ for all n, because otherwise we would have

$$\mu(W_n) > \mu(U_n) \cdot \frac{1}{2^n} > \frac{1}{2^n} \cdot \frac{1}{2^n} = \frac{1}{2^{2n}},$$

a contradiction.

Since A is random, it follows that $\{n \mid A \in U_n\}$ is finite. Thus for all but finitely many n we have $A \notin U_n$, i.e.,

$$\mu(\{Y \mid A \oplus Y \in W_n\}) \leq 1/2^n.$$

Put $V_n^A = \{Y \mid A \oplus Y \in W_n\}$. Then $\mu(V_n^A) \leq 1/2^n$ for all but finitely many n, and V_n^A is uniformly $\Sigma_1^{0,A}$. Moreover $B \subset \bigcap_n V_n^A$, contradicting the assumption that B is random over A. $\square$

This result will be used extensively in the next Lecture. Here is a really pretty application of van Lambalgen's Theorem.

Theorem 4.16: (Miller and Yu [62]) Suppose that A is random and B is n-random. Suppose also that $A \leq_T B$. Then A is n-random.

Proof: (We do $n = 2$.) If B is 2-random, then B is 1-Ω-random (as $\Omega \equiv_T \emptyset'$.) Hence by van Lambalgen's Theorem, $\Omega \oplus B$ is random. Thus, again by van Lambalgen's Theorem, Ω is 1-B-random. But $A \leq_T B$. Hence, Ω is 1-A-random. Hence $\Omega \oplus A$ is random, again by van Lambalgen's Theorem. Thus, A is 1-Ω-random. That is, A is 2-random.

For the case $n + 1$, there is a 1-random Z with $Z \equiv_T \emptyset^{(n)}$ by Kučera's Coding Theorem. If B is n-random then B is n-Z-random. Thus $B \oplus Z$ is random. Hence Z is n-B-random. Hence Z is n-A-random, as $A \leq_T B$. Thus A is n-random. $\square$

Actually, Miller and Yu [62] have also proven for any (not necessarily random Z), any random below a Z-random is itself Z-random. (Of course, this does not use van Lambalgen.)

A nice corollary of van Lambalgen's Theorem and Sacks' Theorem is the following.

Corollary 4.4: (Kautz [33]) Let $n \geq 2$. Then if **a** and **b** are relatively n-random, they form a minimal pair.

Proof: Suppose that $D \leq_T A, B$. Then $A \in \{E : \Phi_e^E = D\}$. By Sacks' Theorem, this set is a Π_2^D-nullset, and hence A is not 2-D-random, and hence not 2-B-random. $\square$

4.8. *Effective 0-1 Laws*

Classically we know that any measurable class of reals closed under finite translations has measure 0 or measure 1. Stillwell's Theorem and other consideration suggest that there should be refined computability-theoretical analogs of this.

The following lemma is the real basis of these results.

Lemma 4.7: (Kučera-Kurtz, [33], Lemma IV.2.1) Let D be a real, and $n \geq 1$. Let T be a Π_n^D class of positive measure. Then T contains a member of every n-D-random degree. Moreover, if A is any n-D-random, then there is some string σ and real B such that $A = \sigma B$ and $B \in T$.

Proof: For simplicity, we do $n = 1$ with D computable, the general case being analogous. We let T be a Π_1^0 class of positive measure, and $S = \overline{T}$, so that there is a c.e. set of strings W with $S = \cup\{[\sigma] : \sigma \in W\}$. Moreover we can assume W is prefix-free.

Choose rational $r < 1$ with $\mu(S) < r$. We define tests. Let $E_0 = S$ and $E_{s+1} = \{\sigma\tau : \sigma \in E_s \land \tau \in W\}$. Then $\mu(E_s) \leq r^s$ by induction on s. Now suppose that for every B with $A = \sigma B$, $A \in S$. Then $A \in \cap_s E_s$, and hence A is not random. Thus for any random B, there is some σ with $\sigma B \in \overline{S} = T$. $\qquad\square$

I remark that this can also be gotten from the lemma for Kučera Coding. The following is an immediate Corollary to this Lemma.

Corollary 4.5: (Kurtz [44], also Kautz [33])

   (i) Every degree invariant Σ_{n+1}^0-class or Π_{n+1}^0 either contains all n-random sets or no n-random sets.
   (ii) In fact the same is true for any such class closed under translations, and such that for all A, if $A \in S$, then for any string σ, $\sigma A \in S$.

Here are some examples of results proven using this fact.

Corollary 4.6: (Kurtz and others)

   (i) The class $\{A : A$ has non-minimal degree $\}$ has measure 1, and includes every 1-random set.
   (ii) The class $\{A \oplus B : A, B$ form a minimal pair $\}$ has measure 1, and includes all 2-random but not every 1-random set.
   (iii) The class $\{A : \deg(A)$ is hyperimmune $\}$ has measure 1 and includes all 2-random but not every 1-random set.

The proofs of (i) and (ii) are done by analyzing the proofs of almost everywhere behaviour. For instance (i) follows since no random real has minimal degree, and the part for 2-randoms follows by Corollary 4.4 We will prove (iii) in Lecture 5, where we will look at measure-theoretical injury priority arguments.

The second part of (ii) is trickier. It uses a result of Kučera.

Theorem 4.17: (Kučera [37]) If A and B are 1-random with $A, B <_T \emptyset'$ then A and B do not form a minimal pair.

Proof: If **a** is random then we have already seen that it is DNC. Choose a low random below $0'$ and using van Lambalgen's Theorem, get another Turing incomparable with it Choose a low random below $0'$ and using van Lambalgen's Theorem, get another Turing incomparable with it. It is a consequence of Kučera's priority free solution to Post's Problem (in [37]) that no pair of Δ_2^0 of FPF degrees form a minimal pair. □

Hirschfeldt, Nies, and Stephan [31] have shown that the degrees below such pairs are K-trivial.

4.9. Omega operators

Unfortunately, I don't have time to discuss the interesting results looking at Ω, or, rather, *Omegas* as operators on Cantor Space. Here we push the possible analogy that Ω looks like the halting problem. The first question was whether it was degree invariant.

In fact there are reals $A =^* B$ such that Ω^A and Ω^B are relatively random!

There are many other very interesting results such as the fact that that Omega operators are lower semicontinuous but not continuous, and moreover, that they are continuous exactly at the 1-generic reals. These results will hopefully be presented by one of the other co-authors at this conference, and we otherwise refer the reader to Downey, Hirschfeldt, Miller and Nies [20] for details.

5. Lecture 4: Calibrating randomness

Our goal is to try to calibrate levels of randomness and the vehicle we will use is comparing initial segment complexities. The idea is that if the initial segment complexity of a real determines whether it is random, then it also should determine *how* random the relevant real is. Naturally, other approaches suggest themselves, supermartingale growth rates, etc. More on this later.

5.1. *Measures of relative randomness and the Kučera-Slaman Theorem*

One of the operators in classical computability theory is the jump operator. In some sense Ω, the halting *probability*, is a kind of analog to $\emptyset'$, the halting *set*. The fact that $\emptyset'$ is computably unique is Myhill's Theorem. Solovay [83] realized that Ω was apparently machine dependent, and proposed an analytic version of m-reducibility that he showed preserved randomness.

Definition 5.1: (Solovay [83]) We say that a real α is *Solovay reducible* to β (or β *dominates* α), $\alpha \leq_S \beta$, iff there is a constant c and a partial computable function f, so that for all $q \in \mathbb{Q}$, with $q < \beta$,

$$c(\beta - q) > \alpha - f(q).$$

The intuition is that a sequence of rationals converging to β can be used to generate one converging to α at the same rate up to a constant factor. Solovay reducibility is particularly relevant to the measure-theoretical analog of the computably enumerable sets, the computably enumerable *reals*, which may be defined as the measures of the domains of prefix-free machines. If $\alpha \leq_S \beta$, then given a sequence of rationals converging to some β and f we can generate a sequence converging to α. We then get the following characterization of Solovay reducibility.

Lemma 5.1: (Calude, Hertling, Khoussainov, Wang [7]) For c.e. reals, $\alpha \leq_S \beta$ iff for all c.e. $q_i \to \beta$ and c.e. $r_i \to \alpha$, there exists a total computable g, and a constant c, such that, for all m,

$$c(\beta - q_m) > \alpha - r_{g(m)}.$$

The proof of this result applies equally well to a β computable sequence $q_i \to \beta$, and hence this lemma means that the ability to compute a close approximation to β (say the first n bits) allows us to compute α to within some close approximation (say 2^{-n+d}) means that the following holds.

Proposition 5.1: (Folklore) For any reals α and β, if $\alpha \leq_S \beta$, then $\alpha \leq_T \beta$.

Since there are only $O(2^{2d})$ many reals within a radius of 2^{-n+d} of a string representing a rational whose dyadic expansion has length n, it follows that $\leq_S$ has the *Solovay Property*.

Proposition 5.2: (Solovay [83]) If $\alpha \leq_S \beta$ then there is a c such that, for all n,

$$K(\alpha \restriction n) \leq K(\beta \restriction n) + c.$$

The same also holds for C in place of K.

Thus we will regard $\leq_S$ as an initial segment *measure of relative randomness*. We can compactly write the facts expressed in Proposition 5.2 as

$$\leq_S \text{ implies } \leq_C, \text{ and,}$$

$$\leq_S \text{ implies } \leq_K.$$

Notice that, by Schnorr's Theorem, if $\Omega \leq_S \beta$, then β is Martin-Löf random.

There is a particularly nice characterization of $\leq_S$ in terms of $+$ for c.e. reals.

Lemma 5.2: (Downey, Hirschfeldt, Nies) $\alpha \leq_S \beta$, with α and β c.e. reals, iff there is a constant c and a c.e. real γ such that $c\beta = \alpha + \gamma$.

The proof is fairly easy. Roughly, the proof works by synchronizing the enumerations so that the approximation to α is "covered" by one for β, (i.e. $\alpha_{s+1} - \alpha_s$ generates a change in $c\alpha$ of the same order. Then we use the amount needed for this covering for α and the excess goes into $\gamma_{s+1} - \gamma_s$.

The formal statement of the lemma being used is the following.

Lemma 5.3: (Calude, Hertling, Khoussainov, Wang, [7]) For c.e. reals, $\alpha \leq_S \beta$ iff for all c.e. $q_i \to \beta$ there exists a total computable g, and a constant c, such that, for all m,

$$c(\beta - q_m) > \alpha - r_{g(m)}.$$

Solovay reducibility is particularly relevant to computably enumerable reals as we soon see. Recall that such reals are the measures of the domains of prefix-free machines. Of course, if M is prefix-free, then $x = \mu(\mathrm{dom}(M))$ is the limit of the following computable increasing sequence of rations.

$$x = \lim_s x_s \text{ where } x_s = \mu(\mathrm{dom}(M_s)) = \sum_{M_s(\sigma)\downarrow} 2^{-|\sigma|}.$$

Equivalently, that is, we can define a c.e. real as the limit of a nondecreasing sequence of rationals. Solovay called a real β Ω-*like* if β was computably enumerable and $\Omega \leq_S \beta$. He remarked that he thought it was very interesting that many of the properties of Ω (which seems machine dependent) also held for Ω-like reals. The first piece of the puzzle was found by Calude, Hertling, Khoussainov, and Wang.

Theorem 5.1: (Calude, Hertling, Khoussainov, Wang [7]) Suppose that β is a c.e. real and that $\Omega \leq_S \beta$. Then β is a halting probability. That is, there is a universal machine $\widehat{U}$ such that $\mu(\text{dom}(\widehat{U})) = \beta$.

Proof: Here we use Lemma 5.3. Suppose that we are given monotone enumerations $\Omega = \lim_s \Omega_s$ and $\beta = \lim_s \beta_s$, and f and c witnessing $\Omega \leq_S \beta$ so that for all rationals $q < \beta$, $f(q) \downarrow < \Omega$ and $2^c(\beta - q) > \Omega - f(q)$. We will build our machine M in stages using KC-axioms.

It is not hard to use c and f to speed up the enumeration of β so that such that for all s, $2^c(\beta_{s+1} - \beta_s) > \Omega_{s+1} - \Omega_s$. Then at stage s, suppose that $U(\sigma) \downarrow$, so that σ enters the domain of U, and hence $\Omega_{s+1} - \Omega_s \geq 2^{-|\sigma|}$. We can assume that exactly on string enters the domain of U. Then $\beta_{s+1} - \beta_s \geq 2^{-(|\sigma|+c)}$. Hence we build M via the KC-axioms $\langle |\sigma| + c, U(\sigma) \rangle$. The result follows. $\qquad\square$

The final piece of the puzzle was provided by the following lovely result of Kučera and Slaman.

Theorem 5.2: (Kučera and Slaman [40]) Suppose that α is random and c.e. Then for all c.e. reals β, $\beta \leq_S \alpha$.

Proof: Suppose that α is random and β is a c.e. real. We need to show that $\beta \leq_S \alpha$. We enumerate a Martin-Löf test $F_n : n \in \omega$ in stages. Let $\alpha_s \to \alpha$ and $\beta_s \to \beta$ be a computable sequence of rationals converging to β monotonically. We assume that $\beta_s < \beta_{s+1}$. At stage s if $\alpha_s \in F_n^s$, do nothing, else put $(\alpha_s, \alpha_s + 2^{-n}(\beta_{s+1} - \beta_{t_s}))$ into F_n^{s+1}, where t_s denotes the last stage we put something into into F_n. One verifies that $\mu(F_n) < 2^{-n}$. Thus the F_n define a Martin-Löf test. As α is random, there is a n such that for all $m \geq n$, $\alpha \notin F_m$. This shows that $\beta \leq_S \alpha$ with constant 2^n. $\qquad\square$

Corollary 5.1: For c.e. reals α the following are equivalent:

(i) α is Martin-Löf random.
(ii) For all c.e. reals, β, for all n, $K(\beta \restriction n) \leq K(\alpha \restriction n) + O(1)$.
(iii) For all c.e. reals, β, for all n, $C(\beta \restriction n) \leq C(\alpha \restriction n) + O(1)$.

(iv) For any version of Ω, for all n, $C(\Omega \restriction n) \leq C(\alpha \restriction n) + O(1)$ and $K(\Omega \restriction n) = K(\alpha \restriction n) + O(1)$.

(v) For all c.e. reals β, $\beta \leq_S \alpha$.

(vi) α is the halting probability of some universal machine.

Whilst we know that all reals have complexity oscillations, the Kučera-Slaman Theorem says that for c.e. random reals, they all happen in the same places.

By all of this, if a real is both random and c.e. it must be Solovay complete and hence Turing complete. Actually, it has to be Turing complete in a very strong way.

Theorem 5.3: (Downey and Hirschfeldt [18]) Suppose that A is a c.e. set and α is a 1-random c.e. real. Then $A \leq_{wtt} \alpha$, and this is true with identity use.

Proof: Given A and α as above we must construct $\Gamma^\alpha = A$, where $\gamma(x) = x$. Fix a universal prefix-free machine U and using Kraft-Chaitin, and the Recursion Theorem, we will be building a part of U, so that if we enumerate a Kraft-Chaitin axiom $\langle 2^n, \sigma \rangle$ we will know that some τ, σ enters U for some τ of length $e + n$. Since α is 1-random we know that for all n, $K_U(\alpha \restriction n) \geq n - c$ for a fixed c, which we will know for the sake of this construction.

Let $\alpha_s \to \alpha$ be a computable sequence converging to α and let $A = \cup_s A_s$. Initially, we will define $\Gamma^{\alpha_s}(x) = 0$ for all x, and maintain this unless x enters $A_{s+1} - A_s$. As usual at such a stage, we would like to change the answer from 0 to 1. To do this we will need $\alpha \restriction x \neq \alpha_s \restriction x$. Should we see a stage $t \geq s$ with $\alpha_t \restriction x \neq \alpha_s \restriction x$ then we can so change the answer. For $x > e + c + 2$, we can force this to happen. We simply enumerate an axiom $\langle 2^{x-c-e-1}, \alpha_s \restriction x \rangle$ into our machine, causing a description of $\alpha_s \restriction x$ of length $x - c - 1$ to enter $U - U_s$, and hence $\alpha \restriction x \neq \alpha_s \restriction x$. We can simply wait for this to happen, correct Γ and move to the next stage. $\square$

Similar methods can be used to prove the following (the first being originally established by Arslanov's Completeness Criterion).

Theorem 5.4: (Kučera [36]) Suppose that A is a random set of c.e. degree. Then A is Turing complete.

Theorem 5.5: (Downey and Hirschfeldt [18]) Suppose that A is a random set of c.e. wtt-degree. Then A is wtt-complete.

Downey, Hirschfeldt and Nies [21], and Downey, Hirschfeldt and LaForte [19] were motivated to look at the structure of c.e. reals under Solovay reducibility. They proved several results, but the structure remains largely unexplored.

Theorem 5.6: (Downey, Hirschfeldt and Nies [21]) The Solovay degrees of c.e. reals forms a distributive upper semilattice, where the operation of join is induced by $+$, arithmetic addition (or multiplication) (namely $[x] \vee [y] \equiv_S [x + y]$.)

Theorem 5.7: (Downey, Hirschfeldt and Nies [21]) If $[\Omega] = \mathbf{a} \vee \mathbf{b}$ then either $[\Omega] = \mathbf{a}$ or $[\Omega] = \mathbf{b}$.

Theorem 5.8: (Downey and Hirschfeldt [18]) There exist c.e. *sets* A and B such that the Solovay degrees of A and B have no infimum in the (global) Solovay degrees.

To prove (i) note that $x, y \leq_S z$ implies there is a triple c, p, q such that $cz = x + p = y + q$. So $2cz = (x + y) + (p + q)$. So $x + y \leq_S z$. Clearly $x, y \leq x + y$. Similarly to prove (ii), roughly, we work as follows. Suppose $z \leq_S x_1 + y_1$. Use Lemma 5.3 to run the enumerations of and cover the $z_{s+1} - z_s$ using bits of $x_{1,s+1} - x_s, y_{s+1} - y_s$.

The proof of density and join inaccessibility are significantly more intricate.

5.2. *The Density Theorem*

Theorem 5.9: (Downey, Hirschfeldt and Nies [21]) The Solovay degrees of c.e. reals are dense. Indeed the following hold.

(i) If $\mathbf{a}$ is incomplete and $\mathbf{b} <_S \mathbf{a}$, then there exist $\mathbf{a_1}|_S\mathbf{a_2}$ such that $\mathbf{b} < \mathbf{a_1}, \mathbf{a_2}$, and $\mathbf{a} = \mathbf{a_1} \vee \mathbf{a_2}$. That is every incomplete degree splits over all lesser ones.

(ii) If $[\Omega] = \mathbf{a} \vee \mathbf{b}$ then either $[\Omega] = \mathbf{a}$ or $[\Omega] = \mathbf{b}$.

Proof: (ii) is a straightforward finite injury argument. (i) is a finite injury argument with some quite novel features. We want to build β^0 and β^1 such that

- $\beta^0, \beta^1 \leq_S \alpha$,
- $\beta^0 + \beta^1 = \alpha$, and

- the following requirement is satisfied for each $e, k \in \omega$ and $i < 2$:

$$R_{i,e,k} : \Phi_e \text{ total} \Rightarrow \exists n(\alpha - \alpha_{\Phi_e(n)} \geq k(\beta^i - \beta^i_n)).$$

There are very interesting timing problems and these can be seen with a two requirement scenario, which we discuss below.

- $R_0 ; \Phi$ total $\Rightarrow \exists n(\alpha - \alpha_{\Phi(n)} \geq h(\beta^0 - \beta^0_n))$
- $R_1 : \Psi$ total $\Rightarrow \exists n(\alpha - \alpha_{\Psi(n)} \geq l(\beta^1 - \beta^1_n))$

We will be measuring whether Φ and Ψ are total and only work when this appears so. Thus, without loss of generality, we will assume that Φ and Ψ are total. The reader should imagine the construction as follows. We have

- two containers, labeled β^0 and β^1, and
- a large funnel, through which bits of α are being poured.

R_0 and R_1 fight for control of the funnel. In particular, bits of α must go into the containers (because we want $\beta^0 + \beta^1 = \alpha$) at the same rate as they go into α (because we want $\beta^0, \beta^1 \leq_S \alpha$). However, each R_i wants to funnel enough of α into β^{1-i} to be satisfied.

As R_0 is stronger, it could potentially put all of α into β^1, but that would leave R_1 unsatisfied. The trouble comes from trying to recognize when enough of α has been put into β^1 so that R_0 is satisfied.

Definition 5.2: R_0 is *satisfied through n at stage s* if $\Phi(n)[s] \downarrow$ and $\alpha_s - \alpha_{\Phi(n)} > k(\beta^0_s - \beta^0_n)$.

To achieve satisfaction, the idea is that R_0 sets a quota for R_1 (how much may be funneled into β^0 from that point on). If the quota is 2^{-m} and R_0 finds that either

- it is unsatisfied or
- the least number through which it is satisfied changes,

then it sets a new quota of $2^{-(m+1)}$ for how much may be funneled into β^0 from that point on.

Lemma 5.4: There is an n through which R_0 is eventually permanently satisfied, that is,

$$\exists n, s \, \forall t > s \, (\alpha_t - \alpha_{\Phi(n)} > k(\beta^0_t - \beta^0_n)).$$

Proof: (of Lemma) Suppose not. Then R_1's quota $\to 0$, so β^0 is computable. Also, $\forall n, s \, \exists t > s \, [\alpha_t - \alpha_{\Phi(n)} \leq k(\beta^0_t - \beta^0_n)]$. So $\forall n \, [\alpha - \alpha_{\Phi(n)} < (k+1)(\beta^0 - \beta^0_n)]$. Thus $\alpha \leq_S \beta^0$ is computable. Contradiction. $\square$

Thus the strategy above yields a method for meeting R_0. At the end of this process, R_0 is permanently satisfied, and R_1 has a final quota 2^{-m} that it is allowed to put into β^0.

Now we hit the crucial problem, precisely where we need incompleteness for α. If R_1 waits until a stage s s.t. $\alpha - \alpha_s < 2^{-m}$ then it can put all of $\alpha - \alpha_s$ into β^0 and R_1 will, in turn, be satisfied.

The problem is that R_1 *cannot tell when such an s arrives*. If R_1 is too quick, it may find itself unsatisfied and unable to do anything about it since it will have used all of its quota *before s* arrives.

The key new idea is that

R_1 *uses Ω as an investment advisor.*

Let s be the stage at which R_1's final quota of 2^{-m} is set. At each stage $t \geq s$, R_1 puts as much of $\alpha_{t+1} - \alpha_t$ into β^0 as possible so that the total amount put into β^0 since stage s *does not exceed* $2^{-m}\Omega_t$. The total amount put into β^0 after stage s is $\leq 2^{-m}\Omega < 2^{-m}$, so the quota is respected. We finish the proof with the following Lemma

Lemma 5.5: There is a stage t after which R_1 is allowed to funnel all of $\alpha - \alpha_t$ into β^0.

Proof: It is enough that $\exists u \geq t \geq s \; \forall v > u \; (2^{-m}(\Omega_v - \Omega_t) \geq \alpha_v - \alpha_t)$. Suppose not. Then $\forall u \geq t \geq s \; \exists v > u \; [\Omega_v - \Omega_t < 2^m(\alpha_v - \alpha_t)]$. Thus $\forall t \geq s \; [\Omega - \Omega_t \leq 2^m(\alpha - \alpha_t)]$. So there is a d s.t. $\forall t \; [\Omega - \Omega_t < d(\alpha - \alpha_t)]$, and hence $\Omega \leq_S \alpha$. Contradiction. $\square$

Downey and Hirschfeldt [18] have shown that the proof can be made to work for any Σ_3^0 measure of relative randomness where $+$ is a join, the 0 degree includes the computable reals, and the top degree is Ω.

5.3. *Other measures of relative randomness*

We remark that there are a number of other interesting measures of relative randomness, aside from $\leq_S$ and the fundamental initial segment ones $\leq_C$ and $\leq_K$. They include

(i) $A \leq_{cl} B$ iff there is a c and a wtt procedure Γ with use $\gamma(n) = n+c$, and $\Gamma^B = A$. (We met this in Theorem 5.3.) If $c = 0$, then this is called *ibT*-reducibility and is the one used by Soare and Csima in differential geometry, such as Soare [81].

(ii) $A \leq_{rK} B$ means that there there is a c such that for all n,
$$K((A \restriction n)|(B \restriction n + c)) = O(1).$$

I don't really have room to discuss all the results but will try to give a flavour. cl-reducibility[g] says that not only is the case that the initial segments converge at more of less the same rate, but there is an *efficient* way to convert the *bits* of B into those of A.

Thus in spite of the fact that the Kučera-Slaman Theorem says that all versions of Ω are the same in terms of their S-degrees, Yu and Ding [97] established the following.

Theorem 5.10: (Yu and Ding [97])

   (i) There is no cl-complete c.e. real.

   (ii) There are two c.e. reals β_0 and β_1 so that there is no c.e. real α with $\beta_0 \leq_{cl} \alpha$ and $\beta_1 \leq_{cl} \alpha$.

The proof roughly works by picking two long intervals $\beta_0 \upharpoonright [n, n+t], \beta_1 \upharpoonright [n, n+t]$, to diagonalize against some α and cl reduction Γ_e with use $n + e$. Initially the reals have $\beta_i(k) = 0$ on this interval. Our procedure will work by alternating between β_0 and β_1 adding $2^{-(n+t)}$ each time, where time here means "expansionary stages."

The key observation of Yu and Ding here is that this simple procedure can drive α to be large. This is proven by induction, the nicest proof being due to Bamparlias and Lewis [3]. The main lemma is to show that α's best strategy is the obvious one of *least effort* (nomenclature of Bamparlias and Lewis [3]) where α changes as far right as it can. You argue that if α' works then so would α given by this least effort strategy. This is proven by induction.

Here is an example which would show why the least effort strategy won't work.

stage 1: $\beta_{0,1} = 0.001$, $\beta_{1,1} = 0$ and $\alpha_1 = 0.001$

stage 2: $\beta_{0,2} = 0.001$, $\beta_{1,2} = 0.001$ and $\alpha_2 = 0.010$

stage 3: $\beta_{0,3} = 0.010$, $\beta_{1,3} = 0.001$ and $\alpha_3 = 0.100$

stage 4: $\beta_{0,4} = 0.010$, $\beta_{1,4} = 0.010$ and $\alpha_4 = 0.110$

stage 5: $\beta_{0,5} = 0.011$, $\beta_{1,5} = 0.010$ and $\alpha_5 = 0.111$

stage 6: $\beta_{0,6} = 0.011$, $\beta_{1,6} = 0.011$ and $\alpha_6 = 1.000$

stage 7: $\beta_{0,7} = 0.100$, $\beta_{1,7} = 0.011$ and $\alpha_7 = 1.100$

stage 8: $\beta_{0,8} = 0.100$, $\beta_{1,8} = 0.100$ and $\alpha_8 = 10.000$

---

[g]This reducibility was originally called sw-reducibility by Downey, Hirschfeldt and LaForte, for "strong weak truth table reducibility", but in keeping with the spirit of recent usage, we have adopted the more sensible notion suggested by Barmpalias and Lewis of *computable Lipschitz reducibility*.

Similar arguments show the following.

Theorem 5.11: (Barmpalias and Lewis [3]) There is a c.e. real α such that for any random c.e. real β, $\alpha \not\leq_{cl} \beta$.

An argument of a different kind shows the following.

Theorem 5.12: (Hirschfeldt, unpubl) There is a real α such that for all random reals β, $\alpha \not\leq_{cl} \beta$.

We have seen that if W is a c.e. set the $W \leq_{cl} \Omega$. Also on the c.e. sets, S and cl coincide. Little else is known about $\leq_{cl}$.

The reducibility rK is implied by either of cl and S. It has many of the best features. By unpublished work of Downey, Greenberg, Hirschfeldt, and Miller it is not implied by $\leq_C$ even on the c.e. reals. It is an appropriate Σ_3^0 reducibility and hence it is dense on the c.e. reals. It implies $\leq_T$. Little else is known about $\leq_{rK}$.

Other reducibilities include $\leq_{Km}$, $\leq_{KM}$ (i.e. supermartingale reducibility), but these are more or less completely unexplored, save the fact that $\leq_{SM}$ is dense on the c.e. reals.

5.4. $\leq_C$ and $\leq_K$

Recall $A \leq_K B$ means that, for all n, $K(A \restriction n) \leq K(B \restriction n) + O(1)$, and similarly $\leq_C$.

Thanks mainly to the work of Miller and Yu, we know some basic facts about the $\leq_K$ degrees of *random reals*. The first thing we discover about $\leq_C$ and $\leq_K$ is that they are not really "reducibilities" though we will confusingly call them such.

Theorem 5.13: (Yu, Ding, Downey [99]) For $Q \in \{K, C\}$, $\{X : X \leq_Q Y\}$ has size $2^{\aleph_0}$ and has members of each degree, whenever Y is random.

The proof is relatively easy. Take a sufficiently computable sparse set X. Then it's complexity is eventually always well below a random real. But any degree can be coded into a really sparse set. The replacement for this theorem is a measure-theoretical one:

Theorem 5.14: (Yu, Ding, Downey [99]) For any real A, $\mu(\{B : B \leq_K A\}) = 0$. Hence there are uncountably many K degrees.

Yu and Ding improved this result to construct $2^{\aleph_0}$ many K-degrees directly ([98]), but this can also be obtained by observing that $\leq_K$ is Borel, and hence a theorem of Silver can be used. Using direct analysis of the the fluctuations of initial segment complexities Yu, Ding, and Downey obtained the following.

Theorem 5.15: (Yu, Ding, Downey [99]) For all c and $n < m$,

$$(\exists^{\infty} k)\left[K(\Omega^{(n)} \restriction k) < K(\Omega^{(m)} \restriction k) - c\right].$$

For $n = 0, m = 1$ Theorem 5.15 was proven by Solovay [83], using totally different methods. Using van Lambalgen's Theorem as a base. more powerful results have recently been obtained by Miller and Yu. Recall that van Lambalgen's Theorem states: *B n-random and A is B-n-random iff* $A \oplus B$ *is n-random.*

Definition 5.3: (Miller and Yu [62]) We say that $\alpha \leq_{vL} \beta$, α is *van Lambalgen*[h] *reducible* to β if for all $x \in 2^{\omega}$, $\alpha \oplus x$ is random implies $\beta \oplus x$ is random.

Theorem 5.16: (Miller and Yu [62]) For all random α, β,

  (i) α n-random and $\alpha \leq_{vL} \beta$ implies β is n-random.
  (ii) If $\alpha \oplus \beta$ is random then α and β have no upper bound in the vL-degrees.
  (iii) If $\alpha \leq_T \beta$ and α is 1-random, then $\beta \leq_{vL} \alpha$.
  (iv) There are random $\alpha \equiv_{vL} \beta$ of different Turing degrees.
  (v) There are no maximal or minimal random vL-degrees, and no join.
  (vi) If $\alpha \oplus \beta$ is random then $\alpha \oplus \beta <_{vL} \alpha, \beta$.
  (vii) The Σ_1^0 theory of the vL-degrees is decidable.

The proofs of most of these are relatively easy once you figure out what to do. For instance, suppose that α n-random and $\alpha \leq_{vL} \beta$. Use Kučera's Theorem that there is a random z with $z \equiv_T \emptyset^{(n-1)}$. Then $\alpha \oplus z$ is random, and hence $\beta \oplus z$ is random and hence β is 1-z-random, that is β is n-random.

Furthermore, Miller and Yu show that $\Omega^{(n)}$ and $\Omega^{(m)}$ have no upper bound in the vL degrees for $n \neq m$. This improves the Yu, Ding, Downey (Theorem 5.15) result above.

---

[h]This is closely related to a relation introduced by Nies: He defined $A \leq_{LR} B$ if for all Z, Z is 1-B-random implies Z is 1-A-random. If A and B are both random then $A \leq_{LR} B$ iff $B \leq_{LR} A$. Note that $\leq_{vL}$ is really only interesting on the randoms, but there are a number of possible extensions which also make sense on the non-randoms. These have yet to be explored.

All of this is filters through an interesting relationship between $\leq_{vL}$ and $\leq_C, \leq_K$.

Lemma 5.6: (Miller and Yu [62]) For random α, β,

(i) Suppose that $\alpha \leq_K \beta$. Then $\alpha \leq_{vL} \beta$.
(ii) Suppose that $\alpha \leq_C \beta$. Then $\alpha \leq_{vL} \beta$.

We state the following for $\leq_K$ but they hold equally for $\leq_C$. (Most of the proofs are in this area work equally well for $\leq_C$ and $\leq_K$. There are occasional cases where the proof needs considerably more work in the $\leq_C$ case, such as the proof of Lemma 5.6(ii) above.)

Corollary 5.2: (Miller and Yu [62])

(i) Suppose that $\alpha \leq_K \beta$, and α is n-random and β is random. Then β is n-random.
(ii) If $\alpha \oplus \beta$ is 1-random, then $\alpha |_K \beta$ and have no upper bound in the K-degrees.
(iii) For all $n \neq m$, the K-degrees of $\Omega^{(n)}$ and $\Omega^{(m)}$ have no upper bound.

Miller and Yu provided a natural way to demonstrate that the vL-degree and the K-degrees differ on randoms. The following result should be compared with Theorem 5.16 (vi).

Theorem 5.17: (Miller and Yu [62]) If $\alpha \oplus \beta$ is 1-random, then $\alpha |_K \alpha \oplus \beta$.

A simple application yields Theorem 5.14, that $\mu\{\beta : \beta \leq_K \alpha\}$ for random α is zero. Namely, if β is 1-α-random, then $\beta \not\leq_{vL} \alpha$ and hence, since $\mu(\{\beta : \beta \text{ is 1-}\alpha\text{-random }\}) = 1$, we get $\mu(\{\beta : \beta \leq_K \alpha\}) = 0$, since $\leq_K$ implies $\leq_{vL}$.

Using new techniques and extensions of the above, Miller has proven the following.

Theorem 5.18: (Miller [58])

(i) If α, β are random, and $\alpha \equiv_K \beta$, then $\alpha' \equiv_{tt} \beta'$. As a consequence, every K-degree of a random real is countable.
(ii) If $\alpha \leq_K \beta$, and α is 3-random, then $\beta \leq_T \alpha \oplus \emptyset'$.

Notice that (ii) implies that the cone of K-degrees above a 3-random is countable. Miller and Yu have constructed a 1-random whose K-upper cone is uncountable. This last result uses their method of constructing K-comparable reals. Its proof uses the following clever lemma. The current proof of Lemma 5.19 is quite difficult.

Theorem 5.19: (Miller and Yu) Suppose that $\sum_n 2^{-f(n)} < \infty$, then there is a 1-random Y with

$$K(Y \upharpoonright n) < n + f(n),$$

for almost all n.

Then to get K-comparable reals, use the result taking $g(n) = K(B \upharpoonright n) - n$ for random B. This function is convergent by the Ample Excess Lemma. Now we use Theorem 5.19 on some convergent function f with $g - f \to \infty$.

We remark that Miller has also constructed an uncountable (non-random) K-degree.

Finally Miller has shown that if α is 3-random then its often useless as an oracle. We will call α *weakly-low for K* if $(\exists^\infty n)[K(n) \le K^\alpha(n) + O(1)]$. Thus in a weakly-low real, the information in α is so useless that it cannot help to compress n. The following result echoes the theme articulated by Stephan that most random reals have little *usable* information in them.

Theorem 5.20: (Miller [58])

  (i) If α is 3-random it is weakly-low.
  (ii) If α is weakly-low, and random, then α is strongly Chaitin random in that

$$(\exists^\infty n)[K(\alpha \upharpoonright n) \ge n + K(n) - O(1)].$$

5.5. *Outside of the randoms*

Little is known about $\le_C$ and $\le_K$ outside of the random reals. On the c.e. reals $\le_C$ and $\le_K$ are appropriate Σ_3^0 measures of relative randomness, and hence the following is true.

Theorem 5.21: (Downey and Hirschfeldt [18]) The C- and K- degrees of c.e. reals form dense uppersemilattices.

We remark that this needs the result of Downey, Hirschfeldt, Nies, Stephan [22] that $+$ is a join. To see this, given $x, y < z$, run the enumerations, and have one z-program if $x_s \upharpoonright n$ stops first, and one if $y_s \upharpoonright n$ first.

Other than this, Frank Stephan has shown that $\leq_C$ implies $\leq_T$ on the c.e. reals. (see [18]). This generalizes earlier theorems of Loveland and of Chaitin.

Theorem 5.22: (Loveland [50]) Suppose that for all n, $C(X \upharpoonright n|n) = O(1)$. Then X is computable.

Theorem 5.23: (Chaitin [8]) $A \leq_C 1^\omega$ iff A is computable.

Proof: (Sketch) All of these use the "Π_1^0 class method" each time. They use the fact that a Π_1^0 class with a finite number of paths has computable computable paths. Take Loveland's Theorem, for instance. Since there are only finitely many programs to consider for the $C(X \upharpoonright n|n) = O(1)$. Knowing these and the maximum hit infinitely often will allow for the construction of the Π_1^0 class.

Chaitin's Theorem uses the same method but needs the combinatorial fact below.

Lemma 5.7: (Chaitin [8]) $|\{\sigma : C(\sigma) \leq C(n) + d \wedge |\sigma| = n\}| = O(2^d)$.

Since we know that between n and 2^n there are C-random lengths with $C(n) = \log n$, we can then apply the Lemma to construct the Π_1^0 class. Finally, Stephan's result is similar, but uses a relativization plus enumeration argument.   □

5.6. $\leq_K$ and $\leq_T$

This naturally leads us to ask whether $\leq_K$ implies $\leq_T$ for c.e. reals. As it turns out the implication fails at the very first place it can.

Definition 5.4: We say that A is K-*trivial* iff there is a constant d such that for all n, $K(A \upharpoonright n) \leq K(n) + d$. We would write $A \in KT(d)$.

Theorem 5.24: (Solovay [83]) There are noncomputable K-trivial reals.

As we will see, such reals can even be c.e. sets. The reader might well ask, "what about the Π_1^0 class method?" The Π_1^0 class method kind of works, except to construct the tree, we don't know the K-complexity of any n, like we do for C. For C, we know "$\log n$" is the answer in long enough interval. (That is, between m and 2^m there is a random length and its C-complexity will be $\log n$.) But in the prefix-free case, the tree is only Δ_2^0 since $\emptyset'$ can construct it. Now the same methods work (this time using the Coding Theorem to bound the width). We get the following.

Theorem 5.25: (Chaitin [9], Zambella [100]) There are only $O(2^d)$ members of $KT(d)$. They are all Δ_2^0.

The reader might wonder with the nice computable bound how many K-trivial reals there are. Let $G(d) = |\{X : X \in KT(d)\}|$. Then there is a crude estimate that $G(d) \leq_T \emptyset'''$. This is the best upper bound known. In unpublished work, Downey, Miller and Yu have shown that $G(d) \not\leq_T \emptyset'$, using the fact that $\sum_d \frac{G(d)}{2^d}$ is convergent. This is all related to the Csima-Montalbán functions. We say that f is a *Csima-Montalbán function* if f is nondecreasing and

$$K(A \upharpoonright n) \leq K(n) + f(n) + O(1)$$

implies that $A \upharpoonright n$ is K-trivial. Such functions can be constructed from $\emptyset'' \oplus G$. We define f to be *weakly* Csima-Montalbán, if we weaken the hypothesis to be that $\liminf_n f(n) \to \infty$. Little is known here. It is not known if the arithmetical complexity of f depends upon the universal machine chosen.

Returning to our story, it is very easy to construct a K-trivial real, using the "cost function" method. This construction has been well-reported, and we note it below.

Let

$$A_{s+1} = A_s \cup \{x : W_{e,s} \cap A_s = \emptyset \wedge x \in W_{e,s}$$

$$\wedge \sum_{x \leq j \leq s} 2^{-K(1^j)[s]} < 2^{-(e+1)}\}.$$

Then A is noncomputable and K-trivial. Work of Nies shows that, basically, this is the *only* way to construct such sets.

The K-trivial reals form a fascinating class of reals and has many unexpected properties. The first such property to be established was the following.

Theorem 5.26: (Downey, Hirschfeldt, Nies, Stephan [22]) If A is K-trivial then A is Turing incomplete.

The proof of this result is far from easy, and uses the "decanter" method. This decanter/golden run method has been refined and elaborated through the deep work of Nies, and Hirschfeldt-Nies, (especially [66, 68]) and later Hirschfeldt-Nies-Stephan to establish a number of amazing results, using "golden run" constructions. For example:

Theorem 5.27:

(i) (Nies) Every K-trivial is tt-bounded by a K-trivial c.e. set.

(ii) (Nies) Every K-low is superlow, and "traceable".

(iii) (Nies) K-trivial = low for randomness, meaning that A-random iff random.

(iv) (Hirschfeldt-Nies) K-trivial=low for K, meaning that $K^A(\sigma) = K(\sigma) + O(1)$, for all σ.

(v) (Nies) K-trivials are closed under T-reducibility and form the only known natural Σ_3^0 ideal in the Turing degrees.

(vi) (Nies) They are bounded above by a low$_2$ degree.

Hirschfeldt, Nies and Stephan [31] have shown that if A and B are Δ_2^0 random then there are K-trivials are T-below both of them. I simply don't have time in these lectures to tell the "trivial story" which is still being worked out.

I remark that there are many open questions about this amazing class. Here is one of my favourites. Is there a low (necessarily non-c.e.) degree above all the K-trivials. The reason that this is of interest is that if such a real could be random then it would be a genuinely new strategy for coding into randoms, as has been pointed out by Kučera.

5.7. *Hausdorff Dimension*

We return to the theme of calibrating randomness. This time we will try another approach. Consider Ω. It is random. What about $a_0 0 a_1 0 \ldots$ where $\Omega = a_0 a_1 \ldots$. Then probably we'd expect it to be "half random."

There is an an effectivization of a well-known theory that makes this correct. A refinement of the class of measure zero sets is given by the theory of Hausdorff Dimension. In 1919 Hausdorff [30] generalized earlier work of Carathéodory to define a notion of an s-dimensional measure to include non-integer values.

The basic idea is that you replace measure by a kind of generalized measure, where $\mu([\sigma])$ is replaced by $2^{-s|\sigma|}$ where $0 < s \leq 1$. With $s = 1$ we get normal Lebesgue measure. For $s < 1$ we get a refinement of measure zero. Very informally, it can be shown that there is some limsup where the s-measure is not zero, and this is called the Hausdorff dimension of the set. One can translate this cover version into a s-gale (a version of martingales) definition in the same way that it is possible to frame Lebesgue measure in terms of martingales.

In any case, for the effective version through the work of Lutz, Mayordomo, Hitchcock, Staiger and others we find that the notion corresponds to $\liminf_n \frac{K(A\restriction n)}{n}$, and can take that as a *working definition* of effective Hausdorff dimension. (This sentence does not do justice to the work done by these and other authors in this area.) With this definition, one can easily see that the thinned version of Ω above gets dimension $\frac{1}{2}$. Actually it is $\frac{1}{2}$-random in the sense of Tadaki [89].

In these lectures I don't have the space to talk about all the nice results developed concerning effective Hausdorff dimension (and effective packing and box counting dimension). I simply want to point out that there are other ways to calibrate randomness. One very nice question here is the following. Is there a degree **a** which has a member of nonzero dimension, and yet bounds no random degree? .

I would like to remark that Lutz [51,52] has even developed a notion of dimension for individual strings. Again we see that it is immediately related to prefix-free complexity. The approach is to replace s-gales by "termgales" which are the analogues of s-gales for terminated strings. (You have a third bet that you have reached the end of the string.) One of the results he has proven is the following.

Theorem 5.28: (Lutz [51,52]) Let d denote the discrete dimension obtained using "termgales" as alluded to above. There is a constant $c \in \mathbb{N}$ such that for all $\sigma \in 2^{<\omega}$,

$$|K(\sigma) - |\sigma| \dim(\sigma)| \le c.$$

I refer the reader to Lutz [51,52], Staiger [84,85], Terwijn [90], Reimann [74,75], Downey, Merkle, Reimann [25], Tadaki [89], and Downey and Hirschfeldt [18], for more details.

6. Lecture 5: Measure-theoretical injury arguments

6.1. *Risking measure*

In the last section we met the effective 0-1 laws. These laws can be exploited to prove various results about random degrees by making constructions which work on sets of positive measure. This basic idea goes back to unpublished work of Martin (as we see below) and of Paris [72,73]. The idea was beautifully exploited by Kurtz in the later chapters of his PhD Thesis [44]. These results tend to give theorems that hold for almost all degrees. With more care, we can calculate exactly the level needed to make the constructions work and deduce that they hold for all, for example, 2-random reals, as in Kautz [33].

6.2. *2-random degrees are hyperimmune*

We begin this section with the proof of Martin's Theorem 6.1 The main idea is called "risking measure," in the words of Kurtz [44], and results in a measure-theoretical version of the priority method. The idea is that if we *fail* to meet some requirement then we will fail to define some object under construction (typically, a Turing functional) on some small measure of Cantor space. However, things are arranged that this is small enough that we only fail on a set whose measure is bounded, and then after the fact argue that the result holds for almost all reals by the effective 0-1 laws.

Theorem 6.1: (Martin, unpubl.) Almost every degree is a hyperimmune degree.

Proof: By the zero-one law it is enough to prove that $\mu(\{A : A$ bounds a hyperimmune degree $\}) > 0$. To achieve this we construct a partial computable functional Ξ so that

$$\mu(\{A : \Xi^A \text{ is total and not dominated by any computable function}\} \geq \frac{1}{2}.$$

For this construction, it is convenient to consider Ξ as a partial computable function from strings to strings. This will be a reduction provided that $\Xi(\nu) \downarrow$, $\nu \preceq \widehat{\nu}$ and $\Xi(\widehat{\nu}) \downarrow$, then $\Xi(\nu) \subseteq \Xi(\widehat{\nu})$. We remark that it is not necessarily true that if $\Xi(\nu) \downarrow$ then $\Xi(\gamma) \downarrow$ for all $\gamma \subseteq \nu$.

On a set of positive measure, we will meet the requirements :

$$R_e : \varphi_e \text{ total } \to \Xi^A \text{ is not majorized by } \varphi_e.$$

Now suppose that we *knew* whether φ_e was total. Then we could proceed as follows. We could begin with λ, the empty string. If φ_0 was not total, then we would need to do nothing else and could proceed to treating R_1. Otherwise, we could select some witness n_0, compute $\varphi_0(n_0)$ and then set $\Xi^{\lambda}(n_0) = \varphi_0(n_0) + 1$. Then, of course, φ_0 cannot majorize Ξ^A for any A. We could then use a new witness for φ_1, etc.

The trouble is that we cannot decide whether φ_0 is total, and of course even for a witness n_0, we can't decide if $\varphi_0(n_0) \downarrow [s]$ for some s.

The idea then is to try to implement this process in such a way that should we fail to meet some R_e then the overall cost (in terms of measure) will be small. (Here small will be $2^{-(e+2)}$.)

For the sake of R_0 we will work as follows. Initially, we will devote the string 00 to trying to force φ_0 not to be total. By this we mean that we first pick a witness $n_0 = n_0^{00}$, for the cone [00] of strings σ extending 00.

We will not define $\Xi^\sigma(n_0)$, or indeed Ξ^σ at all for such σ until we see some stage s where $\varphi_0(n_0^{00}) \downarrow [s]$. At this stage we will then define

$$\Xi^{00}(n_0) = \varphi_0(n_0) + 1.$$

The *point* is that now for any strings σ extending 00, $\Xi^\sigma(n_0) = \varphi_0(n_0) + 1$. (And we would make sure that $\Xi^{00}(m)$ is defined for all $m \leq n_0^{00}$ so as to make Ξ^σ total on a set of a of positive measure.) Hence R_0 is met within the cone [00] no matter how else we define Ξ, provided that it is done in a way consistent with Ξ^{00}.

Notice that if we *fail* to define Ξ^{00}, then n_0^{00} is a witness to the fact that φ_0 is not total. R_0 is therefore met globally. The cost of meeting R_0 this way is that Ξ^A will not be defined on a set of measure $2^{-|00|} = 2^{-2}$.

Now whilst we are devoting [00] to testing R_0, we cannot stop Ξ^ν being defined for strings *not* extending 00. Thus, for instance, by the stage s that we get to define $\Xi^{00}(n_0) \downarrow [s] = \varphi(n_0) + 1$, we might well have defined $\Xi^\sigma(m)$ for various σ extending 01.

The idea is that once we succeed for R_0 in [00] we will process R_0 in the cone [01] (and then [10], then [11], always risking 2^{-2} each time). Thus R_1 will assert control of [01] taking control of all extensions of 01 of length s. There will be 2^{s-2} such extensions $\nu_1 \ldots \nu_{2^{s-2}}$ and note that

$$\sum_{i \leq i \leq 2^{s-2}} 2^{-|\nu_i|} = 2^{-2}.$$

Then the idea is that R_0 will pick a new (large, fresh) witness n_0^{01} which will serve for *all* of the ν_i's. Diagram 1 below might be helpful here.

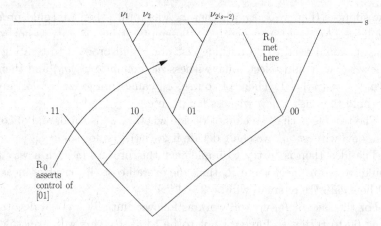

Fig. 1. R_0 changes to the next cone

Then if we ever see a stage $t > s$ such that $\varphi_0(n_0^{01}) \downarrow [t]$, we will be free to define $\Xi^{\nu_i}(n_0^{01}) \downarrow [t] = \varphi_0(n_0^{01}) + 1$, for *all* ν_i simultaneously.

Then R_0 could assert control of $[10]$, via all strings of length t and take a new witness for this cone, etc.

The point is that either R_0 will eventually get stuck on one of $[00], [01], [10], [11]$ at a cost of $\frac{1}{4}$, or we will diagonalize φ_0 in each cone via the cone's witness, in which case R_0 has *no* effect on the measure where Ξ is defined.

The inductive strategies are defined similarly. For R_e choose the collection of strings $\sigma_1, \ldots, \sigma_{2^{2+e}}$, the antichain of strings of length $e + 2$ to take the role of $00, 01, 10, 11$ as above. The goal would be, first process $R - e$ in $[\sigma_1]$ then $[\sigma_2]$, etc. Of course we would let the priorities sort this out. Whilst some R_j is asserting control of some $[\nu] \supset [\sigma_i]$ for $j < e$, we would let R_e assert control of the first available $[\sigma_k]$. In this way, we see that at any stage each R_e has exactly $2^{-(e+2)}$ of measure risked because of it, of $R - e$ is completely met. Thus the overall measure where Ξ is not defined is bounded by $\sum_{e \in \mathbb{N}} 2^{-(e+2)} = \frac{1}{2}$. $\qquad \square$

This result can also be obtained by the following elegant argument of Nies, Stephan, and Terwijn.

Proof: (Nies, Stephan, and Terwijn [69]) Recall that in Lemma 4.3, it is shown that *A is 2-random iff for all computable time bounds g with $g(n) \geq n^2$,*

$$\exists d \exists^\infty n (C^g(A \upharpoonright n) \geq n - d).$$

Define

$$f(k) = \mu n [\exists p_1, \ldots, p_k \leq n](C^g(A \upharpoonright p_i) \geq p_i - d).$$

Then f is A computable. We claim that f is not dominated by any computable function. Suppose that $f(k) \leq h(k)$ for all k. Define the Π_1^0 class

$$P = \{Z : \forall k [\exists p_1, \ldots, p_k \leq h(k)](C^g(Z \upharpoonright p_i) \geq p_i - d)\}.$$

Then $P \neq \emptyset$, and every member of P is 2-random. But then there is a 2-random real that is, for instance, of low degree. $\qquad \square$

6.3. *Almost every degree is CEA*

We use the ideas of the Martin's proof above to establish the following remarkable result of Kurtz, with the observation of Kautz that 2-randomness is enough for the proof.

Theorem 6.2: (Kurtz [44], Kautz [33]) Suppose that A is 2-random. Then A is CEA.

Proof: As in Theorem 6.1, we satisfy things on a set of positive measure. Here, we construct an operator Ξ so that

$$\mu(\{A : \Xi(A) \text{ total and } \Xi(A) \not\leq_T A\}) \geq \frac{1}{4},$$

and A is c.e. in $\Xi(A)$ whenever $\Xi(A)$ is total. The calculation that weak 2-randomness is enough comes from again analyzing the method of satisfaction of the requirements.

To make A c.e. in $\Xi(A)$ we will ask that $n \in A$ iff $\langle n, m \rangle \in \Xi(A)$ for some m. Thus, in fact A is enumeration reducible to $\Xi(A)$. Now whilst doing this we must meet requirements of the form

$$R_e : A \neq \Phi_e^{\Xi(A)},$$

there Γ_e denotes the e-Turing procedure.

Definition 6.1: We say that a string ξ is *acceptable* for a string σ iff

$$\xi(\langle m, n \rangle) = 1 \rightarrow \sigma(n) = 1.$$

Remember we are trying to make A c.e. in Ξ^A. Thus we will always require that $\Xi^\sigma[s]$ is acceptable. We will also try, whenever possible, to use ξ to represent string in the range of $\Xi[s]$.

The principal difficulty with this proof is that it is very difficult for us to actually force

$$A \neq \Phi_e^{\Xi(A)}.$$

The reason is that our opponent is playing Φ_e.

The main idea is the following. In this proof, we will inductively be working above some string β, to which Kurtz assigns a *state* via a colour blue$_e$. This more or less indicates that we are presently happy with the situation up to this string. Now we consider the action in the cone $[\beta]$. As in the previous construction, we will willfully be not defining Ξ on some cone of small measure in the cone $[\beta]$. In this case, the test string will be $\beta^\frown 1^{e+2}$. We will only ever define Ξ^σ for σ extending $\beta^\frown 1^{e+2}$, should it be possible for us to actually perform some kind of diagonalization.

For those reading the account of Kurtz [44] or Kautz [33], this testing is signaled by assigning the string $\beta^\frown 1^{e+2}$ the state red$_e$.

While we are waiting for suitable conditions for diagonalization to occur, we will devote $[\nu_1], \ldots, [\nu_{2^{(e+2)}-1}]$ (the length $e+2$ extensions of β) to the rest of the construction in this cone $[\beta]$. Note that we will only be defining Ξ^τ for τ extending ν_i in this cone, with priority e. In Kurtz's construction, this part of the current inductive module satisfying R_e is given the state/colour yellow$_e$. Diagram 2 below might be helpful here.

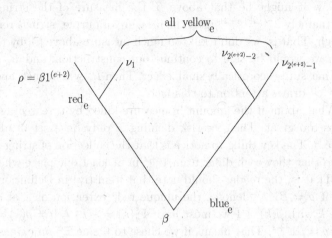

Fig. 2. R_e's basic module

Now, in the background, the construction is running along, gradually defining Ξ^σ for various θ extending ν_i in the cone $[\beta]$.

Question: Under what circumstances would a string θ look bad from R_e's point of view?

Answer: It would look bad if it looked like the initial segment of a real α with $\Phi_e^{\Xi(\alpha)} = \alpha$.

This threatening situation can be detected as by looking at strings which might approximate such reals.

Definition 6.2: We say that a string θ is *threatening* a requirement R_e is there is a yellow$_e$ strings ν extending its blue$_e$ predecessor β with $\nu \preceq \theta$, and a strings ξ acceptable to θ such that

(i) $|\theta| \leq s$
(ii) $|\xi| = s$
(iii) $\Xi^\nu[s] \preceq \xi$
(iv) $\Phi_e^\xi(k) = \nu(k) = 0[s]$ for some k such that $|\beta| < k \leq |\nu|$, and
(v) no initial segment of θ has colour purple$_e$.

Item (v) is a mystery at present, but is simply a minimality condition now to be described, and can presently be ignored. If we see a threatening string θ, then it looks bad for us to continue to define Ξ on strings extending θ. Thus we will do the following. In the construction we will give any string θ which is threatening R_e colour purple$_e$. While this situation holds, we will not define Ξ on any extension of θ.

Now it might be that above β, the measure of the strings coloured red$_e$ (namely $2^{-(|\beta|+e+2)}$) *plus* the measure of purple$_e$ strings remains small enough. That is, we don't kill too much measure above $[\beta]$ by this process. Then we would be safe to continue out construction, and R_e would only have measure-theoretically small effect. Then R_e will be met, and the string $\beta^\frown 1^{(e+2)}$ draws attention to this fact.

What about if the amount of measure killed by threatening strings becomes too great. Then we risk defining Ξ only on a set of measure zero above β. The key thing to notice is that the collection of strings ν_i are chosen so that they each differ from 1^{e+2} in at least one place where they are a 0. That is, the reader should note that item (iv) in Definition 6.2 means that if $\rho = \beta 1^{e+2}$ denotes the unique red$_e$ extension of β at stage s, for some k with $\rho(k) = 1$ we must have $\Phi_e^\xi(k) = \nu(k) = 0 \neq \rho(k) = 1$, by fact that $\rho = \beta 1^{e+2}$. That means if we chose to define Ξ^α on extensions α of ρ to emulate Ξ^ν, then it *cannot be* that $\Phi_e^{\Xi^\alpha} = \alpha$ as it must be wrong on ρ.

For each purple$_e$ string θ, let θ' denote the unique string of length $|\theta|$ extending the red$_e$ string ρ with $\theta(m) = \theta'(m)$ for all $n \geq |\rho|$. The idea is that we will be able to give these θ' the colour green$_e$, and bound their density away from 0, if the density of purple$_e$ strings grows too much.

Specifically, when the density of purple$_e$ strings above the blue$_e$ string β exceeds $2^{-(e+3)}$, there must be some yellow$_e$ string ν such that the density of purple$_e$ strings above ν must also exceed $2^{-(e+3)}$, by the Lebesgue Density Theorem. Let $\{\theta_1, \ldots, \theta_n\}$ list the purple$_e$ strings above ν, where for definiteness ν is chosen lexicographically least. For each θ_i, let ξ_i be the least string which witnesses the threat to R_e according to Definition 6.2.

Thus,

(i) ξ_i is acceptable for θ_i,

(ii) $\Xi^\nu \preceq \xi_i$, and

(iii) $\Phi_e^{\xi_i}(k) = 0 = \theta_i(k)$ for some k with $\rho(k) = 1$.

Now we are in a position to win R_e in the cone above ρ, and on a set of large measure. In the construction, we will now define Ξ to mimic the action of Ξ on the θ_i above ν, so that $\Xi^\nu[s] = \Xi^\rho[s]$. The action is that

(a) Any string extending β loses its colour.

(b) β loses colour blue$_e$.

(c) Each string θ'_i is given colour green$_e$.

(d) We define $\Xi^{\theta'_i} = \xi_i$, which will then force $\Phi_e^{\xi_i}(k) \neq \theta'(k)$ for some k with $|\beta| < k \leq |\nu|$.

Notice that (d) above justifies the use of the colour green$_e$ for the strings θ'_i. Diagram 3 might be helpful here.

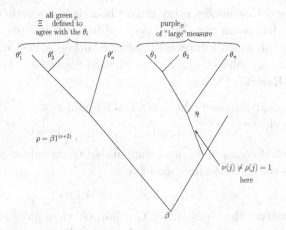

Fig. 3. R_e's acts under purple$_e$ pressure

The reader should note that each time we are forced to use the red$_e$ string ρ, we are guaranteed to succeed on set of measure at least $2^{-(e+5)}$ above β, and hence we will succeed on a set of positive measure. Notice also that since we remove all colours from the purple strings then we will be able to replicate this construction on strings extending those that have lost their colour. Finally, we remark that the fact that we only consider acceptable strings will mean that some subset of A is computably enumerable in Ξ^A. This will be fixed via a "catch up" when we assign blue$_e$ colours. It is also why we chose 1^{e+2} as the extension for β since all strings of interest will remain acceptable, and we can always find an extension to make the enumeration reduction work.

To see that 2-randomness is enough for the result, we have made sure that $\{\Xi^A$ is total $\}$ is a Π_2^0 class of positive measure and contains at least one 2-random real. Thus by the effective 0-1 law, it contains members of every 2-random degree.

Full details can be found in Downey and Hirschfeldt [18], and Kautz [33].

□

In Kautz's Thesis [33], Kautz mistakenly claims that weak 2-randomness is enough. His error is to assume that the effective 0-1 law holds for weak 2 randomness. In fact for hyperimmune free degrees Martin-Löf randomness is the same as weak 2-randomness, as we have already seen.

6.4. *Variations and marginalia*

Kurtz proved that almost every degree bounds a 1-generic degree. He remarks that this result can also be deduced from the theorem above, via the following result which was known to several authors in the 70's. P The proof below is an unpublished one due to Richard Shore (also found in Soare's book [80], Exercise VI, 3.9).

Theorem 6.3: Suppose that A is CEA(B) with $B <_T A$. Then A bounds a 1-generic degree.

Proof: Let $A = \cup_s A_s$ be a B-computable enumeration of A. Let $c(n)$ denote the computation function of A:

$$c(n) = \mu s(A_s \upharpoonright n = A \upharpoonright n).$$

We construct the 1-generic set $G = \lim_s G_s$ via a finite extension argument. Let V_e denote the e-th c.e. set of strings. At stage s we let $G_{s+1} = G_s$ unless (for some least e) we see some extension $\gamma \in V_{e,c(s+1)}$ with $G_s \prec \gamma$. In this latter case, let $G_{s+1} = \gamma$.

Suppose that G is not 1-generic. We claim that $A \leq_T B$. Let e such that G does not meet V_e and every initial segment of G is extended by one in V_e. Let s_0 be a stage after which e has priority. It suffices to compute $c(n)$ from B, since B enumerates A. Using simultaneous induction on s and G_s, for $s > s_0$.

Now assume that G_s is known. Compute a minimal stage t such that some extension γ_s of G_s occurs in $V_{e,t}$. The it must be that $t > c(s+1)$, lest we would act for e. This allows us to compute $c(s+1)$, and hence G_{s+1}.

□

Corollary 6.1: (Kurtz [44])Suppose that **a** is 2-random. Then **a** bounds a 1-generic degree.

Actually, Kurtz proved an even more surprising result in [44]. Recall that we say that a class $\mathcal{C}$ of degrees is *downward dense* below a degree **a** iff for all nonzero **b** < **a** there is a degree **c** ≤ **b** with **c** ∈ $\mathcal{C}$.

Theorem 6.4: (Kurtz [44]) The 1-generic degrees are downward dense below any 2-random degree.

Again the methods of proof are elaborations of the measure risking ones. Roughly speaking, a key ingredient is the de Leeuw, Moore, Shannon, and Shapiro-Sacks result that the cone of degrees above a noncomputable real has measure zero. However, this time the argument is much more delicate, since, in particular, we need to recognize when we are dealing with a computable real where the de Leeuw, Moore, Shannon, and Shapiro-Sacks result might fail. The proof of this result is beyond the scope of the present lecture and we refer the reader to Kurtz [44] and Downey-Hirschfeldt [18].

We remark that there has been a reasonable amount of work pursuing analogs of these results for other classes, such as the generic reals. For instance, Yu Liang [96] has proven an analog of van Lambalgen's Theorem:

Theorem 6.5: (Yu [96]) $x \oplus z$ is n-generic iff x is n-z-generic and z is n-generic.

This allowed Yu to prove a number of interesting results about generics. We can also ask whether reals x can be low for, say, 1-genericity meaning that y is x-generic iff y is generic.

Theorem 6.6: (Greenberg, Miller, Yu [96]) No noncomputable real is low for 1-genericity.

We remark that Miller and Yu have shown that there are reals low for weak genericity. One question is whether the class of low for weakly generic reals is the same class as the computably traceable reals.

References

1. Allender, E., H. Buhrman, M. Koucký, D. van Melkebeek, and D. Ronneburger, *Power from Random Strings,* in *FOCS 2002,* IEEE (2002), 669-678.
2. Allender, E., H. Buhrman, and M. Koucký, *What Can be Efficiently Reduced to the Kolmogorov-Random Strings?,* Annals of Pure and Applied Logic, Vol. 138, No. 1-3 (2006), 2-19.
3. Barmpalias, G., and A. Lewis, *A c.e. real that cannot be sw-computed by any Ω-number,* Notre Dame Journal of Formal Logic, Vol. 47, 2 (2006), 197-209.
4. Calude, C., *Information Theory and Randomness, an Algorithmic Perspective,* Springer-Verlag, Berlin, 1994. Second revised edition 2002.
5. Calude, C., and Coles, R. *Program size complexity of initial segments and domination relation reducibility,* in *Jewels are Forever,* (J. Karhümaki, H. Mauer, G. Paŭn, G. Rozenberg, eds.) Springer-Veralg, New York, 1999, 225-237.

6. Calude, C., Coles, R., Hertling, P., Khoussainov, B., *Degree-theoretic aspects of computably enumerable reals*, in *Models and Computability*, (ed. Cooper and Truss) Cambridge University Press, 1999.

7. Calude, C., Hertling, P., Khoussainov, B., Wang, Y., *Recursively enumerable reals and Chaitin's Ω number*, in STACS '98, Springer Lecture Notes in Computer Science, Vol. 1373, 1998, 596-606.

8. Chaitin, G., *A theory of program size formally identical to information theory*, Journal of the Association for Computing Machinery 22 (1975), pp. 329-340.

9. Chaitin, G. *Information-theoretical characterizations of recursive infinite strings,* Theoretical Computer Science, Vol. 2 (1976), 45-48.

10. Chaitin, G. *Incompleteness theorems for random reals*, Adv. in Appl. Math Vol. 8 (1987), 119-146.

11. Chaitin, G., *Information, Randomness & Incompleteness*, 2nd edition, Series in Computer Science 8, World Scientific, River Edge, NJ, 1990.

12. de Leeuw, K., Moore, E. F., Shannon, C. E., and Shapiro, N. *Computability by probabilistic machines,* Automata studies, pp. 183–212. Annals of mathematics studies, no. 34. Princeton University Press, Princeton, N. J., 1956.

13. Downey, R., *Some recent progress in algorithmic randomness,* in *Proceedings of the 29 th Annual Conference on Mathematical Foundations of Computer Scence, Prague, August 2004* (invited paper) (J. Fiala, V. Koubek and J. Kratochvil eds), Springer-Verlag, Lecture Notes in Computer Science, Vol. 3153 (2004), 42-81.

14. Downey, R., *Some Computability-Theoretical Aspects of Reals and Randomness,* to appear *The Notre Dame Lectures* (P. Cholak, ed) *Lecture Notes in Logic,* Association for Symbolic Logic, 2005.

15. Downey, R., Ding Decheng, Tung Shi Ping, Qiu Yu Hui, Mariko Yasuugi, and Guohua Wu (eds), *Proceedings of the 7th and 8th Asian Logic Conferences,* World Scientific, 2003, viii+471 pages.

16. Downey, R. and E. Griffiths, *On Schnorr randomness*, Journal of Symbolic Logic, Vol. 69 (2) (2004), 533-554.

17. Downey, R., E. Griffiths, and S. Reid, *On Kurtz randomness*, Theoretical Computer Science, Vol. 321 (2004), 249-270.

18. Downey, R., and D. Hirschfeldt, *Algorithmic Randomness and Complexity,* Springer-Verlag Monographs in Computer Science, to appear.

19. Downey, R., D. Hirschfeldt, and G. LaForte, *Undecidability of Solovay degrees of c.e. reals,* to appear, Journal of Comput. and Sys. Sci.

20. Downey, R., D. Hirschfeldt, J. Miller, and A. Nies, *Relativizing Chaitin's halting probability,* Journal of Mathematical Logic, Vol. 5 (2005), 167-192.

21. Downey, R., D. Hirschfeldt, and A. Nies, *Randomness, computability and density* SIAM Journal on Computing, 31 (2002) 1169–1183 (extended abstract in proceedings of STACS 2001 January, 2001. Lecture Notes in Computer Science, Springer-Verlag, (Ed A. Ferriera and H. Reichel, 2001, 195-201)).

22. Downey, R., D. Hirschfeldt, A. Nies, and F. Stephan, *Trivial reals*, extended abstract in *Computability and Complexity in Analysis* Malaga, (Electronic

Notes in Theoretical Computer Science, and proceedings, edited by Brattka, Schröder, Weihrauch, FernUniversität, 294-6/2002, 37-55), July, 2002. Final version appears in [15] 2003, 103-131.

23. Downey, R., D. Hirschfeldt, A. Nies, and S. Terwijn, *Calibrating randomness*, Bulletin Symbolic Logic, Vol. 12 (2006), 411-491.

24. Downey, R., and J. Miller, *A basis theorem for Π_1^0 classes of positive measure and jump inversion for random reals*, Proc. Amer. Math. Soc., Vol. 134 (2006), 283-288.

25. Downey, R., J. Reimann, and W. Merkle, *Schnorr dimension*, preliminary version appeared in *Computability in Europe, 2005*, Springer-Verlag Lecture Notes in Computer Science, Vol. 3526 (2005), 96-105. Final version in *Mathematical Structures in Computer Science*, Vol. 16 (2006), 789-811.

26. Gács, P., *On the symmetry of algorithmic information*, Soviet Mat. Dolk., Vol. 15 (1974), 1477-1480.

27. Gács, P., *On the relation between descriptional complexity and algorithmic probability*, Theoretical Computer Science, Vol. 22 (1983), 71-93.

28. Gács, P., *Every set is reducible to a random one*, Information and Control, Vol. 70, (1986), 186-192.

29. Gaifmann, H., and M. Snir, *Probabilities over rich languages*, J. Symb. Logic, Vol. 47, (1982), 495-548.

30. Hausdorff, F., *Dimension und äußeres Maß*, Mathematische Annalen 79 (1919) 157–179.

31. Hirschfeldt, D, A. Nies, and F. Stephan, *Using random sets as oracles*, Journal of the London Mathematical Society, Vol. 75 (2007), 610-622.

32. Jockusch, C. G. , M. Lerman, R. I. Soare, and R. Solovay, *Recursively enumerable sets modulo iterated jumps and extensions of Arslanov's completeness criterion*, J. Symb. Logic, Vol. 54 (1989), 1288-1323.

33. Kautz, S., *Degrees of Random Sets*, Ph.D. Diss. Cornell, 1991.

34. Kolmogorov, A. N., *Three Approaches to the Quantitative Definition of Information*, in *Problems of Information Transmission (Problemy Peredachi Informatsii)*, 1965, Vol. 1, 1-7.

35. Kraft, L. G., *A device for quantizing, grouping, and coding amplitude modulated pulses*, M.Sc. Thesis, MIT, 1949.

36. Kučera, A., *Measure, Π_1^0 classes, and complete extensions of PA*, in *Springer Lecture Notes in Mathematics*, Vol. 1141, 245-259, Springer-Verlag, 1985.

37. Kučera, A., *An alternative priority-free solution to Post's Problem*, Proceedings, *Mathematical Foundations of Computer Science*, (Eds. J. Gruska, B. Rovan, and J. Wiederman), Lecture Notes in Computer Science, Vol. 233, Springer-Verlag, (1986).

38. Kučera, A., *On the use of diagonally nonrecursive functions*, in *Logic Colloquium, '87*, North-Holland, Amsterdam, 1989, 219-239.

39. Kučera, A., *Randomness and generalizations of fixed point free functions*, in *Recursion Theory Week, Proceedings Oberwolfach 1989*, (K. Ambos-Spies, G. H. Müller and G. E. Sacks, eds.) Springer-Verlag, LNMS 1432 (1990), 245-254.

40. Kučera, A. and T. Slaman, *Randomness and recursive enumerability*, SIAM Journal of Computing, Vol. 31 (2001), 199-211.
41. Kučera, A., and S. Terwijn, *Lowness for the class of random sets*, Journal of Symbolic Logic, vol. 64 (1999), 1396-1402.
42. Kummer, M., *Kolmogorov complexity and instance complexity of recursively enumerable sets*, SIAM Journal of Computing, Vol. 25 (1996), 1123-1143.
43. Kummer, M., *On the complexity of random strings*, (extended abstract) in *STACS, '96*, Springer-Verlag Lecture Notes in Computer Science, Vol. 1046 (11996), 25-36.
44. Kurtz, S., *Randomness and Genericity in the Degrees of Unsolvability*, Ph. D. Thesis, University of Illinois at Urbana, 1981.
45. Levin. L., *On the notion of a random sequence*, Soviet Math. Dokl. 14 (1973) 1413-1416.
46. Levin, L., *Laws of information conservation (non-growth) and aspects of the foundation of probability theory*, Problems Informat. Transmission, Vol. 10 (1974), 206-210.
47. Levin, L., *Measures of complexity of finite objects (axiomatic description)*, Soviet Mathematics Dolkady, Vol. 17 (1976), 552-526.
48. Levy, P., *Theorie de l'Addition des Variables Aleantoires*, Gauthier-Villars, 1937.
49. Li, Ming and Vitanyi, P., *Kolmogorov Complexity and its Applications*, Springer-Verlag, 1993.
50. Loveland, D. *A variant of the Kolmogorov concept of complexity*, Information and Control, Vol. 15 (1969), 510-526.
51. Lutz, J., *The dimensions of individual strings and sequences*, Information and Computation, Vol. 187 (2003), pp. 49-79. (Preliminary version: *Gales and the constructive dimension of individual sequences*, in: U. Montanari, J. D. P. Rolim, E. Welzl (eds.), Proc. 27th International Colloquium on Automata, Languages, and Programming, 902–913, Springer, 2000.)
52. Lutz, J., *Effective fractal dimensions*, Mathematical Logic Quarterly Vol. 51 (2005), pp. 62-72.
53. Martin-Löf, P., *The definition of random sequences*, Information and Control, 9 (1966), 602-619.
54. Martin-Löf, P., *Complexity oscillations in infinite binary sequences*, Z. Wahrscheinlichkeittheorie verw. Gebiete, Vol. 19 (1971), 225-230.
55. Miller, J., *Kolmogorov random reals are 2-random*, Journal of Symbolic Logic,
56. Merkle W., and N. Mihailovic, *On the construction of effective random sets*, in Proceedings MFCS 2002, (ed. Diks and Rytter) Springer-Verlag LNCS 2420.
57. Merkle, W., J. Miller, A. Nies, J. Reimann, and F. Stephan, *Kolmogorov-Loveland randomness and stochasticity*, Annals of Pure and Applied Logic, Vol. 138 (2006), 183-210.
58. Miller, J., *The K-degrees, low for K degrees, and weakly low for K oracles*, in preparation.

59. Miller, J., *Solution to the question of whether* $\{x : K(x) < |x| + K(|x|) - c\}$ *is* Σ_1^0, e-mail communication, 13th January, 2005.

60. Miller, J. *Contrasting plain and prefix-free Kolmogorov complexity*, in preparation.

61. Miller, J., *A hyperimmune-free weakly 2 random real* , e-mail communication, 14th April, 2005.

62. Miller, J., and L. Yu, *On initial segment complexity and degrees of randomness*, to appear, Trans. Amer. Math. Soc.

63. Miller, J., and L. Yu, *Oscillation in the initial segment complexity of random reals*, in preparation.

64. Muchnik, An. A., and S. P. Positelsky, *Kolmogorov entropy in the context of computability theory*, Theor. Comput. Sci., Vol. 271 (2002), 15-35.

65. Muchnik, An. A., A. Semenov, and V. Uspensky, *Mathematical metaphysics of randomness*, Theor. Comput. Sci., Vol. 207(2), (1998), 263-317.

66. Nies, A., *Reals which compute little*, Proceedings of Logic Colloquium 2002, (Eds. Z. Chatizdais, P. Koepke and W. Pohlers), Lecture Notes in Logic, Vol. 27 (2002), 261-275.

67. Nies, A., *Lowness properties and randomness*, Advances in Mathematics, Vol. 197 (2005), 274-305.

68. Nies, A., *Each Low (CR) set is computable*, typeset manuscript, January 2003.

69. Nies, A., F. Stephan and S. A. Terwijn, *Randomness, relativization, and Turing degrees*, Journal of Symbolic Logic, 70(2) (2005), 515-535.

70. Odifreddi, P., *Classical Recursion Theory*, North-Holland, 1990.

71. Odifreddi, P., *Classical Recursion Theory, Vol. 2*, North-Holland, North-Holland, 1999.

72. Paris, J., *Measure and minimal degrees*, Annals of Mathematical Logic, Vol. 11 (1977), 203-216.

73. Paris, J., *Survey of Results on Measure and Degrees*, unpublished notes.

74. Reimann, J, PhD Thesis, University of Heidelberg, in preparation.

75. Reimann, J., *Computability and Dimension*, unpublished notes, University of Heidelberg, 2004

76. G. Sacks, *Degrees of Unsolvability*, Princeton University Press, 1963.

77. Schnorr, C. P., *A unified approach to the definition of a random sequence*, Mathematical Systems Theory, 5 (1971), 246-258.

78. Schnorr, C. P., *Zufälligkeit und Wahrscheinlichkeit*, Springer-Verlag Lecture Notes in Mathematics, Vol. 218, 1971, Springer-Verlag, New York.

79. Schnorr, C. P., *Process complexity and effective random tests*, Journal of Computer and System Sciences, Vol. 7 (1973), 376-388.

80. Soare, R., *Recursively enumerable sets and degrees* (Springer, Berlin, 1987).

81. Soare, R., *Computability Theory and Differential Geometry*, Bulletin Symbolic Logic, Vol. 10, Issue 4 (2004), 457-486.

82. Solomonoff, R., *A formal theory of inductive inference, part 1 and part 2*, Information and Control, 7 (1964), 224-254.

83. Solovay, R., *Draft of paper (or series of papers) on Chaitin's work*, unpublished notes, May, 1975, 215 pages.

84. Staiger, L., *Kolmogorov complexity and Hausdorff dimension*, Information and Computation, Vol. 103 (1993), 159-194.
85. Staiger, L., *A tight upper bound on Kolmogorov complexity and uniformly optimal prediction*, Theory of Computing Sciences, Vol. 31 (1998), 215-229.
86. Staiger, L., *Constructive dimension equals Kolmogorov complexity*, Research Report CDMTCS-210, University of Auckland, January 2003.
87. Stephan, F., *Martin-Löf random sets and PA-complete sets*, ASL Lecture Notes in Logic, Vol. 27 (2006), 342-348.
88. Stillwell, J., *Decidability of "almost all" theory of degrees*, J. Symb. Logic, Vol. 37 (1972), 501-506.
89. Tadaki, K., *A generalization of Chaitin's halting probability Ω and halting self-similar sets*, Hokkaido Math. J., Vol. 32 (2002), 219-253.
90. Terwijn, S. *Computability and Measure*, Ph. D. Thesis, University of Amsterdam, 1998.
91. van Lambalgen, M., *Random Sequences*, Ph. D., Diss. University of Amsterdam, 1987.
92. van Lambalgen, M., *The axiomatization of randomness*, Journal of Symbolic Logic, Vol. 55 (1991), 1143-1167.
93. von Mises, R., *Grundlagen der Wahrscheinlichkeitsrechnung*, Math. Z., Vol. 5 (1919), 52-99.
94. Ville, J., *Étude critique de la concept du collectif*, Gauthier-Villars, 1939.
95. Wang, Y., *Randomness and Complexity*, PhD Diss, University of Heidelberg, 1996.
96. Yu, Liang, *Degrees of generic and random reals*, to appear.
97. Yu Liang and Ding Decheng, *There is no sw-complete c.e. real*, J. Symbolic Logic, Vol. 69 (2004), no. 4, 1163-1170.
98. Yu Liang and Ding Decheng, *The initial segment complexity of random reals*, Proc. Amer. Math. Soc., Vol. 132 (2004), no. 8, 2461–2464.
99. Yu Liang, Ding Decheng, and R. Downey, *The Kolmogorov complexity of the random reals*, Annals of Pure and Applied Logic Volume 129, Issues 1-3, (2004), 163-180.
100. Zambella, D., *On sequences with simple initial segments*, ILLC technical report, ML-1990-05, University of Amsterdam, 1990.
101. Zvonkin A. K., and L.A. Levin, *The complexity of finite objects and the development of concepts of information and randomness by the theory of algorithms*, Russian Math. Surveys, 25(6):83-124 (1970).

GLOBAL PROPERTIES OF THE TURING DEGREES AND THE TURING JUMP

Theodore A. Slaman[*]

Department of Mathematics
University of California, Berkeley
Berkeley, CA 94720-3840, USA
E-mail: slaman@math.berkeley.edu

We present a summary of the lectures delivered to the Institute for Mathematical Sciences, Singapore, during the 2005 Summer School in Mathematical Logic. The lectures covered topics on the global structure of the Turing degrees $\mathscr{D}$, the countability of its automorphism group, and the definability of the Turing jump within $\mathscr{D}$.

1. Introduction

This note summarizes the tutorial delivered to the Institute for Mathematical Sciences, Singapore, during the 2005 Summer School in Mathematical Logic on the structure of the Turing degrees. The tutorial gave a survey on the global structure of the Turing degrees $\mathscr{D}$, the countability of its automorphism group, and the definability of the Turing jump within $\mathscr{D}$.

There is a glaring open problem in this area: Is there a nontrivial automorphism of the Turing degrees? Though such an automorphism was announced in Cooper (1999), the construction given in that paper is yet to be independently verified. In this paper, we regard the problem as not yet solved. The Slaman-Woodin Bi-interpretability Conjecture 5.10, which still seems plausible, is that there is no such automorphism.

Interestingly, we can assemble a considerable amount of information about $Aut(\mathscr{D})$, the automorphism group of $\mathscr{D}$, without knowing whether

---

[*]Slaman was partially supported by the National University of Singapore and by National Science Foundation Grant DMS-0501167. Slaman is also grateful to the Institute for Mathematical Sciences, National University of Singapore, for sponsoring its 2005 program, Computational Prospects of Infinity.

it is trivial. For example, we can prove that it is countable and every element is arithmetically definable. Further, restrictions on $Aut(\mathscr{D})$ lead us to interesting conclusions concerning definability in $\mathscr{D}$.

Even so, the progress that can be made without settling the Bi-interpretability Conjecture only makes the fact that it is open more glaring. With these notes goes the hope that they will spark further interest in this area and eventually a solution to the problems that they leave open.

1.1. Style

In the following text, we will state the results to be proven in logical order. We will summarize the proofs when a few words can convey the reasoning behind them. When that fails, we will try to make the theorem plausible. A complete discussion, including proofs omitted here, can be obtained in the forthcoming paper Slaman and Woodin (2005).

2. The coding lemma and the first order theory of the Turing degrees

Definition 2.1 • $\mathscr{D}$ denotes the partial order of the Turing degrees. $a + b$ denotes the join of two degrees. ($A \oplus B$ denotes the recursive join of two sets.)

• A subset $\mathscr{I}$ of $\mathscr{D}$ is an *ideal* if and only if $\mathscr{I}$ is closed under $\leq_T$ ($x \in \mathscr{I}$ and $y \leq_T x$ implies $y \in \mathscr{I}$) and closed under $+$ ($x \in \mathscr{I}$ and $y \in \mathscr{I}$ implies $x + y \in \mathscr{I}$). A *jump ideal* is closed under the Turing jump ($a \mapsto a'$) as well.

Early work on $\mathscr{D}$ concentrated on its naturally order-theoretic properties. For example, Kleene and Post (1954) showed that every finite partial order is isomorphic to a suborder of $\mathscr{D}$. Sacks (1961) extended this embedding theorem from finite to countable partial orders. Spector (1956) constructed a minimal nonzero degree, which began a long investigation into the structure of the initial segments of $\mathscr{D}$. Some of the high points in that investigation are Lachlan (1968), every countable distributive lattice with a least element is embeddable as an initial segment, Lerman (1971), every finite lattice with a least element is embeddable as an initial segment, and Lachlan and Lebeuf (1976), every countable lattice with a least element is embeddable as an initial segment.

Our focus will be on the logical properties of $\mathscr{D}$. We can view the results in the previous paragraph as steps in deciding the low-level fragments of the first order theory of $\mathscr{D}$. By the Kleene-Post theorem, the existential theory of $\mathscr{D}$ is decidable. An existential statement is true in $\mathscr{D}$ if and only if it is true in some finite partial order of size the length of the sentence. Lerman's theorem can be combined with a strengthening of the Kleene-Post theorem to show that the $\exists\forall$-theory of $\mathscr{D}$ is decidable; see Lerman (1983) or Shore (1978).

One can also come to conclusions concerning undecidability. The theory of distributive lattices is undecidable, Lachlan's theorem yields an interpretation of that theory into the first order theory of $\mathscr{D}$, and so the first order theory of $\mathscr{D}$ is undecidable.

The exact degree of the theory of $\mathscr{D}$ was calculated by Simpson (1977). Simpson showed that the theory of $\mathscr{D}$ is recursively isomorphic to the second order theory of arithmetic. We will obtain a proof of Simpson's theorem as a corollary to the machinery we develop here.

2.1. The coding lemma

Definition 2.2 A countable n-place relation $\mathscr{R}$ on $\mathscr{D}$ is a countable subset of the n-fold Cartesian product of $\mathscr{D}$ with itself. In other words, $\mathscr{R}$ is a countable subset of the set of length n sequences of elements of $\mathscr{D}$.

Theorem 2.3 (The Coding Lemma, Slaman and Woodin (1986))
For every n there is a first order formula $\varphi(x_1,\ldots,x_n,y_1,\ldots,y_m)$ such that for every countable n-place relation $\mathscr{R}$ on $\mathscr{D}$ there is a sequence of degrees $\overrightarrow{p} = (p_1,\ldots,p_m)$ such that for all sequences of degrees $\overrightarrow{d} = (d_1,\ldots,d_n)$,

$$\overrightarrow{d} \in \mathscr{R} \iff \mathscr{D} \models \varphi(\overrightarrow{d},\overrightarrow{p}).$$

By the coding lemma, quantifiers over countable relations on $\mathscr{D}$ can be interpreted in the first order language of $\mathscr{D}$ by quantifying over the parameters used to define these relations. Consequently, the first order theory of $\mathscr{D}$ can interpret all of countable mathematics.

For example, the isomorphism type of the standard model of arithmetic $\mathbb{N}$ is characterized in countable terms. There is a finitely axiomitized theory T such that for any countable model $\mathfrak{M}$ of T, either there is an infinite decreasing sequence in $\mathfrak{M}$ (a countably expressed property) or $\mathfrak{M}$ is isomorphic to $\mathbb{N}$. Similarly, second order quantifiers over a copy of $\mathbb{N}$ are just quantifiers over countable sets. Hence, there is an interpretation of second order arithmetic in the first order theory of $\mathscr{D}$ and Simpson's theorem follows.

One can push the application of the coding lemma further in this direction. Rather than interpreting the second order theory of arithmetic, one can interpret the subsets of the natural numbers, work with them individually, and associate them with the degrees that they represent, to the extent that it is possible to do so. This is a principal theme in what follows.

Finally, we will have a more metamathematical use of the coding lemma. Every first order structure has a countable elementary substructure, typically a Skolem hull of the original. In models of set theory, this fact becomes a reflection property. For example, if a sentence φ is true in L, then it is true in some countable initial segment of L. In the Turing degrees, we will show that global properties of $\mathscr{D}$ can be reflected to countable jump ideals.

In retrospect, the coding lemma should be expected. One can construct sets and directly control everything that is arithmetically definable from them. Consider the Friedberg (1957) jump inversion theorem. Given a set $X \geq_T 0'$, Friedberg constructs another set A such that $A' \equiv_T X$. Though the theory of forcing was only introduced later, Friedberg uses the ingredients Cohen forcing to decide atomic facts about A'. He alternates between meeting the dense sets associated with building a generic real and steps to code atomic facts about X.

One can think of the parameters as sets generically engineered to distinguish between elements of the relation and elements of it's complement. The interactions to be controlled have bounded complexity in the arithmetic hierarchy. Consequently, one can obtain coding parameters which are uniformly arithmetic in any presentation of the relation. A sharp analysis of the coding methods in Slaman and Woodin (2005) gives the following.

Theorem 2.4 *Suppose that there is a presentation of the countable relation $\mathscr{R}$ which is recursive in the set R. There are parameters $\overrightarrow{p}$ which code $\mathscr{R}$ in $\mathscr{D}$ such that the elements of $\overrightarrow{p}$ are recursive in R'.*

Similarly, since the coding only involves arithmetic properties of the parameters, the relation $\mathscr{R}$ is arithmetically definable from the parameters which code it.

Theorem 2.5 (Decoding Theorem) *Suppose that $\overrightarrow{p}$ is a sequence of degrees which lie below y and $\overrightarrow{p}$ codes the relation $\mathscr{R}$. Letting Y be a representative of y, $\mathscr{R}$ has a presentation which is $\Sigma_5^0(Y)$.*

In the particular case of coding a model of arithmetic with a unary predicate, one can do much better than Σ_5^0.

Theorem 2.6 *For any degree x and representative X of x, there are parameters $\overrightarrow{p}$ such that the following conditions hold.*

- $\overrightarrow{p}$ *codes an isomorphic copy of $\mathbb{N}$ with a unary predicate for X.*

- $\overrightarrow{p}$ *is recursive in $x + 0'$.*

In the other direction, suppose that $\overrightarrow{p}$ is a sequence of degrees below y, and $\overrightarrow{p}$ codes an isomorphic copy of $\mathbb{N}$ together with a unary predicate U. As a direct application of the decoding theorem, for Y a representative of y, U is $\Sigma_5^0(Y)$.

3. Properties of automorphisms of $\mathscr{D}$

3.1. Results of Nerode and Shore

We can apply the coding and decoding theorems to obtain some early results of Nerode and Shore on the global properties of $\mathscr{D}$.

Theorem 3.1 (Nerode and Shore (1980)) *Suppose that $\pi : \mathscr{D} \xrightarrow{\sim} \mathscr{D}$. For every degree x, if x is greater than $\pi^{-1}(0')$ then $\pi(x)$ is arithmetic in x.*

Proof: Let Y be a representative of $\pi(x)$. Since $x \geq_T \pi^{-1}(0')$, $Y \geq_T 0'$. By Theorem 2.6, there are parameters $\overrightarrow{p}$ which are recursive in Y such that $\overrightarrow{p}$ codes Y. But then, $\pi(\overrightarrow{p})$ is recursive in x and still codes Y. By the decoding theorem, Y is $\Sigma_5^0(X)$.

Theorem 3.2 (Nerode and Shore (1980)) *Suppose $\pi : \mathscr{D} \xrightarrow{\sim} \mathscr{D}$ is an automorphism of $\mathscr{D}$ and $x \geq_T \pi^{-1}(0')^{(5)} + \pi^{-1}(\pi(0')^{(5)})$. Then, $\pi(x) = x$. Consequently, π is the identity on a cone.*

Proof: Given x above $\pi^{-1}(0')^{(5)}$, fix y_1 and y_2 so that $y_1 \vee y_2 = x$; $\pi(y_1)$ and $\pi(y_2)$ are greater than $0'$; and $y_1^{(5)}$ and $y_2^{(5)}$ are recursive in x. By Theorem 2.6, each $\pi(y_i)$ can compute a sequence of parameters which codes one of its representatives. The preimages of these parameters are recursive in Y_i. By the decoding theorem, representatives of each $\pi(y_i)$ is $\Sigma_5^0(Y_i)$ and hence recursive in X. Thus, $x \geq_T \pi(y_1) \vee \pi(y_2) = \pi(x)$. By symmetry, $\pi(x) \geq_T x$.

The Nerode-Shore theorems pose a challenge. Given an automorphism, where is the base of the cone on which it is the identity? In the 1980's, Jockusch and Shore produced a remarkable sequence of papers on *REA*-operators, with the conclusion that every automorphism of $\mathscr{D}$ is fixed on the cone above the degrees of the arithmetic sets. See Jockusch and Shore (1984). We go a step further and show that every automorphism of $\mathscr{D}$ is fixed on the cone above $0''$.

Our final observation on directly applying the coding lemma is due to Odifreddi and Shore. In brief, the local action of a global automorphism of $\mathscr{D}$ is locally definable.

Theorem 3.3 (Odifreddi and Shore (1991)) *Suppose that π is an automorphism of $\mathscr{D}$ and that $\mathscr{I}$ is an ideal in $\mathscr{D}$ which includes $0'$ such that π restricts to an automorphism of $\mathscr{I}$. For any real I, if there is a presentation of $\mathscr{I}$ which is recursive in I then the restriction of π to $\mathscr{I}$ has a presentation which is arithmetic in I.*

Proof: Code a counting of $\mathscr{I}$ by parameters $\overrightarrow{p}$ which are arithmetic in I. The action of π on $\mathscr{I}$ is determined by the action of π on $\overrightarrow{p}$. Since $0' \in \mathscr{I}$, the Nerode and Shore Theorem implies that $\pi(\overrightarrow{p})$ is arithmetic in I. ∎

4. Slaman and Woodin analysis of *Aut*($\mathscr{D}$)

Until indicated otherwise, we will follow the Slaman and Woodin (2005) analysis of $Aut(\mathscr{D})$.

If $\mathscr{J}$ is an ideal in the Turing degrees (such as $\mathscr{D}$), $\mathscr{I}$ is a countable subideal of $\mathscr{J}$ such that $0' \in \mathscr{I}$ and $\mathscr{J}$ codes a counting of $\mathscr{I}$, and ρ is an automorphism of $\mathscr{J}$ that restricts to an automorphism of $\mathscr{I}$, then $\rho \upharpoonright \mathscr{I}$ is definable from the action of ρ on the parameters which code the counting. Now, suppose that $\rho \upharpoonright \mathscr{I}$ is an automorphism of $\mathscr{I}$ such that for any counting of $\mathscr{I}$, $\rho \upharpoonright \mathscr{I}$ can be extended to an automorphism of an ideal which includes that counting. Then $\rho \upharpoonright \mathscr{I}$ would be definable from that counting. But, if $\rho \upharpoonright \mathscr{I}$ is definable from every counting of $\mathscr{I}$, then $\rho \upharpoonright \mathscr{I}$ is definable from $\mathscr{I}$ itself. If we can apply this analysis to a countable reflection of $\mathscr{D}$, then any automorphism of $\mathscr{D}$ would be definable from the reals, that is be an element of $L[\mathbb{R}]$. We follow just this line of reasoning in this section.

4.1. Persistent automorphisms

Definition 4.1 An automorphism ρ of a countable ideal $\mathscr{I}$ is *persistent* if and only if for every degree x there is a countable ideal $\mathscr{I}_1$ such that the following conditions hold.

- $x \in \mathscr{I}_1$ and $\mathscr{I} \subseteq \mathscr{I}_1$.

- There is an automorphism ρ_1 of $\mathscr{I}_1$ such that the restriction of ρ_1 to $\mathscr{I}$ is equal to ρ.

We will show that ρ is persistent if and only if ρ extends to a automorphism of $\mathscr{D}$. One direction of the equivalence is obvious.

Theorem 4.2 *Suppose that* $\pi : \mathscr{D} \xrightarrow{\sim} \mathscr{D}$. *For any countable ideal* $\mathscr{I}$, *if* π *restricts to an automorphism* $\pi \upharpoonright \mathscr{I}$ *of* $\mathscr{I}$ *then* $\pi \upharpoonright \mathscr{I}$ *is persistent.*

Thus, if $\mathscr{D}$ is not rigid, then there is a nontrivial persistent automorphism of some countable ideal in $\mathscr{D}$.

Theorem 4.3 *Suppose that* $\rho : \mathscr{I} \xrightarrow{\sim} \mathscr{I}$, *that* $\mathscr{J}$ *is a jump ideal contained in* $\mathscr{I}$, *and that* $\rho(0') \vee \rho^{-1}(0') \in \mathscr{J}$. *Then* $\rho \upharpoonright \mathscr{J}$ *is an automorphism of* $\mathscr{J}$.

Proof: The theorem follows from the effective coding and decoding theorems. If $x \in \mathscr{J}$, then $\rho(x)$ is arithmetic in $x \vee \rho^{-1}(0')$, which is also in $\mathscr{J}$. ∎

Corollary 4.4 *Suppose that* $\mathscr{I}$ *is a countable ideal such that* $0'$ *is an element of* $\mathscr{I}$ *and suppose that* ρ *is a persistent automorphism of* $\mathscr{I}$. *For any countable jump ideal* $\mathscr{J}$ *extending* $\mathscr{I}$, ρ *extends to an automorphism of* $\mathscr{J}$.

To prove the corollary, extend ρ to an automorphism of a countable ideal containing some upper bound of $\mathscr{J}$ and apply Theorem 4.3.

Theorem 4.5 *Suppose that* $\mathscr{I}$ *is a countable ideal in* $\mathscr{D}$ *such that* $0'$ *is an element of* $\mathscr{I}$. *Suppose that there is a presentation of* $\mathscr{I}$ *which is recursive in* I. *Finally, suppose that* $\mathscr{J}$ *is a jump ideal which includes the degree of* I *and* ρ *is an automorphism of* $\mathscr{J}$ *that restricts to an automorphism of* $\mathscr{I}$. *Then, the restriction* $\rho \upharpoonright \mathscr{I}$ *of* ρ *to* $\mathscr{I}$ *has a presentation which is arithmetic in* I.

Proof: There is a code $\overrightarrow{p}$ for a counting of $\mathscr{I}$ which is arithmetic in I and hence an element of the subideal of $\mathscr{J}$ consisting of the degrees of sets which are arithmetic in I. By Theorem 4.3, ρ restricts to an automorphism of this subideal. Consequently, $\rho(\overrightarrow{p})$ is arithmetic in I. Now, apply the Odifreddi-Shore argument of Theorem 3.3. ■

Corollary 4.6 *Suppose that $\mathscr{I}$ is a countable ideal and $0'$ is an element of $\mathscr{I}$. If ρ is a persistent automorphism of $\mathscr{I}$, then ρ is arithmetically definable in any presentation of $\mathscr{I}$.*

Consequently, persistent automorphisms of $\mathscr{I}$ are locally presented and there are at most countably many of them.

4.2. Persistently extending persistent automorphisms

Theorem 4.7 *Suppose that $\mathscr{I}$ is a countable ideal and $0'$ is an element of $\mathscr{I}$. Suppose that ρ is a persistent automorphism of $\mathscr{I}$. For any countable jump ideal $\mathscr{J}$ which extends $\mathscr{I}$, ρ extends to a persistent automorphism of $\mathscr{J}$.*

Proof: Suppose that $\mathscr{J}$ were a countable jump ideal such that there is no persistent automorphism of $\mathscr{J}$ which extends ρ. Let J compute a presentation of $\mathscr{J}$. Choose x_e so that the eth arithmetic in J extension of ρ to $\mathscr{J}$ cannot be extended further to include x_e. Let x bound the x_e's. By its persistence, extend ρ to an automorphism ρ_1 of the jump ideal generated by x. Then, $\rho_1 \restriction \mathscr{J}$ is arithmetic in J, contradiction. ■

We now draw some conclusions about the complexity of ρ's being persistent.

Theorem 4.8 *The property I is a representation of a countable ideal $\mathscr{I}$, $0' \in \mathscr{I}$, and R is a presentation of a persistent automorphism ρ of $\mathscr{I}$ is Π_1^1.*

Proof: ρ is persistent if and only if for every presentation J of a countable jump ideal $\mathscr{J}$ extending $\mathscr{I}$, there is an arithmetic in J extension of ρ to $\mathscr{J}$. This property is Π_1^1. ■

Corollary 4.9 *The properties R is a presentation of a persistent automorphism and There is a countable map $\rho : \mathscr{I} \xrightarrow{\sim} \mathscr{I}$ such that $0' \in \mathscr{I}$, ρ is persistent, and ρ is not equal to the identity are absolute between well-founded models of ZFC.*

Proof: These properties are Π_1^1 and Σ_2^1, respectively. The corollary then follows from Shoenfield (1961) Absoluteness. ∎

4.3. Persistence and reflection

Let T be the fragment of *ZFC* in which we include only the instances of replacement and comprehension in which the defining formula is Σ_1.

Definition 4.10 Suppose that $\mathfrak{M} = (M, \in^{\mathfrak{M}})$ is a model of T.

1. $\mathfrak{M}$ is an *ω-model* if and only if $\mathbb{N}^{\mathfrak{M}}$ is isomorphic to the standard model of arithmetic.

2. $\mathfrak{M}$ is *well-founded* if and only if the binary relation $\in^{\mathfrak{M}}$ is well-founded. That is to say that there is no infinite sequence $(m_i : i \in \mathbb{N})$ of elements of $\mathfrak{M}$ such that for all i, $m_{i+1} \in^{\mathfrak{M}} m_i$.

Theorem 4.11 *Suppose that $\mathfrak{M}$ is an ω-model of T. Let $\mathscr{I}$ be an element of $\mathfrak{M}$ such that*

$$\mathfrak{M} \models \mathscr{I} \text{ is a countable ideal in } \mathscr{D} \text{ such that } 0' \in \mathscr{I}.$$

Then, every persistent automorphism of $\mathscr{I}$ is also an element of $\mathfrak{M}$.

Proof: $\mathfrak{M}$ is closed under arithmetic definability. ∎

Corollary 4.12 *Suppose that $\mathfrak{M}$ is an ω-model of T and that ρ and $\mathscr{I}$ are elements of $\mathfrak{M}$ such that $0' \in \mathscr{I}$, $\rho : \mathscr{I} \xrightarrow{\sim} \mathscr{I}$, and $\mathscr{I}$ is countable in $\mathfrak{M}$. Then,*

$$\rho \text{ is persistent} \Longrightarrow \mathfrak{M} \models \rho \text{ is persistent}.$$

Proof: Persistent automorphisms extend to larger countable jump ideals persistently. Hence, these extensions belong to $\mathfrak{M}$. ∎

4.4. Generic persistence

We now extend the notion of persistence to uncountable ideals. In what follows, V is the universe of sets and G is a V-generic filter for some partial order in V.

Definition 4.13 Suppose that $\mathscr{I}$ is an ideal in $\mathscr{D}$ and ρ is an automorphism of $\mathscr{I}$. We say that ρ is *generically persistent* if there is a generic extension $V[G]$ of V in which $\mathscr{I}$ is countable and ρ is persistent.

Theorem 4.14 *Suppose that* $\rho : \mathscr{I} \xrightarrow{\sim} \mathscr{I}$ *is generically persistent. If* $V[G]$ *is a generic extension of* V *in which* $\mathscr{I}$ *is countable, then* ρ *is persistent in* $V[G]$.

Proof: Generics for any two forcings can be realized simultaneously. By absoluteness, persistence is evaluated consistently in the two generic extensions.   ∎

Theorem 4.15 *Suppose that* $\pi : \mathscr{D} \xrightarrow{\sim} \mathscr{D}$. *Then,* π *is generically persistent.*

Proof: If not, then the failure of π to be generically persistent would reflect to a countable well-founded model $\mathfrak{M}$. One could then add a generic counting of $\mathscr{D}^{\mathfrak{M}}$ to $\mathfrak{M}$ and obtain $\mathfrak{M}[G]$ in which $\pi \restriction \mathscr{D}^{\mathfrak{M}}$ is not persistent. This would contradict the persistence of $\pi \restriction \mathscr{D}^{\mathfrak{M}}$ and Corollary 4.12.   ∎

Theorem 4.16 *Suppose that* $V[G]$ *is a generic extension of* V. *Suppose that* π *is an element of* $V[G]$ *which maps the Turing degrees in* V *automorphically to itself (that is,* $\pi : \mathscr{D}^V \xrightarrow{\sim} \mathscr{D}^V$). *If* π *is generically persistent in* $V[G]$, *then* π *is an element of* $L(\mathbb{R}^V)$. *That is,* π *is constructible from the set of reals in* V.

Proof: π is generically persistent, so π is arithmetically definable relative to any $V[G]$-generic counting of $\mathscr{D}^V$. Consequently, π must belong to the ground model for such countings, namely $L(\mathbb{R}^V)$.   ∎

Theorem 4.17 *Suppose that* $\mathscr{I}$ *is a countable ideal in* $\mathscr{D}$, $0'$ *is an element of* $\mathscr{I}$, *and* $\rho : \mathscr{I} \xrightarrow{\sim} \mathscr{I}$ *is persistent. Then* ρ *can be extended to an automorphism* $\pi : \mathscr{D} \xrightarrow{\sim} \mathscr{D}$.

Proof: ρ can be persistently extended to $\mathscr{D}^V$ in a generic extension of V. By Theorem 4.16, this extension belongs to $L[\mathbb{R}^V]$ and hence to V.   ∎

Corollary 4.18 *The statement* There is a non-trivial automorphism of the Turing degrees *is equivalent to a* Σ_2^1 *statement. It is therefore absolute between well-founded models of ZFC.*

Theorem 4.19 *Let* π *be an automorphism of* $\mathscr{D}$. *Suppose that* $V[G]$ *is a generic extension of* V. *Then, there is an extension of* π *in* $V[G]$ *to an automorphism of* $\mathscr{D}^{V[G]}$, *the Turing degrees in* $V[G]$.

Proof: There is a persistent extension π_1 of π in any generic extension of $V[G]$ in which $\mathscr{D}^{V[G]}$ is countable. This π_1 belongs to $V[G]$.   ∎

4.5. Definability of automorphisms of $\mathscr{D}$

Definition 4.20 Given two functions $\tau : \mathscr{D} \to \mathscr{D}$ and $t : 2^\omega \to 2^\omega$, we say that t *represents* τ if for every degree x and every set X in x, the Turing degree of $t(X)$ is equal $\tau(x)$.

We will analyze the behavior of an automorphism of $\mathscr{D}$ in terms of the action of its extensions on the degrees of the generic reals.

Theorem 4.21 *Suppose that* $\pi : \mathscr{D} \tilde{\to} \mathscr{D}$. *There is a countable family* $\vec{D}$ *of dense open subsets of* $2^{<\omega}$ *such that* π *is represented by a continuous function* f *on the set of* $\vec{D}$-*generic reals.*

Proof: The proof has several steps, which we will sketch. We use $\Pi(Z)$ to denote a representative of $\pi(degree(Z))$.

1. Let $V[\mathscr{G}]$ be a generic extension of V obtained by adding ω_1-many Cohen reals and let π_1 be an extension of π to $\mathscr{D}^{V[\mathscr{G}]}$.

2. Since $\pi_1 \in L[\mathbb{R}^{V[\mathscr{G}]}]$, fix $X \in \mathbb{R}^{V[\mathscr{G}]}$ so that π_1 is ordinal definable from X in $L[\mathbb{R}^{V[\mathscr{G}]}]$. Work in $V[X]$ and note that $V[\mathscr{G}]$ is a generic extension of $V[X]$ obtained by adding ω_1-many Cohen reals. (The forcing factors.)

3. Consider a set G, of degree g, which is Cohen generic over $V[X]$. $\pi_1(g)$ is arithmetically definable relative to g and $\pi^{-1}(0')$. We can find an e and a k such that it is forced that $\pi_1(g)$ is represented by $\{e\}((G \oplus \Pi^{-1}(\emptyset'))^{(k)})$. Since G is Cohen generic, we can assume that e has the form $\{e\}(G \oplus \Pi^{-1}(\emptyset')^{(k)})$. Thus, π_1 is continuously represented on the set of $V[X]$-generic reals.

4. We make an aside to exploit a phenomenon first observed by Jockusch and Posner (1981). For any $\vec{D}$, the $\vec{D}$-generic degrees generate $\mathscr{D}$ under meet and join. We fix a mechanism by which this coding can be realized.

Let G and Y be given. Let G_{even} and G_{odd} be the even and odd parts of G. Construct $\mathbb{C}(Y, G)$ by injecting the values of Y into G_{even} at those places where G_{odd} is not zero. That is, we shuffle the bitstreams of G_{even} and Y like a deck of playing cards and use G_{odd} to determine the points at which the cards in the Y half of the deck are inserted between the cards in the G_{even} half of the deck.

Lemma 4.22 *If G_{odd} is infinite, then $\mathbb{C}(Y, G) \oplus G \equiv_T Y \oplus G$.*

Lemma 4.23 *For any dense open subset of $2^{<\omega}$, D, there is a dense open set D^*, such that for all D^*-generic G and all Y, $\mathbb{C}(Y, G)$ is D-generic. In particular, for all G, Y, and Z, if G is generic over $V[Z]$, then so is $\mathbb{C}(Y, G)$.*

Definition 4.24 For $Y \in 2^\omega$, let (Y) denote the set $\{Z : Z \leq_T Y\}$.

Let Y be given with Turing degree y, and let G_1 and G_2 be mutually Cohen generic over $V[X \oplus Y]$. We can write the ideal generated by Y as the meet of joins of generic ideals.

$$(\mathbb{C}(Y, G_1) \oplus G_1) \cap (\mathbb{C}(Y, G_2) \oplus G_2) = (Y \oplus G_1) \cap (Y \oplus G_2)$$
$$= (Y)$$

Thus, as Jockusch and Posner observed, the degrees of the generic sets generate the Turing degrees under meet and join.

5. The previous equality is preserved by π_1, as represented on generic reals.

$\{Z : \text{the degree of } Z \text{ belongs to } (\pi_1(y))\} =$
$$\left(\{e\}(\mathbb{C}(Y, G_1) \oplus \Pi^{-1}(\emptyset')^{(k)}) \oplus \{e\}(G_1 \oplus \Pi^{-1}(\emptyset')^{(k)})\right)$$
$$\bigcap \left(\{e\}(\mathbb{C}(Y, G_2) \oplus \Pi^{-1}(\emptyset')^{(k)}) \oplus \{e\}(G_2 \oplus \Pi^{-1}(\emptyset')^{(k)})\right)$$

When Y is also generic:

$$\left(\{e\}(Y \oplus \Pi^{-1}(\emptyset')^{(k)})\right) =$$
$$\left(\{e\}(\mathbb{C}(Y, G_1) \oplus \Pi^{-1}(\emptyset')^{(k)}) \oplus \{e\}(G_1 \oplus \Pi^{-1}(\emptyset')^{(k)})\right)$$
$$\bigcap \left(\{e\}(\mathbb{C}(Y, G_2) \oplus \Pi^{-1}(\emptyset')^{(k)}) \oplus \{e\}(G_2 \oplus \Pi^{-1}(\emptyset')^{(k)})\right)$$

This exhibits the desired representation of π on generic reals Y. ∎

Sharper results will follow, but we can obtain some preliminary information concerning the definability of π from what we already know.

Corollary 4.25 *Suppose $\pi : \mathscr{D} \overset{\sim}{\to} \mathscr{D}$. Then π has a Borel representation; in fact, π has a representation that is arithmetic in the real parameter $\Pi^{-1}(\emptyset')$.*

Next in this line is the proof that for any Z and any sufficiently generic real G, $(G \oplus Z)'' \geq_T \Pi(Z)$. The proof uses two facts. First, for any countable collection of dense open sets $\overrightarrow{D}$, there is another collection $\overrightarrow{D}^*$ such that if G^* is $\overrightarrow{D}^*$-generic then there is a $\overrightarrow{D}$-generic G with $\Pi(G^*) \geq_T G$. Second, for any set X, there is an X-recursive partial order P such that the degrees of any sufficiently generic sets for P form parameters to code X. The coding is sufficiently effective that X is recursive in the double-jump of any upper bound on these parameters. We leave the details of the proof to Slaman and Woodin (2005).

Theorem 4.26 *Suppose $\pi : \mathscr{D} \overset{\sim}{\to} \mathscr{D}$. For every $z \in \mathscr{D}$, $z'' \geq_T \pi(z)$.*

Proof: Let Z be given. Fix $\overrightarrow{D}$ so that for any $\overrightarrow{D}$-generic G, $\Pi(Z) \oplus G$ can compute a generic set for the partial order to produce parameters which code $\Pi(Z)$. Fix $\overrightarrow{D}^*$ so that if G^* is $\overrightarrow{D}^*$-generic, then $\Pi(G^*)$ computes a $\overrightarrow{D}$-generic. Let G^* be $\overrightarrow{D}^*$-generic and let G be a $\overrightarrow{D}$-generic recursive in $\Pi(G^*)$. The coding is preserved by π, so we may conclude that $degree(\Pi(Z)) \leq_T \pi^{-1}(degree(\Pi(Z) \oplus G))''$. Hence $\Pi(Z) \leq_T (Z \oplus G^*)''$, and so $\Pi(Z) \leq_T Z'' \oplus G^*$. G^* was any sufficiently generic, so $\Pi(Z) \leq_T Z''$. ∎

Corollary 4.27 *Suppose $\pi : \mathscr{D} \overset{\sim}{\to} \mathscr{D}$. For any 2-generic set G,*

$$degree(G) \vee 0'' \geq_T \pi(degree(G)).$$

Theorem 4.28 *Suppose that $\pi : \mathscr{D} \overset{\sim}{\to} \mathscr{D}$.*

- *For all $x \in \mathscr{D}$, $x \vee 0'' \geq_T \pi(x)$.*

- *For all $x \in \mathscr{D}$, if $x \geq_T 0''$ then $x = \pi(x)$.*

Proof: A degree above $0''$ can be written as a join of 2-generic degrees. ∎

Theorem 4.29 *Suppose that $\pi : \mathscr{D} \overset{\sim}{\to} \mathscr{D}$.*

- *There is a recursive functional $\{e\}$ such that for all G, if G is 5-generic, then $\pi(degree(G))$ is represented by $\{e\}(G, \emptyset'')$.*

- *There is an arithmetic function $F : 2^\omega \to 2^\omega$ such that for all $X \in 2^\omega$, $\pi(degree(X))$ is represented by $F(X)$.*

Proof: Replay the proof of Theorem 4.21, using the new information that $\pi(degree(G))$ is recursive in G''. Conclude that there is a fixed reduction which works for all 5-generic G's. Since the 5-generics generate $\mathscr{D}$, the representation on 5-generics propagates to an arithmetic representation everywhere. ∎

Theorem 4.30 $Aut(\mathscr{D})$ *is countable.*

Theorem 4.31 *If g is 5-generic and $\pi : \mathscr{D} \overset{\sim}{\to} \mathscr{D}$, then π is determined by its action on g.*

Proof: If G is 5-generic, then $\{e\}(G, \emptyset'') \equiv_T G$ if and only if the same is true for all 5-generics. ∎

4.6. Invariance of the double jump

The efficient coding that lies behind the proof of Theorem 4.26 can be sharpened not just to produce parameters that code z but rather to produce parameters that code z''.

Theorem 4.32 *For every $Z \subseteq \omega$, there is a countable family of dense open sets $\vec{D}$ such that such that for all $\vec{D}$-generic G, $\pi(degree(Z \oplus G))'' \geq_T degree(Z'')$*

Theorem 4.33 *Suppose that $\pi : \mathscr{D} \overset{\sim}{\to} \mathscr{D}$. For all $z \in \mathscr{D}$, $z'' = \pi(z)''$.*

Theorem 4.34 *The relation $y = x''$ is invariant under π.*

Proof: Suppose that $y = x''$. Since $y \geq_T 0''$, $\pi(y) = y$. By the previous theorem, $x'' = \pi(x)''$. Consequently, $\pi(y) = \pi(x)''$. By the same argument applied to π^{-1}, if $\pi(y) = \pi(x)''$ then $y = x''$. ∎

5. Definability in $\mathscr{D}$

5.1. Bi-interpretability

Definition 5.1 An *assignment* of reals consists of

- A countable ω-model $\mathfrak{M}$ of T $(T = \Sigma_1\text{-}ZFC)$.

- A function f and a countable ideal $\mathscr{I}$ in $\mathscr{D}$ such that $f : \mathscr{D}^{\mathfrak{M}} \to \mathscr{I}$ surjectively and for all x and y in $\mathscr{D}^{\mathfrak{M}}$, $\mathfrak{M} \models x \geq_T y$ if and only if $f(x) \geq_T f(y)$ in $\mathscr{I}$.

An assignment of reals to an ideal $\mathscr{I}$ is a representation of an isomorphism between the Turing degrees of the reals in an ω-model $\mathfrak{M}$ and the elements of $\mathscr{I}$. We can work with countable assignments within $\mathscr{D}$, via the coding lemma, and we can investigate which assignments extend, just as we did with persistent automorphism.

Definition 5.2 For assignments $(\mathfrak{M}_0, f_0, \mathscr{I}_0)$ and $(\mathfrak{M}_1, f_1, \mathscr{I}_1)$, $(\mathfrak{M}_1, f_1, \mathscr{I}_1)$ *extends* $(\mathfrak{M}_0, f_0, \mathscr{I}_0)$ if and only if

- $\mathscr{D}^{\mathfrak{M}_0} \subseteq \mathscr{D}^{\mathfrak{M}_1}$,

- $\mathscr{I}_0 \subseteq \mathscr{I}_1$,

- and $f_1 \upharpoonright \mathscr{D}^{\mathfrak{M}_0} = f_0$.

Definition 5.3 An assignment $(\mathfrak{M}_0, f_0, \mathscr{I}_0)$ is *extendable* if

$$\forall z_1 \exists (\mathfrak{M}_1, f_1, \mathscr{I}_1) \left[\begin{array}{l} (\mathfrak{M}_1, f_1, \mathscr{I}_1) \text{ extends } (\mathfrak{M}_0, f_0, \mathscr{I}_0)), z_1 \in \mathscr{I}_1, \text{ and} \\ \forall z_2 \exists (\mathfrak{M}_2, f_2, \mathscr{I}_2) \\ \left(\begin{array}{l} (\mathfrak{M}_2, f_2, \mathscr{I}_2) \text{ extends } (\mathfrak{M}_1, f_1, \mathscr{I}_1), z_2 \in \mathscr{I}_2, \text{ and} \\ \forall z_3 \exists (\mathfrak{M}_3, f_3, \mathscr{I}_3) \left[\begin{array}{l} (\mathfrak{M}_3, f_3, \mathscr{I}_3) \text{ extends} \\ (\mathfrak{M}_2, f_2, \mathscr{I}_2) \text{ and } z_3 \in \mathscr{I}_3 \end{array} \right] \end{array} \right) \end{array} \right]$$

Theorem 5.4 *If* $(\mathfrak{M}, f, \mathscr{I})$ *is an extendable assignment, then there is a* $\pi : \mathscr{D} \xrightarrow{\sim} \mathscr{D}$ *such that for all* $x \in \mathscr{D}^{\mathfrak{M}}$, $\pi(x) = f(x)$.

Proof: One can compare ideals $\mathscr{D}^{\mathfrak{M}}$ and $\mathscr{I}$ by considering the sets that are coded within them. Sets coded in the range $\mathscr{I}$ belong to the domain $\mathfrak{M}$. Sets in the domain which together with $0'$ can only code elements of $\mathfrak{M}$ must belong to $\mathfrak{M}$.

One shows that if $(\mathfrak{M}, f, \mathscr{I})$ is an extendable assignment, then $f : \mathscr{D}^{\mathfrak{M}} \to \mathscr{I}$ extends to a persistent automorphism of a larger ideal. Hence, it extends to an automorphism of $\mathscr{D}$. ∎

Theorem 5.5 *If* g *is the Turing degree of an arithmetically definable 5-generic set, then the relation* $R(\vec{c}, d)$ *given by*

$$R(\vec{c}, d) \iff \vec{c} \text{ codes a real } D \text{ and } D \text{ has degree } d$$

is definable in $\mathscr{D}$ *from* g.

This is the internal realization of the previous result that every automorphism is determined by its action on g.

Corollary 5.6 *Suppose that R is a relation on $\mathcal{D}$. The following conditions are equivalent.*

- *R is induced by a projective, degree invariant relation R_{2^ω} on 2^ω.*

- *R is definable in $\mathcal{D}$ using parameters.*

Proof: $\overrightarrow{x}$ satisfies R if and only if there is a correct assignment of representatives to degrees such that $f(degree(\overrightarrow{Y})) = \overrightarrow{x}$, and $\overrightarrow{Y}$ satisfies R_{2^ω}. (The correctness of the assignment is defined using the arithmetic 5-generic of the previous theorem.) ∎

Theorem 5.7 *Suppose that R is a relation on $\mathcal{D}$. The following conditions are equivalent.*

- *R is induced by a relation R_{2^ω} on 2^ω such that the following conditions hold.*

  - *R_{2^ω} is definable in second order arithmetic and degree invariant.*
  - *R_{2^ω} is preserved by $Aut(\mathcal{D})$.*

- *R is definable in $\mathcal{D}$.*

Proof: $\overrightarrow{x}$ satisfies R if and only if there is an extendable assignment such that $f(degree(\overrightarrow{Y})) = \overrightarrow{x}$ and $\overrightarrow{Y}$ satisfies R_{2^ω}. ∎

Definition 5.8 *$\mathcal{D}$ is bi-interpretable with second order arithmetic if and only if the relation on $\overrightarrow{c}$ and d given by*

$$R(\overrightarrow{c}, d) \iff \overrightarrow{c} \text{ codes a real } D \text{ and } D \text{ has degree } d$$

is definable in $\mathcal{D}$.

Theorem 5.9 *The following are equivalent.*

- *$\mathcal{D}$ is bi-interpretable with second order arithmetic.*

- *$\mathcal{D}$ is rigid.*

Conjecture 5.10 (Slaman and Woodin (2005)) $\mathscr{D}$ *is bi-interpretable with second order arithmetic.*

The Bi-interpretability Conjecture, if true, reduces all the logical questions that one could ask of $\mathscr{D}$ to the exact same questions about second order arithmetic. The structures would be logically identical, though presented in different first order languages.

6. The Turing jump

Theorem 6.1 *The function* $x \mapsto x''$ *is definable in* $\mathscr{D}$.

Proof: We have already shown that the relation $y = x''$ is invariant under all automorphisms of $\mathscr{D}$. It is clearly degree invariant and definable in second order arithmetic. Therefore, it is definable in $\mathscr{D}$.   ∎

We now turn to showing that $x \mapsto x'$ is definable in $\mathscr{D}$. This is an account of work appearing in Shore and Slaman (1999).

We show that (Δ_2^0), the ideal of Δ_2^0 degrees, is definable in $\mathscr{D}$. Our definition is based on the following join theorem for the double-jump.

Theorem 6.2 (Shore and Slaman, 1999) *For* $A \in 2^\omega$, *the following conditions are equivalent.*

- *A is not recursive in $0'$.*

- *There is a $G \in 2^\omega$ such that $A \oplus G \geq_T G''$.*

Theorem 6.2 is an extension of the Posner and Robinson (1981) Theorem that for every nonrecursive degree A there is a G such that $A \oplus G \geq_T G'$. The proof uses a notion of forcing introduced by Kumabe and Slaman.

By Theorem 6.2, (Δ_2^0) is definable in terms of order, join, and the double jump. Consequently, it is definable in $\mathscr{D}$.

Theorem 6.3 *The functions* $a \mapsto (\Delta_2^0(a))$ *and* $a \mapsto a'$ *are definable in* $\mathscr{D}$∎

Proof: By relativizing the previous theorem. For each degree a and each d greater than or equal to a, d is not Δ_2^0 relative to a if and only if there is an x greater than or equal to a such that $d + x \geq_T x''$. Again, the double jump is definable in $\mathscr{D}$, and this equivalence provides first order definitions as required.

Recently, Shore (2007) has produced an alternate proof of Theorem 6.3. Shore's proof replaces the definition of the double jump obtained from the analysis that we have given for $Aut(\mathscr{D})$ with one that is more arithmetically based. He then argues as here that the definability of the double jump implies the definability of the Turing jump.

6.1. Recursive enumerability

We have already stated the Bi-interpretability Conjecture, which we regard as the central question concerning the global structure of the Turing degrees. As we have seen with the Turing jump, specialized definability results are possible even without settling the conjecture. We suggest the following.

Question 6.4 *Is the relation y is recursively enumerable relative to x definable in $\mathscr{D}$?*

A positive answer would follow from a proof of the Bi-interpretability Conjecture. Conceivably, it could also lead to a proof.

References

Cooper, S. B. (1999). Upper cones as automorphism bases. *Siberian Advances in Math.* 9(3), 1–61.

Friedberg, R. M. (1957). A criterion for completeness of degrees of unsolvability. *J. Symbolic Logic* 22, 159–160.

Jockusch, Jr., C. G. and D. Posner (1981). Automorphism bases for degrees of unsolvability. *Israel J. Math.* 40, 150–164.

Jockusch, Jr., C. G. and R. A. Shore (1984). Pseudo-jump operators II: Transfinite iterations, hierarchies, and minimal covers. *J. Symbolic Logic* 49, 1205–1236.

Kleene, S. C. and E. L. Post (1954). The upper semi-lattice of degrees of recursive unsolvability. *Ann. of Math.* 59, 379–407.

Lachlan, A. H. (1968). Distributive initial segments of the degrees of unsolvability. *Z. Math. Logik Grundlag. Math.* 14, 457–472.

Lachlan, A. H. and R. Lebeuf (1976). Countable initial segments of the degrees. *J. Symbolic Logic* 41, 289–300.

Lerman, M. (1971). Initial segments of the degrees of unsolvability. *Ann. of Math.* 93, 311–389.

Lerman, M. (1983). *Degrees of Unsolvability.* Perspectives in Mathematical Logic. Heidelberg: Springer–Verlag. 307 pages.

Nerode, A. and R. A. Shore (1980). Reducibility orderings: theories, definability and automorphisms. *Ann. Math. Logic 18*, 61–89.

Odifreddi, P. and R. Shore (1991). Global properties of local structures of degrees. *Boll. Un. Mat. Ital. B (7) 5*(1), 97–120.

Posner, D. B. and R. W. Robinson (1981). Degrees joining to **0′**. *J. Symbolic Logic 46*(4), 714–722.

Sacks, G. E. (1961). On suborderings of degrees of recursive unsolvability. *Z. Math. Logik Grundlag. Math. 17*, 46–56.

Shoenfield, J. R. (1961). The problem of predicativity. In *Essays on the Foundations of Mathematics*, pp. 132–139. Hebrew University, Jerusalem: Magnes Press.

Shore, R. A. (1978). On the ∀∃-sentences of α-recursion theory. In R. O. G. J. Fenstad and G. E. Sacks (Eds.), *Generalized Recursion Theory II*, Volume 94 of *Stud. Logic Foundations Math.*, Amsterdam, pp. 331–354. North–Holland Publishing Co.

Shore, R. A. (2007). Direct and local definitions of the Turing jump. preprint.

Shore, R. A. and T. A. Slaman (1999). Defining the Turing jump. *Mathematical Research Letters 6*, 711–722.

Simpson, S. G. (1977). First order theory of the degrees of recursive unsolvability. *Ann. of Math. 105*, 121–139.

Slaman, T. A. and W. H. Woodin (1986). Definability in the Turing degrees. *Illinois J. Math. 30*(2), 320–334.

Slaman, T. A. and W. H. Woodin (2005). Definability in degree structures. Unpublished.

Spector, C. (1956). On degrees of recursive unsolvability. *Ann. of Math. (2) 64*, 581–592.

SET THEORY TUTORIALS

DERIVED MODELS ASSOCIATED TO MICE

John R. Steel

Department of Mathematics, University of California, Berkeley
717 Evans Hall #3840, Berkeley, CA 94720-3840, USA
E-mail: steel@math.berkeley.edu

We shall present some results on the the properties of derived models of mice, and on the existence of mice with large derived models. The basic plan of the paper is:

Sections 1–5. Introduction. Preliminary definitions and background material. The iteration independence of the derived model associated to a mouse. The mouse set conjecture in derived models associated to mice.

Sections 6–8 The Solovay sequence in derived models associated to mice. A method of "translating away" extenders overlapping Woodin cardinals in mice.

Sections 9–14. Some partial results on capturing Σ_1^2 truths in AD^+ models by mice. (That is, partial results on the mouse set conjecture.)

Section 15 A theorem of Woodin on the consistency strength of $\mathsf{AD}^+ + \theta_0 < \theta$.

Section 16 The global mouse set conjecture implies its local version.

Section 17 Capturing sets of reals in $\mathbb{R}$-mice.

1. Introduction

In this paper, we shall present some results on the the properties of derived models of mice, and on the existence of mice with large derived models.[1]

In order to motivate what follows, let us recall some well-known Holy Grails of inner model theory:

---

[1] The paper began as a set of notes which accompanied some lectures at the workshop *Computational Prospects of Infinity*, held at the Institute for Mathematical Sciences of the National University of Singapore, in June 2005. The author wishes to thank the conference organizers and the Institute for their gracious hospitality.

Conjecture 1.0.1 *Each of the following statements implies the existence of an* $\omega_1 + 1$-*iterable mouse with a superstrong cardinal:*

(1) There is a supercompact cardinal,

(2) There is a strongly compact cardinal,

(3) PFA,

(4) CH *holds, and there is a homogeneous pre-saturated ideal on* ω_1,

(5) $AD_\mathbb{R}$ *holds, and* θ *is regular.*

This list could be lengthened substantially. In all cases, we can prove the statement in question implies that there is an iterable mouse with infinitely many Woodin cardinals. In cases (2)-(5), we cannot yet produce a mouse with a Woodin limit of Woodin cardinals. (Case (1) is special; here the best partial results are those of [13], which reach a bit past a Woodin limit of Woodins.)

The partial results we have in cases (2)-(5) come from the *core model induction* method. In this method (due to Woodin), one uses core model theory to construct new mice, while using descriptive set theory to keep track of the degree of correctness and the complexity of the iteration strategies of the mice one has constructed. See [4], [32], and [27] for examples of core model inductions. We suspect that even in case (1), one will not be able to go significantly beyond where we are now without bringing in core model induction ideas.

Core model inductions reaching infinitely many Woodins, or further, rely on a key step in which one passes from some kind of *hybrid mouse* to an ordinary $L[\vec{E}]$ mouse. The hybrid mouse $\mathcal{H}$ will have a small number of Woodin cardinals, but these cardinals will be Woodin with respect to predicates coding up a lot of information. The $L[\vec{E}]$ mouse $\mathcal{M}$ is built by a full background extender construction inside $\mathcal{H}$, and may have all sorts of Woodin cardinals, even cardinals strong past a Woodin, and presumably even superstrong cardinals. We sometimes call the ordinary $L[\vec{E}]$ mice *ms-mice*; they are mice in the sense of [19], except that we allow them to be relativised by putting some arbitrary transitive set y at the bottom, in which case we speak of an ms-mouse *over y*. (See e.g. [27].) The following conjecture is the basic test problem for our ability to translate hybrid mice into ordinary ms-mice.

Conjecture 1.0.2 ((Mouse Set Conjectures (MSC))) *Assume* AD^+, *and that there is no ω_1-iteration strategy for a mouse with a superstrong cardinal; then for any countable transitive set y,*

(I) *if $x \subseteq y$, and x is ordinal definable from parameters in $y \cup \{y\}$, then there is an ω_1-iterable ms-mouse $\mathcal{M}$ over y such that $x \in \mathcal{M}$, and*

(II) *if $\exists A \subseteq \mathbb{R}(HC, \in, A) \models \varphi[y]$, then there is an ω_1-iterable ms-mouse $\mathcal{M}$ over y, and a λ such that*

$$\mathcal{M} \models \lambda \text{ is a limit of Woodin cardinals,}$$

and

$$\mathcal{M} \models \exists A \in \mathrm{Hom}_{<\lambda}((HC, \in, A) \models \varphi[y]).$$

MSC2 is equivalent to asserting that the Σ_1^2 fact $\exists A \subseteq \mathbb{R}(HC, \in, A) \models \varphi[y]$ holds in the derived model of $\mathcal{M}$ below λ. It is easy to show that MSC2 implies MSC1, and not too hard to show that MSC1 implies MSC2. We shall therefore use MSC to stand for either MSC1 or MSC2, in situations where we don't need to make a distinction.

Clearly, the conclusion of MSC1 is an equivalence: if x is in an iterable mouse over y, then $x \in \mathrm{OD}(y \cup \{y\})$. In MSC2, one seems to need to add something about the iteration strategy for $\mathcal{M}$ in order to conclude that any Σ_1^2 statement true in the derived model of $\mathcal{M}$ is actually true. This can be done in a natural way. (It is enough that the strategy be "good" in the sense described at the beginning of section 3.)

MSC is known for a certain initial segment of the Wadge hierarchy. Woodin has shown:

Theorem 1.1 (Woodin, late 90's) MSC *holds if the hypothesis is strengthened to* AD^+ *plus there is no iteration strategy for an ms-mouse $\mathcal{M}$ such that for some λ,*

$$\mathcal{M} \models \lambda \text{ is a limit of Woodin cardinals },$$

and

$$\{\kappa \mid \kappa \text{ is } < \lambda\text{-strong in } \mathcal{M}\} \text{ has order type } \lambda.$$

The proof builds on Woodin's proof of MSC under the stronger hypothesis that there is no iteration strategy for a nontame mouse. That proof is written up in [23]. There are many additional difficulties in dealing with ms-mice having extenders overlapping Woodin cardinals, however. An ω_1 iteration strategy for a countable premouse is essentially a set of reals, so the hypothesis of Theorem 1.1 asserting that there is no ω_1 iteration strategy for a mouse of the type amounts to a restriction to the initial segment of the Wadge hierarchy below the Wadge-least such strategy. A roughly equivalent way to put this restriction is: there is no proper initial segment Γ of the Wadge hierarchy such that $L(\Gamma, \mathbb{R}) \models \mathsf{AD}_{\mathbb{R}} + \mathsf{DC}$.

More recently, Itay Neeman and the author have shown

Theorem 1.2 (Neeman-Steel, 2004) MSC *holds if the hypothesis is strengthened to* AD^+ *plus there is no iteration strategy for an ms-mouse* $\mathcal{M}$ *such that for some* λ,

$$\mathcal{M} \models \lambda \text{ is a limit of Woodin cardinals,}$$

and for some $\kappa < \lambda$,

$$\mathcal{M} \models \kappa \text{ is } < \lambda \text{ strong, and a limit of } < \lambda\text{-strongs.}$$

Although the large cardinal reached here is only slightly beyond the large cardinals reached in 1.1, the proof has some new ideas, and seems somewhat simpler, and closer to Woodin's original argument in the tame mouse case. In sections 9 through 14, we shall give these arguments in the special case that V is the minimal model of $\mathsf{AD}^+ + \theta_0 < \theta$. Pretty much all the ideas occur in this case. In section 15, we use this work to re-prove a theorem of Woodin which identifies the consistency strength of $\mathsf{AD}^+ + \theta_0 < \theta$.

The mouse set conjectures ask us to construct ms-mice with given derived models. This leads naturally to the question: *what can one say about the derived model of a mouse?* (See [25] for an exposition of the basic facts about the derived model construction.) This will be the focus of sections 3 through 8. We shall show that for certain "tractable" $\mathcal{M}, \lambda$ such that $\mathcal{M}$ is a mouse and λ is a limit of Woodins in $\mathcal{M}$, the derived model $\mathcal{D}(\mathcal{M}, \lambda)$ of $\mathcal{M}$ at λ satisfies MSC. We shall show that for these tractable pairs $(\mathcal{M}, \lambda)$, there is a canonical derived model of an iterate of $\mathcal{M}$ whose reals are precisely the reals in V. Finally, we shall investigate the *Solovay sequence* $\langle \theta_\alpha \rangle$ of various $\mathcal{D}(\mathcal{M}, \lambda)$. For example, letting $M^\sharp_{\text{wlim}}$ be the minimal active mouse with a Woodin limit of Woodin cardinals, we shall show

Theorem 1.3 (Closson, Steel) *Let* λ *be the Woodin limit of Woodins in* $M^\sharp_{\text{wlim}}$; *then*

$$\mathcal{D}(M^\sharp_{\text{wlim}}, \lambda) \models \theta = \theta_\theta.$$

We do not know what the cofinality of θ is in the model of 1.3. Possibly θ is regular here, in which case $AD_\mathbb{R}$ plus "θ is regular" is much weaker than conjecture 1.0.1(5) would have it.

In section 16, we combine ideas from the two parts of the paper in order to show that MSC implies a reasonably fine local strengthening of itself. In section 17, we show that MSC implies that every $OD(\mathbb{R})$ set of reals is in a countably iterable mouse over $\mathbb{R}$.

This paper is far from self-contained. We have gathered together a fair amount of background material for the reader's convenience, but in order to keep this document to a reasonable size, we have also simply pointed to the literature at many points. Unfortunately, some of the key results have not been written up in full anywhere. We have sketched proofs of such results, or given references for parts of them, when it was feasible to do so.

2. Some background and preliminaries

2.1. *Homogeneously Suslin sets*

The good sets of reals, from the point of view of descriptive set theory, are the homogeneously Suslin sets.

Definition 2.1 *A homogeneity system with support Z is a function $\bar{\mu}$ such that for all $s, t \in \omega^{<\omega}$,*

1. μ_t is a countably complete ultrafilter concentrating on $Z^{\mathrm{dom}(t)}$, and

2. $s \subseteq t \Rightarrow \mu_t$ projects to μ_s.

If all μ_t are κ-complete, then we say $\bar{\mu}$ is κ-complete.

Definition 2.2 *If $\bar{\mu}$ is a homogeneity system, then for $x \in \omega^{<\omega}$,*

$$x \in S_{\bar{\mu}} \Leftrightarrow \text{ the tower of measures } \langle \mu_{x\restriction n} \mid n < \omega \rangle$$
$$\text{is countably complete.}$$

Let

$$\mathrm{Hom}_\kappa = \{ S_{\bar{\mu}} \mid \bar{\mu} \text{ is } \kappa\text{-complete} \},$$
$$\mathrm{Hom}_{<\lambda} = \bigcap_{\kappa < \lambda} \mathrm{Hom}_\kappa,$$
$$\mathrm{Hom}_\infty = \bigcap_{\kappa \in \mathrm{OR}} \mathrm{Hom}_\kappa.$$

A set of reals is *homogeneously Suslin* just in case it is in Hom_κ, for some κ. Although not literally stated in the paper, one of the main results of Martin [7] is that every homogeneously Suslin set is determined. There are no interesting homogeneously Suslin sets unless there are measurable cardinals. For the most part, we shall be working under the assumption that there are infinitely many Woodin cardinals, in which case one has:

Theorem 2.3 *Let λ be a limit of Woodin cardinals; then*

(a) $\text{Hom}_{<\lambda}$ is closed under complements and real quantification ([10]),

(b) every $\text{Hom}_{<\lambda}$ set has a $\text{Hom}_{<\lambda}$ scale ([25]).

Feng, Magidor, and Woodin ([1]) introduced an important notion which turns out to be equivalent to homogeneity in the presence of Woodin cardinals.

Definition 2.4 *Let T and T^* be trees on $\omega \times X$ and $\omega \times Y$ respectively; then T and T^* are κ-absolute complements iff whenever G is V-generic over a poset of size $\leq \kappa$, then $V[G] \models p[T] = \mathbb{R} \setminus p[T^*]$. We say that $A \subseteq \mathbb{R}$ is κ-universally Baire (or κ-UB) iff $A = p[T]$ for some tree T for which there is a κ-absolute complement.*

The results of [9], [30], and [10] show that if λ is a limit of Woodin cardinals, then the $\text{Hom}_{<\lambda}$ sets are precisely the $< \lambda$-universally Baire sets. (See [25].) The theorem and this equivalence also hold for $\lambda = \infty = \text{OR}$.

Stronger large cardinal hypotheses imply stronger closure properties of $\text{Hom}_{<\lambda}$. For example, Woodin has shown that if λ is a limit of Woodin cardinals, and there is a measurable cardinal above λ, then for any $A \in \text{Hom}_{<\lambda}$, $A^\sharp \in \text{Hom}_{<\lambda}$. However, no large cardinal hypothesis is known to imply that sets of reals definable using quantification over $\text{Hom}_{<\lambda}$ must be $\text{Hom}_{<\lambda}$, or even good in some weaker sense. The inner model theory for such a large cardinal hypothesis would be significantly different from the one we have now, as we explain in the next subsection.

2.2. Hom_∞ *iteration strategies*

Let $\mathcal{M}$ be a *ms-premouse*: a model of the form $(J_\alpha[\vec{E}], \in, \vec{E}, F)$, where $\vec{E}^\frown F$ is a fine extender sequence. ([19].) In the *iteration game* $G(\mathcal{M}, \theta)$, I and II cooperate to build an iteration tree $\mathcal{T}$ on $\mathcal{M}$:

- I extends $\mathcal{T}$ at successor steps by applying an extender from the current model to some, possibly earlier, model. The resulting ultrapower is the new current model.

- At limit steps II picks a cofinal branch b, and $\lim_{\alpha \in b} \mathcal{M}_\alpha$ becomes the current model.

II wins if, after θ moves, no illlfounded model has been produced. See [19] for a more complete description of $G(\mathcal{M}, \theta)$.

Definition 2.5 *A θ-iteration strategy for $\mathcal{M}$ is a winning strategy for II in $G(\mathcal{M}, \theta)$. $\mathcal{M}$ is θ-iterable if there is such an iteration strategy.*

Notice that if $\mathcal{M}$ is countable, then an ω_1-iteration strategy for $\mathcal{M}$ is essentially a set of reals. "Good" $\mathcal{M}$ (the "standard" ones), have ω_1-iteration strategies which are Hom_∞. This implies θ-iterability, for all θ.

For $L[\vec{E}]$-models in the region where we have a theory (can prove iterability),

$$x <^{L[\vec{E}]} y \Leftrightarrow \exists \mathcal{M}(\mathcal{M} \text{ has a } \mathrm{Hom}_\infty \text{ iteration strategy}$$
$$\text{and } \mathcal{M} \models x <^{L[\vec{E}]} y).$$

This can be used to show that the large cardinal hypotheses true in these models are compatible with there being a wellorder of $\mathbb{R}$ in $L(\mathbb{R}, \mathrm{Hom}_\infty)$.

2.3. *The derived model*

Given $A \in \mathrm{Hom}_\kappa$, as witnessed by both $\bar{\mu}$ and $\bar{\nu}$, and G which is V-generic for a poset of size $< \kappa$, we have

$$(S_{\bar{\mu}})^{V[G]} = (S_{\bar{\nu}})^{V[G]}.$$

So we write $A^{V[G]}$ for the common value of all $(S_{\bar{\mu}})^{V[G]}$ such that $A = S_{\bar{\mu}}$.

Now let λ be a limit of Woodin cardinals, and G be V-generic for $\mathrm{Col}(\omega, < \lambda)$, and set

$$\mathbb{R}_G^* = \bigcup_{\alpha < \lambda} \mathbb{R} \cap V[G \restriction \alpha],$$
$$\mathrm{Hom}_G^* = \{A^* \mid \exists \alpha < \lambda (A \in \mathrm{Hom}_{<\lambda}^{V[G \restriction \alpha]})\}$$

where $A^* = \bigcup_{\beta < \lambda} A^{V[G \restriction \beta]}$. Note that one has

$$\mathrm{Hom}_G^* = \{p[T] \cap \mathbb{R}_G^* \mid \exists \alpha < \lambda (V[G \restriction \alpha] \models T \text{ has an } < \lambda\text{-absolute complement})\}.$$

Theorem 2.6 (Derived model theorem, Woodin 1987)

(1) $L(\mathbb{R}^*, \mathrm{Hom}^*) \models \mathsf{AD}^+$.

(2) $\mathrm{Hom}^* = \{A \subseteq \mathbb{R}^* \mid A, \mathbb{R} \setminus A \text{ have scales in } L(\mathbb{R}^*, \mathrm{Hom}^*)\}$.

If $\mathcal{M} \models \lambda$ is a limit of Woodins, and a reasonable fragment of ZFC, then we write

$$D(\mathcal{M}, \lambda)$$

for the associated derived model. This is analogous to the V^P notation, since we have not specified a generic. Since the forcing is homogeneous, the theory of $D(\mathcal{M}, \lambda)$ does not depend on a generic. We shall speak of "realizations" of $D(\mathcal{M}, \lambda)$ if we have specified a generic G. Note that only $\mathbb{R}_G^*$, not all of G, is needed to realize $D(\mathcal{M}, \lambda)$.

Remark 2.7 This is the "old" derived model; the model is always of the form $L(\Gamma, \mathbb{R})$, where Γ is the class of its Suslin-co-Suslin sets. Not all models of AD^+ have this form. Woodin has shown that the Suslin-co-Suslin sets of any AD^+ model can be realized as the Suslin-co-Suslin sets of a derived model, however, and given a variant construction which hits all models of AD^+ plus $V = L(P(\mathbb{R}))$ in full.

If $\mathcal{M} = M_\omega$, the minimal mouse with ω Woodins, then letting λ be their sup, $D(\mathcal{M}, \lambda) = L(\mathbb{R}^*)$. On the other hand, for stronger $\mathcal{M}$, $D(\mathcal{M}, \lambda)$ may be larger.

Definition 2.8 *(AD^+.) For $A \subseteq \mathbb{R}$, $\theta(A)$ is the least ordinal α such that there is no surjection of $\mathbb{R}$ onto α which is ordinal definable from A and a real. We set*

$$\theta_0 = \theta(\emptyset),$$
$$\theta_{\alpha+1} = \theta(A), \text{ for any (all) } A \text{ of Wadge rank } \theta_\alpha,$$
$$\theta_\lambda = \bigcup_{\alpha < \lambda} \theta_\alpha.$$

$\theta_{\alpha+1}$ is defined iff $\theta_\alpha < \Theta$. Note $\theta(A) < \Theta$ iff there is some $B \subseteq \mathbb{R}$ such that $B \notin \mathrm{OD}(\mathbb{R} \cup \{A\})$. In this case, $\theta(A)$ is the least Wadge rank of such a B.

Theorem 2.9 (Woodin, mid 80's) *Assume AD^+, and suppose A and $\mathbb{R} \setminus A$ admit scales; then*

(a) All $\Sigma_1^2(A)$ sets of reals admit scales, and

(b) All $\Pi_1^2(A)$ sets admit scales iff $\theta(A) < \Theta$.

Theorem 2.10 (Martin, Woodin, mid 80's) *Assume* AD^+; *then the following are equivalent:*

(1) $\mathsf{AD}_{\mathbb{R}}$,

(2) Every set of reals admits a scale,

(3) $\Theta = \theta_\lambda$, *for some limit* λ.

Sadly, the proofs of these theorems have never fully appeared. There is a good deal on consequences of AD^+ in [29]. There is most of a proof that Scale(Σ_1^2) holds in derived models in [25], section 7.

Theorem 2.11 (Woodin, 1988, 2000) *Suppose λ is a limit of Woodins, and $L(\mathbb{R}^*, \mathrm{Hom}^*)$ is a derived model at λ; then*

(a) $\exists \kappa < \lambda(\kappa$ is $< \lambda$-strong$) \Rightarrow L(\mathbb{R}^*, \mathrm{Hom}^*) \models \theta_0 < \Theta$.

(b) λ is a limit of κ which are $< \lambda$-strong $\Rightarrow L(\mathbb{R}^*, \mathrm{Hom}^*) \models \Theta = \theta_\alpha$, for some limit α.

(c) λ is an inaccessible limit of κ which are $< \lambda$-strong $\Rightarrow L(\mathbb{R}^*, \mathrm{Hom}^*) \models \Theta = \theta_\alpha$, where $\mathrm{cof}(\alpha) \geq \omega_1$.

(d) $\exists \kappa < \lambda(\kappa$ is λ-supercompact$) \Rightarrow L(\mathbb{R}^*, \mathrm{Hom}^*) \models \mathsf{AD}_{\mathbb{R}} + \Theta$ is regular.

Remarks.

(i) Parts (a)-(c) are proved in [25]. Part (d) has not been written up.

(ii) The hypotheses of (a)-(c) follow from λ being a Woodin limit of Woodin cardinals.

(iii) The hypothesis of (b), that there is a λ which is a limit of Woodins and of $< \lambda$-strongs, is called the $\mathsf{AD}_{\mathbb{R}}$ *hypothesis*. The author has recently shown the $\mathsf{AD}_{\mathbb{R}}$ hypothesis is equiconsistent with $\mathsf{AD}_{\mathbb{R}}$. One direction is (b) above.

(iv) There is a big jump going from (c) to (d). How strong is $\mathsf{AD}_{\mathbb{R}} + \Theta$ regular?

(v) There are other ways of calibrating the strength of determinacy models. One can look at the complexity of the games of length ω_1 which are determined in the model. One can look at the complexity of the mice which have iteration strategies in the model.

2.4. *Iterations to make* $\mathbb{R}^V = \mathbb{R}^*$

Let $\mathcal{M}$ be countable and (ω_1+1)-iterable, with λ a limit of Woodin cardinals of $\mathcal{M}$. Working in $V^{\mathrm{Col}(\omega,\mathbb{R})}$, we can form an $\mathbb{R}$-*genericity iteration* (of $\mathcal{M}$, below λ), that is, a sequence

$$I = \langle \mathcal{T}_n \mid n < \omega \rangle$$

such that the $\mathcal{T}_n$ are iteration trees whose composition

$$\mathcal{T} = \oplus_n \mathcal{T}_n$$

is a normal, nondropping iteration tree on $\mathcal{M}$, with

$$\mathcal{M}_\infty^I = \lim{}_n \mathcal{M}_n^I,$$

the direct limit along the main branch of $\mathcal{T}$ (where $\mathcal{M}_n^I$ is the base model of $\mathcal{T}_n$, and the last model of $\mathcal{T}_{n-1}$ if $n > 0$), being such that $\mathbb{R}^V$ is the reals of a symmetric collapse over $\mathcal{M}_\infty^I$ below λ_∞^I, the image of λ. We write

$$\mathrm{Hom}_I^* = \bigcup\{p[T] \cap \mathbb{R}^V \mid \exists x \in \mathbb{R}^V\,(\mathcal{M}_\infty^I \models T \text{ is } < \lambda \text{ absolutely complemented})\},$$

and

$$D(\mathcal{M}_\infty^I, \lambda_\infty^I) = L(\mathbb{R}^V, \mathrm{Hom}_I^*)$$

for the derived model of $\mathcal{M}_\infty^I$ at λ_∞^I whose set of reals is $\mathbb{R}^* = \mathbb{R}^V$.

2.5. *Premice over a set*

We shall be most interested in ms premice over countable, transitive sets a which are *self-wellordered*, where a is self-wellordered iff there is a wellorder of a which is rudimentary in a. If $\mathcal{P}$ is a premouse over such an a, it satisfies AC, and the usual Los theorem for fine-structural ultrapowers, etc. We will sometimes speak of a mouse over a when $a \in \mathrm{HC}$ is not transitive. In this case, we are really speaking of a mouse over the transitive closure of $a \cup \{a\}$.

Nevertheless, the notion of an a-premouse makes sense for any transitive set a, self-wellordered or not, and we shall phrase some definitions in this generality. The reader should see [24] for a discussion of the elementary properties of such premice, in the representative special case that $a = \mathrm{HC}$.

The main thing is that if $\mathcal{M}$ is an a-premouse with top extender E, and $f\colon (a \times \xi) \to E_b$ with $\xi < \mathrm{crit}(E)$ and $f \in \mathcal{M}$, then $\bigcap \mathrm{ran}(f) \in E_b$. This implies that i_E is the identity on $a \cup \{a\}$, and that we have Los' theorem for Σ_n ultrapowers, whenever $\mathrm{crit}(E) < \rho_n^{\mathcal{M}} = \text{least } \rho$ such that there is a new $\Sigma_n^{\mathcal{M}}$ subset of $a \times \rho$. These properties imply that if g is $\mathrm{Col}(\omega, a)$-generic over $\mathcal{M}$, then $\mathcal{M}[g]$ can be regarded as an ordinary premouse over the swo $\langle a, g \rangle$. This last fact summarizes what it is to be an a-premouse: you become an ordinary premouse when a wellorder of a is added generically.

3. Iteration independence for derived models of mice

In this section we consider derived models of ms-mice.

An $\mathbb{R}$-genericity iteration I of a countable $\mathcal{M}$ cannot belong to V, unless λ happens to be measurable in $\mathcal{M}$. Nevertheless, in many interesting cases, the derived model $D(\mathcal{M}_\infty^I, \lambda_\infty^I)$ is in V.

We need a minor technical strengthening of iterability in order to state this result precisely. Let us call an $\omega_1 + 1$ iteration strategy Σ for $\mathcal{M}$ *good* iff whenever $\mathcal{T}$ is a countable normal iteration tree played by Σ with last model $\mathcal{P}$, and $\beta < \mathrm{lh}(\mathcal{T})$, then the phalanx obtained from $\Phi(\mathcal{T} \restriction (\beta + 1))$ obtained by replacing its last model $\mathcal{M}_\beta^{\mathcal{T}}$ with $\mathcal{P}$ is such that Ψ is $\omega_1 + 1$-iterable. All our iterability proofs give good strategies, but we do not see how to show that every $\omega_1 + 1$, or even $(\omega, \omega_1 + 1)$, iteration strategy is good.

Proposition 3.0.1 *Let $\mathcal{M}$ be ω-sound and project to ω, and let Σ be a good $\omega_1 + 1$ iteration strategy for $\mathcal{M}$. Let I be an $\mathbb{R}$-genericity iteration of $\mathcal{M}$ below λ such that I is played according to Σ; then every set in Hom_I^* is projective in $\Sigma \restriction HC$. In particular, $\mathrm{Hom}_I^* \subseteq V$.*

Proof. Let $A \in \mathrm{Hom}_I^*$, as witnessed by $T \in \mathcal{M}_\infty^I[x]$, where $x \in \mathbb{R}^V$. Let $S \in \mathcal{M}_\infty^I[x]$ absolutely complement T for forcings of size $< \lambda_\infty^I$. Let $I = \langle \mathcal{T}_n \mid n < \omega \rangle$, and let $n < \omega$ be large enough that, letting

$$\mathcal{N} = \text{ last model of } \mathcal{T}_n \, ,$$

and

$$\pi \colon \mathcal{N} \to \mathcal{M}_\infty^I$$

be the canonical embedding, we have that for $\kappa = \mathrm{crit}(\pi)$,

$$x \in \mathcal{N}[g],$$

where $g \in V$ is $\mathcal{N}$-generic over $\mathrm{Col}(\omega, \eta)$, for some $\eta < \kappa$. Note that we can lift π to an embedding of $\mathcal{N}[g]$ into $\mathcal{M}^I_\infty[g]$, which we also call π. We assume that n has been chosen large enough that the g-terms for T and S have pre-images in $\mathcal{N}$, and let then

$$\pi((\bar{T}, \bar{S})) = (T, S).$$

Working now in V, we can define A as follows: for $y \in \mathbb{R}$,

$$y \in A \Leftrightarrow \exists \mathcal{U} \in \mathrm{HC}((\oplus_{k \leq n} T_k) \oplus \mathcal{U} \text{ is a normal iteration tree}$$
$$\text{on } \mathcal{M} \text{ played by } \Sigma, \text{ and } y \in p[i^{\mathcal{U}}(\bar{T})].$$

On the right hand side of this equivalence, "$i^{\mathcal{U}}$" stands for the lift of the canonical embedding of $\mathcal{U}$, so that $i^{\mathcal{U}} : \mathcal{N}[g] \to \mathcal{P}[g]$, where $\mathcal{P}$ is the last model of $\mathcal{U}$.

The $\Rightarrow$ direction of the equivalence follows by taking $\mathcal{U} = \oplus_{n < k \leq m} T_k$, for m sufficiently large. For the other direction, suppose $\mathcal{U}$ is as on the right hand side, and toward contradiction, that $y \notin A$. We can then find an iteration tree $\mathcal{W}$ on $\mathcal{N}$ such that $\oplus_{k \leq n} T_k \oplus \mathcal{W}$ is a normal tree by Σ with $y \in p[i^{\mathcal{W}}(\bar{S})]$, where $i^{\mathcal{W}}$ is the lift of the canonical embedding of $\mathcal{W}$. (Take $\mathcal{W} = \oplus_{n < k \leq m} T_k$, for m sufficiently large.) Let $\mathcal{P}$ and $\mathcal{Q}$ be the last models of $\mathcal{U}$ and $\mathcal{W}$ respectively, and let Ψ and Γ be the phalanxes obtained from $\Phi((\oplus_{k \leq n} T_k))$ by replacing its last model $\mathcal{N}$ with $\mathcal{P}$ and $\mathcal{Q}$, respectively. Since Σ is good, Ψ and Γ are $\omega_1 + 1$ iterable, so we can coiterate them.

Standard arguments using the fact that $\mathcal{M}$ is ω-sound and projects to ω show that the last model on the two sides in the comparison of Ψ and Γ is the same, call it $\mathcal{R}$, and is above $\mathcal{P}$ and $\mathcal{Q}$ respectively, and the branches $\mathcal{P}$-to-$\mathcal{R}$ and $\mathcal{Q}$-to-$\mathcal{R}$ do not drop. Thus we have

$$\mathcal{N} \xrightarrow{i^{\mathcal{U}}} \mathcal{P} \xrightarrow{j} \mathcal{R}$$

and

$$\mathcal{N} \xrightarrow{i^{\mathcal{W}}} \mathcal{Q} \xrightarrow{k} \mathcal{R},$$

where j and k are given by the comparison of Ψ and Γ. Let

$$\sigma : \mathcal{M} \to \mathcal{N}$$

be given by $\oplus_{k \leq n} T_k$, and let ν be the sup of the generators of extenders in this tree used on $\mathcal{M}$-to-$\mathcal{N}$. Every element of $\mathcal{N}$ is definable from points in $\mathrm{ran}(\sigma)$ and ordinals $< \nu$. On the other hand, $j \circ i^{\mathcal{U}}$ and $k \circ i^{\mathcal{W}}$ are the identity on ν, and agree on $\mathrm{ran}(\sigma)$. Thus

$$j \circ i^{\mathcal{U}} = k \circ i^{\mathcal{W}}.$$

But then $y \in p[i^{\mathcal{U}}(\bar{T})]$, so $y \in p[j(i^{\mathcal{U}}(\bar{T}))]$, so $y \in p[k(i^{\mathcal{W}}(\bar{T}))]$. On the other hand, $y \in p[i^{\mathcal{W}}(\bar{S})]$, so $y \in p[k(i^{\mathcal{W}}(\bar{S}))]$. Thus $k(i^{\mathcal{W}}(\bar{T}))$ and $k(i^{\mathcal{W}}(\bar{S}))$ do not have disjoint projections, although it is satisfied by $\mathcal{R}[g]$ that their projections are disjoint. This contradicts the wellfoundedness of $\mathcal{R}[g]$. □

Remark 3.1 (i) Under sufficiently strong hypotheses, one can expect that the iteration strategy Σ of 3.0.1 will be Hom_∞, in which case we get that Hom_I^* is a Wadge initial segment of Hom_∞.

(ii) It is easy to see that the strategy Σ is not itself in Hom_I^*, as otherwise $\mathcal{M}$ would be ordinal definable over $D(\mathcal{M}_\infty^I, \lambda_\infty^I)$, hence in $\mathcal{M}_\infty^I$, hence in $\mathcal{M}$. Given that Σ is Hom_∞, we conjecture that if $\mathcal{M}$ is a sharp mouse which reconstructs itself below λ (see 3.3, 3.4 below), and λ is a limit of Woodin cardinals in $\mathcal{M}$ such that every extender on the $\mathcal{M}$-sequence except the last one has length $< \lambda$, then Σ is in any scaled pointclass closed under $\exists^{\mathbb{R}}$ and $\neg$ which properly includes Hom_I^*. This amounts to asserting that any new Suslin sets beyond Hom_I^* bring with them new Σ_1^2 facts, and that a local MSC holds just beyond Hom_I^*. One needs to assume that $\mathcal{M}$ is a sharp mouse here.

(iii) Let $\mathcal{T}$ be a non-dropping iteration tree on $\mathcal{M}$ played by Σ, with last model $\mathcal{N}$, and let $\eta < \delta$ be $\mathcal{N}$-cardinals, with $\nu(E_\alpha)^{\mathcal{T}} < \eta$ for all α, so that $\mathcal{N}$ is η-sound. We write $\Sigma|(\mathcal{T}, [\eta, \delta])$ for the restriction of Σ to iteration trees on $\mathcal{N}|\delta$ which use only extenders with critical point $> \eta$, and do not drop anywhere. (Such trees are normal continuations of $\mathcal{T}$.) We call $\Sigma|(\mathcal{T}, [\eta, \delta])$ a *window-based fragment* of Σ. We call the window $[\eta, \delta]$ *short* if there are no Woodin cardinals of $\mathcal{N}$ strictly between η and δ, and in this case call $\Sigma|(\mathcal{T}, [\eta, \delta])$ a *small fragment* of Σ. The proof of 3.0.1 then shows that every set in Hom_I^* is projective in some small fragment of Σ. In some cases, we can show that all the small fragments of Σ are in Hom_I^*, so that they are Wadge cofinal, and use this to compute the cofinality of θ in $D(\mathcal{M}, \lambda)$. See 7.3. We do not know whether this is true in general. In particular, we do not know whether if Σ is the unique ω_1 strategy for the sharp of the minimal model with a Woodin limit of Woodins, then all small fragments of Σ are in the associated derived model.

(iv) In general, the *initial segment based* fragments of Σ, i.e., the fragments based on windows of the form (empty tree, $[0, \delta]$), are not all in

$D(\mathcal{M}_\infty^I, \lambda_\infty^I)$. For suppose the cofinality of λ is not measurable in $\mathcal{M}$, so all iteration maps are continuous at λ. Suppose also $D(\mathcal{M}, \lambda) \models$ every set of reals is Suslin. By 3.0.1, if all the initial segment based fragments of Σ were in $D(\mathcal{M}_\infty^I, \lambda_\infty^I)$, they would be Wadge cofinal in Hom_I^*, which would imply that the Wadge ordinal of Hom_I^* has V-cofinality ω. In fact, the V-cofinality of this ordinal is often uncountable, as for example in the case of 2.11(c).

Given that for any $\mathbb{R}$-genericity iteration I, Hom_I^* is in V, it is natural to conjecture that in fact Hom_I^* is independent of I. We do not see how to prove that in full generality, but we can get reasonably close.

There is a natural partial order for producing $\mathbb{R}$-genericity iterations of $\mathcal{M}$.

Definition 3.2 *Let $\mathcal{M}$ be a countable mouse, and $\mathcal{M} \models \lambda$ is a limit of Woodin cardinals. Let Σ be an $\omega_1 + 1$ iteration strategy for $\mathcal{M}$. We let $\mathcal{I}(\mathcal{M}, \lambda, \Sigma)$ be the set of all finite sequences*

$$p = \langle \mathcal{T}_0, ..., \mathcal{T}_n \rangle$$

such that

(a) *$\mathcal{T}_0$ is an iteration tree on $\mathcal{M}$, and for $0 < k \leq n$, $\mathcal{T}_k$ is an iteration tree on the last model of $\mathcal{T}_{k-1}$, and $\mathcal{T}_n$ has a last model, and*

(b) *the composition $\oplus_{k \leq n} \mathcal{T}_k$ is a normal tree on $\mathcal{M}$ played according to Σ, with empty drop-set, and*

(c) *letting $i \colon \mathcal{M} \to \mathcal{N}$ be the canonical embedding of $\mathcal{M}$ into the last model of $\mathcal{T}_n$, and δ the sup of the lengths of the extenders used in $\oplus_{k \leq n} \mathcal{T}_k$, we have $\delta < i(\lambda)$.*

We regard $\mathcal{I}$ as a partial order, under reverse inclusion.

Clause (c) of 3.2 guarantees that every $p \in \mathcal{I}$ has a proper extension in $\mathcal{I}$. It is not hard to see that if G is V-generic over $\mathcal{I}(\mathcal{M}, \lambda, \Sigma)$, then $I = \bigcup G$ is an $\mathbb{R}$-genericity iteration of $\mathcal{M}$. We call such an I an $\mathcal{I}(\mathcal{M}, \lambda, \Sigma)$-*generic* ($\mathbb{R}$-genericity) iteration. We are interested in the case $\mathcal{M}$ is ω-sound and projects to ω, so that Σ is determined by $\mathcal{M}$, and we can write $\mathcal{I}(\mathcal{M}, \lambda)$ for $\mathcal{I}(\mathcal{M}, \lambda, \Sigma)$.

If $p = \langle \mathcal{T}_0, ..., \mathcal{T}_n \rangle \in \mathcal{I}(\mathcal{M}, \lambda, \Sigma)$, then we shall write $\mathcal{T}_i(p) = \mathcal{T}_i$, $\mathcal{M}_0(p) = \mathcal{M}$ and $\mathcal{M}_{i+1}(p) = $ the last model of $\mathcal{T}_i$, $\mathcal{M}_p = \mathcal{M}_{n+1}(p)$, $i_{k,l}^p$ for the canonical embedding of $\mathcal{M}_k(p)$ into $\mathcal{M}_l(p)$, λ_p for $i^p_{0,n+1}(\lambda)$, and δ_p for the sup of the lengths of the extenders used in the $\mathcal{T}_k$, for $k \leq n$.

We shall show that for certain natural $(\mathcal{M}, \lambda)$, any two $\mathcal{I}(\mathcal{M}, \lambda)$-generic iterations give rise to the same derived model. The $\mathcal{M}$ we have in mind here are mice like $M^\sharp_\omega$ (the sharp of the minimal model with ω Woodins), $M^\sharp_{\mathrm{adr}}$ (the sharp of the minimal model of the $\mathsf{AD}_\mathbb{R}$ hypothesis), and $M^\sharp_{\mathrm{wlim}}$ (the sharp of the minimal model with a Woodin limit of Woodins). The main feature of these mice which lets our argument go through is that they reconstruct themselves below the relevant λ. There are some other conditions we seem to need as well. We are by no means sure that the definitions we are about to give abstract the most general hypotheses on $\mathcal{M}$ and λ yielding derived-model invariance.

Definition 3.3 *We call a premouse $\mathcal{M}$ a sharp mouse iff*

(a) $\mathcal{M}$ is active, and for some $\alpha < \mathrm{crit}(\dot{F}^{\mathcal{M}})$, $\dot{E}^{\mathcal{M}}_\eta = \emptyset$ for all $\eta \geq \alpha$,

(b) there is a sentence φ such that for $\kappa = \mathrm{crit}(\dot{F}^{\mathcal{M}})$, $\mathcal{M}|\kappa \models \varphi$, but whenever $\mu = \mathrm{crit}(\dot{E}^{\mathcal{M}}_\eta)$ for some η, then $\mathcal{M}|\mu \not\models \varphi$, and

(c) $\mathcal{M}$ is $\omega_1 + 1$-iterable.

If α is as in part (a), then (a) says that $\mathcal{M}$ is essentially the sharp of $\mathcal{M}|\alpha$. Part (b) is a φ-minimality condition, and it implies that $\mathcal{M}$ projects to ω. If $\mathcal{M}$ is also ω-sound, the iteration strategy asserted to exist in (c) is unique.

The *reduct* of a sharp mouse $\mathcal{M}$ is $\mathcal{M}\|o(\mathcal{M})$, that is, $\mathcal{M}$ with its last extender removed.

Definition 3.4 *Let $\mathcal{M}$ be an ω-sound sharp mouse, and λ a limit cardinal in $\mathcal{M}$. We say that $\mathcal{M}$ reconstructs itself below λ iff whenever $\mathcal{N}$ is a countable, non-dropping iterate of $\mathcal{M}$ by its unique $\omega_1 + 1$ iteration strategy, with λ^* the image of λ, and $\kappa < \lambda^*$, and $\mathcal{P}$ is the extender model to which the maximal plus-1 certified ms-array over κ of $\mathcal{N}$ converges (see [21]), then there is a sharp mouse $\mathcal{Q}$ whose reduct is $\mathcal{P}$, and a Σ_1-elementary $\pi \colon \mathcal{M} \to \mathcal{Q}$ such that $\pi(\lambda) = \lambda^*$.*

The maximal plus-1 certified ms-array of $\mathcal{N}$ over κ is the output of a K^c-construction which uses all possible reasonably strong extenders from the $\mathcal{N}$-sequence with critical point $> \kappa$ as background extenders; see [21].

Definition 3.5 *A tractable pair is a pair* $\langle \mathcal{M}, \lambda \rangle$ *such that*

(a) $\mathcal{M}$ *is an ω-sound sharp mouse,*

(b) $\mathcal{M} \models$ *"λ is a limit of Woodin cardinals, and* $\mathrm{cof}(\lambda)$ *is not measurable",* *and*

(c) $\mathcal{M}$ *reconstructs itself below* λ.

Some (probable) examples of tractable pairs are

(1) $M_\omega^\sharp$, with λ the sup of its Woodin cardinals,

(2) $M_{\mathrm{adr}}^\sharp$, with λ the sup of its Woodin cardinals, or with λ the sup of the first ω Woodin cardinals of $M_{\mathrm{adr}}^\sharp$,

(3) $M_{\mathrm{dc}}^\sharp$, the sharp of the minimal model with a cardinal μ which is a limit of Woodins, and such that the set of $\kappa < \mu$ which are $< \mu$-strong has order type μ, with $\lambda = \mu$, or with λ the sup of the first ω Woodins,

(4) $M_{\mathrm{inlim}}^\sharp$, the sharp of the minimal model with a cardinal which is inaccessible, and a limit of Woodins, and a limit of $< \lambda$-strong cardinals, with λ the sup of all its Woodins, or with λ the sup of its first ω Woodins,

(5) $M_{\mathrm{wlim}}^\sharp$, with λ the Woodin limit, or with λ the sup of the first ω Woodins.

The author believes that tractability can be verified in each case using [21]. This is easy to see, except in (2)-(5) with λ the sup of the first ω Woodin cardinals of the mouse in question. The author has not gone through all the details in those cases.

We believe that $M_{\mathrm{adr}}^\sharp$ is the least mouse with a derived model satisfying $\mathrm{AD}_{\mathbb{R}}$, and $M_{\mathrm{dc}}^\sharp$ is the least mouse with a derived model satisfying $\mathrm{AD}_{\mathbb{R}}$ plus DC (or equivalently, $\theta = \theta_{\omega_1}$). This should come out of Woodin's proof that $\mathrm{AD}_{\mathbb{R}}$ is equiconsistent with the $\mathrm{AD}_{\mathbb{R}}$ hypothesis, and his parallel result for $\mathrm{AD}_{\mathbb{R}} + \mathrm{DC}$ versus the $\mathrm{AD}_{\mathbb{R}} + \mathrm{DC}$ hypothesis, but we have not gone through the details. We shall show in section 8 that the derived model of $M_{\mathrm{wlim}}^\sharp$ at its top Woodin satisfies $\mathrm{AD}_{\mathbb{R}}$ plus "$\theta = \theta_\theta$".

Theorem 3.6 *Let* $\langle \mathcal{M}, \lambda \rangle$ *be a tractable pair, and let I and J be* $\mathcal{I}(\mathcal{M}, \lambda)$-*generic iterations; then* $D(\mathcal{M}_\infty^I, \lambda_\infty^I) = D(\mathcal{M}_\infty^J, \lambda_\infty^J)$.

Proof. Fix $\mathcal{M}$ and λ, and let Σ be the unique $\omega_1 + 1$ iteration strategy for $\mathcal{M}$. Let $\mathcal{I} = \mathcal{I}(\mathcal{M}, \lambda)$.

Given $A \subset \mathbb{R}$, let us say (p, g) *captures* A iff $p \in \mathcal{I}$, and g is $\mathcal{M}_p$-generic over $\mathrm{Col}(\omega, \delta_p)$, and there are trees $S, T \in \mathcal{M}_p[g]$ such that

$$\mathcal{M}_p[g] \models S, T \text{ are } < \lambda_p\text{-absolute complements,}$$

and

$$A = \bigcup \{ p[i^{\mathcal{U}}(S)] \mid p^\frown \langle \mathcal{U} \rangle \in \mathcal{I} \}.$$

Here $i^{\mathcal{U}}$ is the embedding along the main branch of $\mathcal{U}$. By the proof of 3.0.1, it is enough to prove

Lemma 3.7 *Suppose (p, g) captures A; then there are densely many $r \in \mathcal{I}$ such that for some h, (r, h) captures A.*

Proof. Let (p, g) capture A, and let $q \in \mathcal{I}$. We seek an $r \leq q$ and h such that (r, h) captures A. The idea is just to reconstruct inside $\mathcal{M}_q$, above δ_q, a model $\mathcal{R}$ into which $\mathcal{M}$ embeds. We can then iterate $\mathcal{M}_q$ above δ_q so that we get an $r \leq q$, with $\mathcal{M}_p$ embedding into the image $\mathcal{S}$ of $\mathcal{R}$ under this iteration. Our universally Baire representation of A over $\mathcal{M}_p$ then lifts to $\mathcal{S}$, and then to the background universe $\mathcal{M}_r$ for $\mathcal{S}$, as desired.

In order to lift the universally Baire representations, we must work with their associated extender normal forms. Recall that an extender normal form is a map

$$\mathcal{E} = (s \mapsto \mathcal{E}_s), \text{ for } s \in \omega^{<\omega},$$

where each $\mathcal{E}_s$ is an alternating chain with $2 \cdot \mathrm{dom}(s)$ many models, and $\mathcal{E}_s$ extends $\mathcal{E}_t$ whenever s extends t. For $x \in \omega^\omega$, we write $\mathcal{E}_x^e$ and $\mathcal{E}_x^o$ for the direct limits along the even and odd branches of the infinite alternating chain associated to x. If $\mathcal{E}$ is an extender normal form which uses only extenders with critical point $> \kappa$, and whenever G is V-generic for a poset of size $< \kappa$,

$$V[G] \models \forall x (\text{ exactly one of } \mathcal{E}_x^e \text{ and } \mathcal{E}_x^o \text{ is wellfounded,})$$

then we call $\mathcal{E}$ a κ-*ENF*. We say that $\mathcal{E}$ *represents* C iff $C \subseteq \mathbb{R}$, and for all $x \in \mathbb{R}$,

$$x \in C \Leftrightarrow \mathcal{E}_x^e \text{ is wellfounded.}$$

It is worth noting the following absoluteness property of such representations: if S, T are κ-absolute complements, and $\mathcal{E}$ is a κ-ENF which represents $p[S]$ in V, then whenever G is V-generic for a poset of size $\leq \kappa$, we have that $\mathcal{E}$ represents $p[S]$ in $V[G]$. (This uses the uniqueness of wellfounded branches in $V[G]$ which we have built into our definition of κ-ENF.)

We say that $\mathcal{E}$ is an $\mathcal{N}$-*based* κ-*ENF in* $\mathcal{N}[g]$ iff $\mathcal{N}$ is a premouse, and g is $\mathcal{N}$-generic over some poset of size $< \kappa$ in $\mathcal{N}$, and $\mathcal{N}[g] \models \mathcal{E}$ is a κ-ENF, and $\mathcal{E}$ uses only extenders from the sequence of $\mathcal{N}$ and its images whose sup-of-generators is a cardinal in the model in which they appear. The following carries over a basic fact about ENF's to this fine-structural setting.

Sublemma 3.7.1 *[[25],[10]] Let $\mathcal{N}$ be a premouse, and $\mathcal{N} \models$ ZFC+ " λ is a limit of cardinals which are Woodin via extenders on my sequence". Let g be $\mathcal{N}$-generic over some poset of size $< \lambda$; then working inside $\mathcal{N}[g]$, the following are equivalent, for any set C of reals:*

(a) C is $< \lambda$-universally Baire,

(b) for any $\kappa < \lambda$ there is an $\mathcal{N}$-based κ-ENF $\mathcal{E}$ which represents C.

We continue with the proof of 3.7. By the sublemma, $\mathcal{M}_p[g]$ satisfies the statement that there is a sequence $\langle \mathcal{E}(\kappa) \mid \kappa < \lambda_p \rangle$ such that for each κ, $\mathcal{E}(\kappa)$ is an $\mathcal{M}_p$-based κ-ENF representing $p[S]$. Fix such an $\langle \mathcal{E}(\kappa) \mid \kappa < \lambda_p \rangle$. Using a g-name for this sequence, we can fix a sequence $\langle X_\kappa \mid \kappa < \lambda_p \rangle \in \mathcal{M}_p$ such that for all $\kappa < \lambda_p$,

(a) $\mathcal{M}_p \models |X_\kappa| \leq \delta_p$, and

(b) for all $s \in \omega^{<\omega}$, $\mathcal{E}_s(\kappa) \in X_\kappa$.

Now let $\mathcal{N}_q$ be the version of $\mathcal{M}$ reconstructed inside $\mathcal{M}_q$, with all critical points above δ_q. Thus we have a Σ_1 elementary

$$\pi: \mathcal{M} \to \mathcal{N}_q \subseteq \mathcal{M}_q$$

such that $\pi(\lambda_p) = \lambda_q$. Using the fact that λ does not have measurable cofinality, and a simple comparison argument, one can show

$$\operatorname{ran}(\pi) \text{ is cofinal in } \lambda_q.$$

This was the reason for our restriction on the cofinality of λ. We leave the details to the reader.

We can use π to lift the iteration from $\mathcal{M}$ to $\mathcal{M}_p$ to an iteration tree on $\mathcal{N}_q$, and thence via the background-extender iterability proof to an iteration tree on $\mathcal{M}_q$ which is above δ_q. This gives us a condition $t \leq q$ in $\mathcal{I}$, and a Σ_1 elementary

$$\sigma \colon \mathcal{M}_p \to \mathcal{N}_t \subseteq \mathcal{M}_t,$$

where $\mathcal{N}_t$ is a sharp mouse whose reduct is the limit model of a plus-1 certified ms-array in the sense of $\mathcal{M}_t$.

Let h_0 be $\mathcal{M}_t$-generic over $\mathrm{Col}(\omega, \sigma(\delta_p))$. Letting

$$Y = \sigma(\langle X_\kappa \mid \kappa < \lambda_p \rangle),$$

we have that Y_μ is countable in $\mathcal{M}_t[h_0]$ for all $\mu < \lambda_t$, and $Y_{\sigma(\mu)}$ covers $\sigma``X_\mu$, for all $\mu < \lambda_p$. Fix a sequence $\langle f_\mu \mid \mu < \lambda_t \rangle$ n $\mathcal{M}_t[h_0]$ such that f_μ maps ω onto Y_μ, for all μ. Finally, fix a real z such that for all $\mu < \lambda_p$, the map

$$s \mapsto f^{-1}_{\sigma(\mu)}(\sigma(\mathcal{E}_s(\mu)),$$

which is essentially a real, is recursive in z.

Now let δ be a Woodin cardinal of $\mathcal{M}_t[h_0]$ such that $\sup(\sigma(\delta_p), \delta_t) < \delta < \lambda_t$. We do a genericity iteration of $\mathcal{M}_t[h_0]$ via an $\mathcal{M}_t$-based tree which is above $\sup(\sigma(\delta_p), \delta_t)$, and below δ, and obtain thereby a condition $r \leq t$ in $\mathcal{I}$, and a generic h on $\mathrm{Col}(\omega, \delta_r)$, so that $h_0, z \in \mathcal{M}_r[h]$. Let

$$\tau \colon \mathcal{M}_t \to \mathcal{M}_r$$

be the natural map, and let

$$\psi = \tau \circ \sigma \colon \mathcal{M}_p \to \mathcal{N}_r,$$

where $\mathcal{N}_r = \tau(\mathcal{N}_t)$.

Claim. (r, h) captures A.

Proof. Recall that there is a $\nu < \lambda_r$ such that in $\mathcal{M}_r[h_0]$, any ν-homogeneous set of reals is $< \lambda_r$-universally Baire. Fix such a $\nu = \nu_0$. Now pick $\mu < \lambda_p$ such that

(a) $\nu_0 < \psi(\mu)$,

(b) $\gamma_0 < \mu$, where γ_0 is the least Woodin cardinal γ of $\mathcal{M}_p$ such that $\delta_r < \psi(\gamma)$, and

(c) there is a Woodin cardinal κ_0 of $\mathcal{M}_r$ such that $\psi(\gamma_0) < \kappa_0 < \psi(\mu)$.

Since λ_r is a limit of Woodin cardinals in $\mathcal{M}_r$, and $\mathrm{ran}(\psi)$ is cofinal in λ_r, we can easily find such a μ, γ_0, and κ_0.

Let

$$\mathcal{E}_s = \mathcal{E}_s(\mu),$$

and

$$\mathcal{F}_s = \psi(\mathcal{E}_s),$$

and

$$\mathcal{F}_s^* = \text{ alternating chain on } \mathcal{M}_r[h] \text{ induced by } \mathcal{F}_s,$$

for all $s \in \omega^{<\omega}$. Here we mean that $\mathcal{F}_s^*$ arises from $\mathcal{F}_s$ as in the iterability proof, using the background extenders provided by the K^c-construction in $\mathcal{M}_r$ whose output is $\mathcal{N}_r$.

We need the following small extension of the theorem 3.3 of [21] on UBH in extender models. We omit the proof.

Lemma 3.8 *Let* $\mathcal{P} \models$ ZFC *be fully iterable mouse, and let* G *be generic over* $\mathcal{P}$ *for a poset of size* $< \kappa$ *in* $\mathcal{P}$; *then there are no* $\mathcal{T}, b, c \in \mathcal{P}[G]$ *such that*

$\mathcal{P}[G] \models \mathcal{T}$ *is a* $\mathcal{P}$-*based plus-2 iteration tree on* $\mathcal{P}[G]$ *of limit length,*

with all critical points of extenders in $\mathcal{T}$ *above* κ, *and* $D^{\mathcal{T}} = \emptyset$, *and*

$\mathcal{P}[G] \models b$ *and* c *are distinct, cofinal, wellfounded branches of* $\mathcal{T}$.

It follows immediately from lemma 3.8 that $\mathcal{M}_r[h] \models \mathcal{F}$ is a $\psi(\mu)$-ENF. We therefore have trees $U, W \in \mathcal{M}_r[h]$ such that

$$\mathcal{M}_r[h] \models U, W \text{ are } < \lambda_r\text{-absolute complements},$$

and

$$\mathcal{M}_r[h] \models p[U] = \{x \mid (\mathcal{F}^*)_x^e \text{ is wellfounded}\}.$$

We claim that U, W witness that (r, h) captures A. For that, fix $x \in \mathbb{R}$. It is enough to show that if $x \in A$, then there is a condition $u \leq r$ such that $x \in p[i_{r,u}(U)]$, and if $x \notin A$, then there is a condition $u \leq r$ such that $x \in p[i_{r,u}(W)]$. So assume $x \in A$; the proof when $x \notin A$ is completely parallel.

Let $a \leq p$ come from iterating $\mathcal{M}_p$ above δ_p, and below γ_0, so as to make x generic over $\mathcal{M}_a[g]$ for the image of the extender algebra at γ_0. We have then that $x \in p[i_{p,a}(S)]$, and thus working in $\mathcal{M}_a[g][x]$, we can conclude that $i_{p,a}(\mathcal{E})^o_x$ is illfounded. Let $b < r$ come from lifting the genericity iteration from $\mathcal{M}_p$ to $\mathcal{N}_r$ using ψ, then further lifting it to $\mathcal{M}_r$ using the iterability proof for background-certified models. It is important here to note that the iteration strategy Σ on normal extensions of $\mathcal{T}_p$ is unique, so that it agrees with the strategy of lifting to $\mathcal{M}_r$ as above, and using Σ to pick branches of the evolving tree on $\mathcal{M}_r$. This shows that the lifting process succeeds, and does give us a condition in $b \in \mathcal{I}$.

Let $\phi \colon \mathcal{M}_a \to \mathcal{N}_b$ be the natural lifting map, where $\mathcal{N}_b = i_{r,b}(\mathcal{N}_r)$. We have from the commutativity of the copy maps that

$$\phi \circ i_{p,a} = i_{r,b} \circ \psi,$$

and from this we get that

$$\phi(i_{p,a}(\mathcal{E})) = i_{r,b}(\mathcal{F}).$$

Here we understand the identity just displayed by letting the embeddings act on $\mathcal{E}$ "pointwise", for ϕ is not actually defined on $\mathcal{M}_a[g]$. (This was the reason we moved to extender normal forms in the first place.) We have then that

$$(i_{r,b}(\mathcal{F}^*))^o_x \text{ is illfounded.}$$

Finally, let $u \le b$ come from a genericity iteration of $\mathcal{M}_b[h]$ above δ_b and below $i_{r,b}(\kappa_0)$ which makes x generic over $\mathcal{M}_u[h]$ for the extender algebra at $i_{r,u}(\kappa_0)$. Clearly, we still have

$$(i_{r,u}(\mathcal{F}^*))^0_x \text{ is illfounded.}$$

But note that x is generic over $\mathcal{M}_u[h]$ for a poset of size $< i_{r,u}(\psi(\mu))$, and that the universally Baire representation $i_{r,u}(U)$ is satisfied to be equivalent to $i_{r,u}(\mathcal{F})$ in $\mathcal{M}_u[h]$, and that this equivalence is absolute for forcing of size $< i_{r,u}(\psi(\mu))$ over $\mathcal{M}_u[h]$. It follows that $x \in p[i_{r,u}(U)]$, as desired.

This completes the proof of the claim, and hence the proof of lemma 3.7, and hence the proof of theorem 3.6. $\qquad\qquad\square$

4. Mouse operators and jump operators

The sharp mice we introduced earlier all relativise: e.g., we can form $M^\sharp_{\mathrm{adr}}(y)$ for any transitive set y. In this way, we obtain a *mouse operator*. In this section we give a general notion of mouse operator which is useful in the AD context, and derive some elementary consequences of AD concerning mouse operators. We then record some special properties of the operators one gets by relativising sharp mice.

Definition 4.1 *A premouse jump operator is a function $x \mapsto \mathcal{M}(x)$, defined on a Turing cone of reals x, such that for each x in its domain*

(1) $\mathcal{M}(x)$ is a ω-sound ms-premouse over x which projects to ω, and

(2) for all $y \equiv_T x$, $\mathcal{M}(y)$ is the canonical re-arrangement of $\mathcal{M}(x)$ as a premouse over y.

A mouse jump operator is a premouse jump operator $\mathcal{M}$ such that $\mathcal{M}(x)$ is ω_1-iterable, for all $x \in \mathrm{dom}(\mathcal{M})$.

Identifying $\mathcal{M}(x)$ with the real $\mathrm{Th}^{\mathcal{M}(x)}(x)$, we see that any premouse jump operator is uniformly Turing-invariant, and hence a jump operator in the sense of [28]. We shall use the notation and results of that paper. In particular, $\leq_m$ is the prewellorder of jump operators defined by Martin.

The next lemma is implicit in [28].

Lemma 4.2 *Assume AD^+ and $\mathsf{MSC1}$, and let J be a jump operator; then*

(1) there is a premouse jump operator $\mathcal{N}$ such that $J \equiv_m \mathcal{N}$, and

(2) if J is $\mathbf{\Delta}_1^2$, then there is a mouse jump operator $\mathcal{N}$ such that $J \equiv_m \mathcal{N}$.

Proof. We first prove (2). Let J be $\Delta_1^2(z)$ and its domain contain the cone above z. For $x \geq_T z$, let

$$\mathcal{P}(x) = \text{ least } x\text{-mouse } \mathcal{M} \text{ such that } J(x) \in \mathcal{M}.$$

Since $J(x)$ is $\mathrm{OD}(x)$, there is such an x-mouse by $\mathsf{MSC1}$. Identifying $\mathcal{P}(x)$ with its theory, it is easy to see that $\mathcal{P}$ is a mouse jump operator, and $J \leq_m \mathcal{P}$.

It follows from the proof of Theorem 3 of [28] that if $\mathcal{Q}$ is a mouse jump operator, and $H \leq_m \mathcal{Q}$, then there is a mouse jump operator $\mathcal{N}$ such that $J \equiv_m \mathcal{N}$. (This was observed by M. Rudominer and the author.) This proves (2).

Part (1) follows from part (2) and Woodin's basis theorem for $\mathbf{\Sigma}_1^2$ (see 9.10, part (b)). For notice that it is a projective statement about J to say that it is a counterexample to (1). Thus if there is any counterexample J to (1), there is a $\mathbf{\Delta}_1^2$ counterexample J to (1). But then J is a counterexample to (2). $\qquad\square$

A minor variant of a proposition from [18] shows that every mouse jump operator is preserved by Cohen forcing on a cone:

Lemma 4.3 *Assume* AD, *and let* $\mathcal{M}$ *be a mouse jump operator; then for a cone of reals* r, *we have that whenever* g *is Cohen generic over* $\mathcal{M}(r)$, *then* $\mathcal{M}(r)[g]$, *when regarded as a mouse over* $\langle r, g \rangle$, *is equal to* $\mathcal{M}(\langle r, g \rangle)$.

Proof. Let $f(r) = o(\mathcal{M}(r))$. f is a Turing invariant, ordinal-valued function. It is easy to see that any such function is order-preserving on a cone, so that for a cone of r, $f(r) \leq f(s)$ whenever $r \leq_T s$. Since there is at most one sound $\langle r, g \rangle$- mouse projecting to ω of a given height, it is enough to see that for a cone of r, $f(r) = f(\langle r, g \rangle)$ whenever g is Cohen generic over $\mathcal{M}(r)$. Suppose not, and let r_0 be the base of a cone where $f(r) < f(\langle r, g \rangle)$ holds. Let $g : (\omega \times \omega) \to 2$ be Cohen generic with respect to all $\mathrm{OD}(r_0)$ dense sets, and let $r_i = \langle r_0, g \restriction (\omega \setminus i) \times \omega$. Clearly, $r_i \equiv_T \langle r_{i+1}, g \rangle$. for some g Cohen over $\mathrm{OD}(r_{i+1})$. But $\mathcal{M}(r_{i+1}) \subseteq \mathrm{OD}(r_{i+1})$, so $f(r_{i+1}) < f(r_i)$ for all i, a contradiction. $\qquad\square$

We shall show that every mouse jump operator has a canonical extension to a operator defined on a cone of countable transitive sets, and take mouse operators to just be these canonical extensions. The extension is obtained simply by looking at generic enumerations of the countable transitive set.

Definition 4.4 *An HC-cone is a set of the form* $c = \{b \mid b$ *is countable and transitive, and* $V_\omega \cup \{x\} \subseteq\}$, *where* x *is a real. We say in this case that* x *is a base for the cone* c.

Theorem 4.5 *Assume* AD, *and let* $\mathcal{M}_0$ *be a mouse jump operator which is defined, and preserved by Cohen forcing in the sense of 4.3, on the Turing cone above* x; *then for each* a *in the HC-cone with base* x, *there is a unique* ω-*sound mouse* $\mathcal{P}$ *over* a *such that such that whenever* g *is* $\mathrm{Col}(\omega, a)$-*generic over* $\mathcal{P}$, *and* r_g *is some real coding* $\langle a, g \rangle$ *in a natural way (so that* $r_0 \leq_T r$), *we have*

$$\mathcal{P}[g] = \mathcal{M}_0(r_g).$$

Proof. (Sketch.) Fix g which is $\mathrm{OD}(a)$-generic over $\mathrm{Col}(\omega, a)$. It is easy to construct by induction on levels a mouse $\mathcal{P}^g$ over a such that $\mathcal{P}^g[g] = \mathcal{M}_0(r_g)$, by simply restricting the extender sequence of $\mathcal{M}_0(r_g)$ to the model one is building. See [24] or [15] for the details of this construction. But if g and h agree at all but finitely many arguments, we have $r_g \equiv_T r_h$, so $\mathcal{M}_0(r_g)$ is $\mathcal{M}_0(r_h)$ up to a trivial re-arrangement, which implies $\mathcal{P}^g = \mathcal{P}^h$. It follows that $\mathcal{P}^g$ is constant on a comeager set of g, and we let $\mathcal{P}$ be this constant value. We then have that $\mathcal{P}$ is a sound mouse over a projecting to a, and for comeager many g on $\mathrm{Col}(\omega, a)$, $\mathcal{P}[g]$ is $\mathcal{M}_0(r_g)$, when re-arranged as a mouse over r_g.

Now let h be an arbitrary $\mathrm{Col}(\omega, a)$-generic over $\mathcal{P}$. Let g be $\mathrm{Col}(\omega, a)$-generic over $\mathrm{OD}(r_h)$, and in the comeager set where $\mathcal{P}[g] = \mathcal{M}_0(r_g)$ (re-arranged). Let $\mathcal{Q}(r_h)$ be the re-arrangement of $\mathcal{P}[h]$ as a mouse over r_h. We have

$$\mathcal{Q}(r_h)[g] = \mathcal{P}[h][g] = \mathcal{P}[g][h] = \mathcal{M}_0(r_g)[h] = \mathcal{M}_0(\langle r_g, r_h \rangle),$$

using 4.3 for the last equality. So

$$\mathcal{Q}(r_h)[g] = \mathcal{M}_0(r_h)[g]$$

by 4.3, which then implies $\mathcal{Q}(r_h) = \mathcal{M}_0(r_h)$, because the two are sound mice over r_h projecting to ω and having the same ordinals. This shows there is a mouse $\mathcal{P}$ as claimed. It is unique because it projects to a, and is a-sound. □

It is easy to see that the mouse $\mathcal{P}$ of the claim is unique, so we may set $\mathcal{M}(a) = \mathcal{P}$. We leave it to the reader to check that $\mathcal{M}(V_\omega \cup \{r\}) = \mathcal{M}_0(r)$ when r is a real $\geq_T r_0$. For arbitrary $a \in \mathrm{HC}$, let us write $\mathcal{M}(a)$ for $\mathcal{M}(V_\omega \cup \mathrm{TC}\,(a \cup \{a\}))$. We have then that $\mathcal{M}$ extends $\mathcal{M}_0$.

Definition 4.6 *Let $\mathcal{M}_0$ be a premouse jump operator defined on all reals $r \geq_T r_0$. We say $\mathcal{M}_0$ is HC-extendible iff for any a in the HC-cone with base r_0, there is an a-premouse $\mathcal{P}$ such that whenever g is $\mathrm{Col}(\omega, a)$-generic over $\mathcal{P}$, and r_g is some real coding $\langle a, g \rangle$ in a natural way (so that $r_0 \leq_T r$), we have*

(1) $\mathcal{P}[g] = \mathcal{M}_0(r_g)$, and

(2) $\mathcal{P}$ is obtained from $\mathcal{M}_0(r_g)$ via the contruction in the proof of 4.5.

If $\mathcal{M}_0$ is HC-extendible, then letting $\mathcal{M}(a)$ be the unique $\mathcal{P}$ as in (1) and (2), we call $\mathcal{M}$ the HC-extension of $\mathcal{M}_0$.

We have then by 4.5 that every mouse jump operator is HC-extendible to some HC-cone. This strengthens part (2) of 4.2, and we can use it to strengthen part (1) as well:

Lemma 4.7 *Assume* AD$^+$ *and* MSC1, *and let* J *be a jump operator; then there is an HC-extendible premouse jump operator* $\mathcal{N}$ *such that* $J \equiv_m \mathcal{N}$.

Proof. Again, this follows from Woodin's basis theorem for $\boldsymbol{\Sigma}_1^2$. If there is any counterexample, there is a $\boldsymbol{\Delta}_1^2$ counterexample J to (1). But by 4.2(2) and 4.5, there is no $\boldsymbol{\Delta}_1^2$ counterexample. $\qquad\square$

Definition 4.8 $\mathcal{M}$ *is a premouse operator iff* $\mathcal{M}$ *is the HC-extension of some premouse jump operator.* $\mathcal{M}$ *is a mouse operator iff it is the HC-extension of some mouse jump operator.*

One useful property a premouse operator might have is *condensation*.

Definition 4.9 *Let* $\mathcal{M}$ *be a premouse operator defined on the HC-cone with base* x. *We say* $\mathcal{M}$ *condenses well (or has condensation) iff whenever* $\pi\colon \mathcal{P} \to \mathcal{M}(a)$ *is* Σ_1 *elementary, and* $TC\,(x \cup \{x\}) \subseteq \mathrm{ran}(\pi)$, *then* $\mathcal{P} = \mathcal{M}(\pi^{-1}(a))$.

It is not the case that every mouse operator condenses well on some HC-cone. Operators determined by Π_2 theories do, however.

Definition 4.10 *A* Π_2 *mouse (jump) operator is a mouse (jump) operator* $\mathcal{M}$ *such that for some set* T *of* Π_2 *sentences in the language of relativised premice, for all* $x \in \mathrm{dom}(\mathcal{M})$, $\mathcal{M}(x)$ *is the least* $\omega_1 + 1$-*iterable ms-mouse* $\mathcal{P}$ *over* x *such that* $\mathcal{P} \models T$.

Lemma 4.11 *Let* $\mathcal{M}_0$ *be a* Π_2 *mouse jump operator, and* $\mathcal{M}$ *its unique HC-extension; then* $\mathcal{M}$ *is a* Π_2 *mouse operator.*

Lemma 4.12 *Every* Π_2 *mouse operator condenses well.*

We now turn to the special case of sharp mice.

The sharp mice we introduced earlier all relativise: e.g., we can form $M_{\mathrm{adr}}^{\sharp}(y)$ for any transitive set y. Similarly, the λ's which were part of our tractable pairs relativise. We now abstract some properties of the resulting operators:

Definition 4.13 *A sharp-mouse operator is a mouse operator* $\mathcal{M}$ *such that there is a sentence* φ *in the language of relativised premice such that for all* $y \in \mathrm{dom}(\mathcal{M})$

(a) $\mathcal{M}(y)$ is active, and for some $\alpha < \mathrm{crit}(\dot{F}^{\mathcal{M}(y)})$, $\dot{E}_\eta^{\mathcal{M}(y)} = \emptyset$ for all $\eta \geq \alpha$,

(b) for $\kappa = \mathrm{crit}(\dot{F}^{\mathcal{M}(y)})$, $\mathcal{M}(y)|\kappa \models \varphi$, but whenever $\mu = \mathrm{crit}(\dot{E}_\eta^{\mathcal{M}(y)})$ for some η, then $\mathcal{M}(y)|\mu \not\models \varphi$.

Note that if $\mathcal{M}$ is a sharp-mouse operator, then $\mathcal{M}(y)$ is sound and $\omega_1 + 1$-iterable, for all $y \in \mathrm{dom}(\mathcal{M})$.

Definition 4.14 *Let $\mathcal{M}$ be an sharp mouse operator, and $\lambda = (y \mapsto \lambda(y))$ be such that for all $y \in HC$, $\lambda(y)$ is a limit cardinal in $\mathcal{M}(y)$. We say that $\mathcal{M}$ reconstructs itself below λ iff whenever $y \in HC$ and $\mathcal{N}(y)$ is a countable, non-dropping iterate of $\mathcal{M}(y)$ by its unique $\omega_1 + 1$ iteration strategy, with λ^* the image of $\lambda(y)$, and $\kappa < \lambda^*$, and z is $\mathcal{N}(y)$-generic over a poset of size $< \kappa$ in $\mathcal{N}(y)$ and $\mathcal{P}(z)$ is the extender model to which the maximal plus-1 certified ms-array over κ of $\mathcal{N}(y)[z]$ converges (see [21]), then there is a sharp mouse $\mathcal{Q}(z)$ whose reduct is $\mathcal{P}(z)$, and a Σ_1-elementary $\pi\colon \mathcal{M}(z) \to \mathcal{Q}(z)$ such that $\pi(\lambda(z)) \leq \lambda^*$.*

Definition 4.15 *A tractable operator is a pair $(\mathcal{M}, \lambda)$ such that*

(a) *$\mathcal{M}$ is a sharp-mouse operator,*

(b) *$\lambda = (y \mapsto \lambda(y))$ is such that $\lambda(y)$ is definable from y over $\mathcal{M}(y)$, uniformly in y,*

(c) *for all y, $\mathcal{M}(y) \models$ "$\lambda(y)$ is a limit of Woodin cardinals, and $\mathrm{cof}(\lambda(y))$ is not measurable",*

(d) *$\mathcal{M}$ reconstructs itself below λ.*

With these definitions, we can extend 3.6 to:

Theorem 4.16 *Let $\langle \mathcal{M}, \lambda \rangle$ be a tractable operator, let $x, y \in HC$, and let I and J be $\mathcal{I}(\mathcal{M}(x), \lambda(x))$ and $\mathcal{I}(\mathcal{M}(y), \lambda(y))$ generic iterations; then $D(\mathcal{M}(x)_\infty^I, \lambda_\infty^I) = D(\mathcal{M}(y)_\infty^J, \lambda_\infty^J)$.*

Later we shall need a further strengthening of 3.6, in which the relativised mice are *hybrid mice* of the form $\mathcal{M}^\Sigma(y)$, where Σ is an appropriate set of reals in the derived model of $\mathcal{M}(y)$.

5. The mouse set conjecture in $D(\mathcal{M}, \lambda)$

If $\mathcal{M}$ is a mouse, then $\mathcal{M}$ itself "captures" all reals which are OD in its derived model. So the following result is not surprising:

Theorem 5.1 *Let $(\mathcal{M}, \lambda)$ be a tractable operator, and suppose D is a realization of $D(\mathcal{M}(z), \lambda)$, and y is transitive and countable in D, and $x \subseteq y$. Then the following are equivalent:*

(a) $D \models x \in OD(y \cup \{y\})$,

(b) $x \in \mathcal{M}(y)$,

(c) $D \models x$ is in an ω_1-iterable mouse over y.

Proof. (a) $\Rightarrow$ (b): We easily get that $D^* \models x \in \mathrm{OD}(y \cup \{y\})$, where $D^* = D(\mathcal{M}(z)_\infty^I)$ for an $\mathcal{I}(\mathcal{M}(z), \lambda(z))$-generic I. But then $D^* = D(\mathcal{M}(y)_\infty^J)$ for an $\mathcal{I}(\mathcal{M}(y), \lambda(y))$-generic J. Since the forcing is homogeneous, $x \in \mathcal{M}(y)_\infty^J$. Therefore $x \in \mathcal{M}(y)$.

(b) $\Rightarrow$ (c) (*Sketch*): It is enough to show that if $\mathcal{P}$ is a proper initial segment of $\mathcal{M}(y)$, and $\mathcal{P}$ projects to y, then $\mathcal{P}$ has an ω_1-iteration strategy in $D(\mathcal{M}(y), \lambda)$. (One can then pass to $D = D(\mathcal{M}(z), \lambda(z))$ by 4.16.) For that, we show that the unique strategy Σ for $\mathcal{P}$ has a μ-UB code in $\mathcal{M}(y)$, for any $\mu < \lambda(y)$. Fix $\mu < \lambda(y)$. Because $\mathcal{M}(y)$ reconstructs itself below $\lambda(y)$, the full background extender construction $\langle \mathcal{C}_k(\mathcal{N}_\eta) \rangle$ of $\mathcal{M}(y)$, done with all critical points $> \mu$ (and using μ-closed extenders from the sequence of $\mathcal{M}(y)$ as backgrounds), reaches $\mathcal{P} = \mathcal{C}_k(\mathcal{N}_\eta)$. Let $\mathcal{M}(y)|\xi \models \mathcal{P} = \mathcal{C}_k(\mathcal{N}_\eta) \wedge \mathsf{ZFC}^-$. By the results of section 12 of [12], Σ is induced by lifting trees on $\mathcal{P}$ to trees on $\mathcal{M}(y)$, then following the unique strategy for $\mathcal{M}(y)$. The lifting process is uniformly definable over $\mathcal{M}(y)[g]$, for any size $< \mu$ generic g over $\mathcal{M}(y)$, and this definition has the generic absoluteness required to obtain a μ-UB code. The key to the generic absoluteness is the fact that **UBH** holds in ms-mice, or more precisely, the sharper version of this fact stated as Theorem 3.4 in [21]. The iteration trees on $\mathcal{P}$ being lifted have size less than the closure of the background extenders, so by **UBH** they are continuously illfounded off the branches they choose, which guarantees both the existence-in-$\mathcal{M}(y)[g]$ and the absolute definabilty of the branches chosen by Σ.

[Here is a bit more detail. Let φ be the formula which defines Σ over $< \mu$-generic extensions of $\mathcal{M}(y)|\xi$ via this lifting process. It is enough to show that club many hulls of $\mathcal{M}(y)|\xi$ are generically correct. So let

$\pi\colon N \to \mathcal{M}(y)$ be elementary, where N is transitive, with $\pi \restriction (\mathcal{P} \cup \{\mathcal{P}\}) =$ identity. Let g be $< \pi^{-1}(\mu)$-generic over N. Let $\mathcal{T}$ be an iteration tree satisfying φ over $N[g]$. We need to see $\mathcal{T}$ is by Σ. Since Σ is the unique strategy for $\mathcal{P}$, it suffices to see that the phalanx $\Phi(\mathcal{T})$ is iterable. Let $\mathcal{U}$ be the lift of $\mathcal{T}$ to N, as in [12]. It suffices to see $\Phi(\mathcal{U})$ is iterable. But $\Phi(\mathcal{U})$ is continuously illfounded off the branches it chooses, and therefore must be according to any iteration strategy for N.]

(c) $\Rightarrow$ (b): This is one of the elementary corollaries of the comparison lemma. (Note ω_1-iterability implies $(\omega_1 + 1)$-iterability, granted AD.)

$\qquad\qquad\qquad\qquad\qquad\qquad\qquad\qquad\qquad\qquad\qquad\qquad\qquad$ $\square$

The independence of the derived model $D(\mathcal{M}_\infty^I)$ for *arbitrary* $\mathbb{R}$-genericity iterations (as opposed to $\mathcal{I}$-generic ones) is perhaps not so important. However, 5.1 does give us something there.

Corollary 5.2 *Let* $(\mathcal{M}, \lambda)$ *be a tractable operator, and let* D_0 *and* D_1 *be derived models associated to* $\mathbb{R}$-genericity iterations of $\mathcal{M}(x)$ *and* $\mathcal{M}(y)$. *Then* $(\Sigma_1^2)^{D_0} = (\Sigma_1^2)^{D_1}$. *Thus if* $D_0 \models \theta = \theta_0$, *then* $D_0 = D_1$.

The proof is an easy consequence of the fact that, assuming AD^+, $x \in OD(y)$ is a Σ_1^2-complete relation on reals. (Note also $D_0 \equiv D_1$ by 3.6, and since we are talking about "old" derived models, if $\theta_0 = \theta$ in such a model, then it is of the form $L(\mathbb{R} \cup \{U\})$, for U a Σ_1^2- complete set.)

We also get a sort of converse to 2.11(a), in the case the ground model is a mouse.

Corollary 5.3 *Let* $(\mathcal{M}, \lambda)$ *be a tractable operator, and suppose that for some* x, $\mathcal{M}(x) \models \lambda(x)$ *is a limit of cutpoints. Then for all* x, $\mathcal{M}(x) \models \lambda(x)$ *is a limit of cutpoints, moreover and derived model* $D(\mathcal{M}(x), \lambda(x))$ *satisfies* $\theta_0 = \theta$.

Proof. Let D be a realization of $D(\mathcal{M}(x), \lambda(x))$, ans suppose toward contradiction that $D \models \theta_0 < \theta$. AD^+ and $\theta_0 < \theta$ together imply that the relation on reals $z \notin OD(y)$ can be uniformized by a Suslin-co-Suslin function. Let $f(y) = z$ be such a function in the sense of D. Because the cutpoints of $\mathcal{M}(x)$ are cofinal below $\lambda(x)$, we can find such a cutpoint η and a g which is $\mathcal{M}(x)$-generic for $\mathrm{Col}(\omega, \eta)$ such that $f = p[T]$, for some tree T in $\mathcal{M}(x)[g]$. Let y be a real in $\mathcal{M}(x)[g]$ which codes $\langle \mathcal{M}(x)|\eta, g \rangle$ is some simple way, and let $z = f(y)$. Because $T \in \mathcal{M}(x)[g]$, we get $z \in \mathcal{M}(x)[g]$. But η is a cutpoint, so $\mathcal{M}[x][g]$ can be re-arranged as a mouse over y. It is not hard to see that this mouse must be $\trianglelefteq \mathcal{M}(y)$, so $z \in \mathcal{M}(y)$, so $z \in \mathrm{OD}(y)$ in $\mathcal{D}$, a contradiction. $\qquad\square$

Remark 5.4 The proof of Corollary 5.2 shows a bit more. Let $(\mathcal{M}, \lambda)$ and $\mathcal{M}, \gamma)$ be tractable operators, with the same mouse operator $\mathcal{M}$, but different limits of Woodins below which our $\mathbb{R}$-genericity iterations are being done. Let D_0 and D_1 be derived models associated to $\mathbb{R}$-genericity iterations of $\mathcal{M}(x)$ below $\lambda(x)$ and $\mathcal{M}(y)$ below $\gamma(y)$. Then $(\Sigma_1^2)^{D_0} = (\Sigma_1^2)^{D_1}$. Thus if $D_0 \models \theta = \theta_0$, then D_0 is a Wadge initial segment of D_1. This is what happens if $\lambda(x)$ is the sup of the first ω Woodin cardinals of $\mathcal{M}(x)$, by 5.3. D_0 is then just $L(U, \mathbb{R})$, where U is the universal Σ_1^2 set of D_1.

6. The Solovay sequence in $D(\mathcal{M}, \lambda)$

We are not sure how to properly state or prove the results of this section in the case of arbitrary tractable operators. We can handle the specific operators introduced above: $M_{\mathrm{adr}}^\sharp, M_{\mathrm{dc}}^\sharp, M_{\mathrm{inlim}}^\sharp, M_{\mathrm{wlim}}^\sharp$, with λ the sup of the Woodin cardinals of the mouse in each case. We can handle other tractable operators like these, but have not abstracted a good definition. So for now, let us call those four operators *paradigmatic*. Woodin's 2.11 implies that for our paradigmatic $(\mathcal{M}, \lambda)$, the derived model $D(\mathcal{M}, \lambda)$ satisfies $\mathsf{AD}_\mathbb{R}$, or equivalently, $\theta = \theta_\xi$ for some limit ordinal ξ. In fact, one can give a somewhat simpler proof of 2.11 in this case, one which gives more information as to what sets of reals sit at the θ_α's in the Wadge hierarchy of the derived model. To begin with

Theorem 6.1 *Let $(\mathcal{M}, \lambda)$ be paradigmatic, and let I be $\mathcal{I}(\mathcal{M}(x), \lambda(x))$-generic, and $D = D(\mathcal{M}_\infty^I, \lambda_\infty^I)$. Then the function*

$$z \mapsto \mathcal{M}(z), \text{ for } z \in \mathbb{R}$$

is in D, and has Wadge rank θ_0 in D.

Proof. We may as well take $x = \emptyset$. Let us write $\mathcal{M} = \mathcal{M}(\emptyset)$, $\lambda = \lambda(\emptyset)$. Let κ be the least $< \lambda$-strong cardinal, and δ the least Woodin cardinal $> \kappa$, in $\mathcal{M}$. By 3.6, we may assume that the first normal tree in I is a genericity iteration below δ, with all critical points above κ, giving rise to

$$i_{0,1} : M_0^I \to M_1^I,$$

in such a way that

$$\mathcal{M} = M_0^I \in M_1^I[g],$$

with g being $\mathrm{Col}(\omega, i_{0,1}(\delta))$-generic over M_1^I. We can also assume that $\mathrm{crit}(i_{1,\infty}) > i_{0,1}(\delta)$.

Claim 1. The function $z \mapsto \mathcal{M}(z)$, for $z \in \mathrm{HC}^V$, is in the symmetric model $M_\infty^I(\mathbb{R}^V)$.

Proof. Note that we can extend $i_{1,\infty}$ to act on $M_1[g]$. Fix $z \in \mathrm{HC}^V$; we show informally how to compute $\mathcal{M}(z)$ inside $M_\infty(\mathbb{R}^V)$. First, let $\alpha < \lambda_\infty$ be an inaccessible cardinal of M_∞ large enough that z is generic over $M_\infty[g]|\alpha$, and $i_{0,1}(\delta) < \alpha$. Let E be an extender on the M_∞ sequence such that $\kappa = \mathrm{crit}(E)$ and $\alpha \le \mathrm{lh}(E)$. In $M_\infty[g]$ we have $\mathcal{M}$ *as a set*, and hence we can form $\mathrm{Ult}(\mathcal{M}, E)$ as a set. Since $\mathcal{M}|\kappa^{+\mathcal{M}} = M_\infty|\kappa^{+M_\infty}$, we have that

$$\mathrm{Ult}(\mathcal{M}, E)|i_E^{\mathcal{M}}(\kappa) = \mathrm{Ult}(M_\infty, E)|i_E^{M_\infty}(\kappa),$$

and thus

$$\mathrm{Ult}(\mathcal{M}, E)|\alpha = M_\infty|\alpha$$

by coherence. Thus z is generic over $\mathrm{Ult}(\mathcal{M}, E)$ for a partial order of size $< i_E(\lambda)$. Since the $(\mathcal{M}, \lambda)$ operator is tractable, we can now use the modified background-extender construction of [21] to re-build $\mathcal{M}(z)$ from $\mathrm{Ult}(\mathcal{M}, E)[z]$. Once again, the key is that we get $\mathcal{M}(z)$ *as a set* in $M_\infty[g][z]$. $\square$

Claim 2. The function $F = z \mapsto \mathcal{M}(z)$, for $z \in \mathbb{R}^V$, is in $D(M_\infty, \lambda_\infty)$; in fact, it has a universally Baire code in $M_\infty[g]$.

Proof. Working in $M_\infty[g]$, where we have $\mathcal{M}$, let α be any inaccessible cardinal as in claim 1. Let E be an extender on the M_∞-sequence such that $\kappa = \mathrm{crit}(E)$ and $\alpha < \mathrm{lh}(E)$. Let φ be the formula defining F on size $< \alpha$ generic extensions of $M_\infty[g]$ from $\mathcal{M}$ and E which is implicit in the proof of claim 1.

Let ξ be sufficiently large. Since any tractable operator condenses to itself, we get that in $M_\infty[g]$, there are club many countable $X \prec M_\infty|\xi$ such that if N is the transitive collapse of X, and H is N-generic for a poset of size $<$ the collapse of α, then φ defines $F \restriction N[H]$ over $N[H]$ from $\mathcal{M}$ and the collapse of E. (I.e., there are club many generically correct X.) As is well known, this gives us an α-universally Baire code for $F \restriction M_\infty[g]$

in $M_\infty[g]$. It is easy to see that the code (T, T^*) has the property that $p[T] \cap M_\infty[g][H] = F \cap M_\infty[g][H]$ for all size $< \alpha$ generic H. ⊓

Claim 3. The function $F = z \mapsto \mathcal{M}(z)$, for $z \in \mathbb{R}^V$, has Wadge rank θ_0 in $D(M_\infty, \lambda_\infty)$.

Proof. Suppose F were $OD(z)$ in $D(M_\infty, \lambda_\infty)$, where $z \in \mathbb{R}$. By 5.1, we then get $F(z) \in \mathcal{M}(z)$, a contradiction. Thus F has Wadge rank at least θ_0.

Let $\mathcal{M}(z)^-$ be the proper class model obtained by iterating away the last extender of $\mathcal{M}(z)$, and put

$$F_n(z) = \text{type of 1st } n \text{ indiscernibles over } \mathcal{M}(z)^-,$$

for $n < \omega$ and $z \in \mathbb{R}$. It is clear that F is Wadge equivalent to the join of the F_n's. (This is actually just a matter of definition; strictly speaking, $F(z)$ isn't even a real. It is coded by its 1st order theory, however, and this theory is easily intercomputable with the join of the $F_n(z)$.) It is enough then to show that each F_n is $OD(\mathbb{R})$ in $D(M_\infty, \lambda_\infty)$.

For this, note first that F_n is uniformly Turing invariant, that is, it is a jump operator. Also, $F_n(z) \in \mathcal{M}(z)$ for all z, so we can set

$$g(z) = \text{least } \alpha \text{ such that } F_n(z) \in \mathcal{M}(z)|\alpha,$$

and

$$G(z) = \Sigma_1 \text{ theory of } p \text{ in } \mathcal{M}(z)|g(z),$$

where p is the first standard parameter of $\mathcal{M}(z)|g(z)$. Clearly $z \equiv_T y \Rightarrow g(z) = g(y)$, so g induces a function from the Turing degrees $\mathcal{D}$ to OR, which has an ordinal rank γ in $OR^{\mathcal{D}}$ mod the Martin measure. Also, G is a jump operator, so by the prewellordering of jump operators (see [28]) we have a recursively pointed perfect set P and an e such that $\forall x \in P$, $F_n(x) = \{e\}^{G(x)}$. Clearly γ and P determine the value of $F_n(x)$ on a pointed perfect set's worth of x. However, it is then easy to see that this determines the value of $F_n(x)$ at all x. Thus F_n is definable over $D(M_\infty, \lambda_\infty)$ from γ and P. □

The claims complete our proof. □

Remark 6.2 Something close to 6.1 was first proved by Woodin, by another method.

We can produce mouse operators sitting at the higher θ_α in $D(M_\infty^I, \lambda_\infty^I)$ by *nesting* the $\mathcal{M}$-operator at various depths. For the mouse operators $\mathcal{M}^\alpha$

we define this way, $\mathcal{M}^\alpha(x)$ will only be defined on the cone of x for which α is countable in the first admissible set over x. This is because we want $\mathcal{M}^\alpha(x)$ to project to x.

Definition 6.3 *Let $\mathcal{M}$ be a premouse operator, and let $\mathcal{P}$ be a premouse, with $\lambda \leq o(\mathcal{P})$. We say that $\mathcal{P}$ is $\mathcal{M}$-closed below λ iff whenever $\xi < \lambda$ and ξ is a cutpoint of $\mathcal{P}$, then $\mathcal{M}(\mathcal{P}|\xi) \trianglelefteq \mathcal{P}$.*

Notice that for each of our paradigmatic operators $(\mathcal{M}, \lambda)$ we have a large cardinal hypothesis φ such that λ is the first η such that the truth of φ in $\mathcal{M}$ is witnessed by the *total* extenders from the $\mathcal{M}|\eta$-sequence. Let us call φ the *hypothesis associated to* $(\mathcal{M}, \lambda)$.

Definition 6.4 *Let $(\mathcal{M}, \lambda)$ be paradigmatic, with associated hypothesis φ. For $\alpha < \omega_1$, we define the mouse operator $\mathcal{M}^\alpha$ by setting $\mathcal{M}^0 = \mathcal{M}$, and for x such that α is countable in the least admissible set over x:*

$$\mathcal{M}^{\alpha+1}(x) = \text{minimal active } x\text{-mouse } \mathcal{P} \text{ such that}$$
$$\exists \gamma < \text{crit}(\dot{F}^\mathcal{P}) \ (\mathcal{P} \text{ is } \mathcal{M}^\alpha\text{- closed below } \gamma$$
$$\text{and } \mathcal{P} \models \varphi, \text{ as witnessed by total extenders of } \mathcal{P}|\gamma).$$

For η a limit ordinal which is countable in the least admissible set over x:

$$\mathcal{M}^\eta(x) = \bigcup_{\alpha < \eta} \mathcal{M}^\alpha(x).$$

Theorem 6.5 *Let $(\mathcal{M}, \lambda)$ be paradigmatic, and $D = D(\mathcal{M}^I_\infty, \lambda^I_\infty)$ where I is $\mathcal{I}(\mathcal{M}, \lambda)$- generic. Then*

(a) if $\mathcal{M} = M^\sharp_{\text{adr}}$, then for all $n < \omega$, the $\mathcal{M}^n$-operator is in D and has Wadge rank θ_n in D;

(b) if $\mathcal{M} = M^\sharp_{\text{dc}}, M^\sharp_{\text{inlim}}$, or $M^\sharp_{\text{wlim}}$, then for all $\alpha < \omega_1$, the $\mathcal{M}^\alpha$ operator is in D, and has Wadge rank θ_α in D.

Proof sketch. We show that (a) and (b) hold when $\alpha = n = 1$, and leave the rest to the reader.

First, 3.6 generalizes in the following way.

Lemma 6.6 *Let I and J be generic $\mathbb{R}$-genericity iterations of $\mathcal{M}(x)$ below $\lambda(x)$ and $\mathcal{M}^1(y)$ below $\lambda^1(y)$, respectively. Then $D(\mathcal{M}(x)^I_\infty, \lambda(x)^I_\infty) = D(\mathcal{M}^1(y)^J_\infty, \lambda^1(y)^J_\infty)$.*

The proof is like that of 3.6. The main additional point is that an appropriate generic extension of $\mathcal{M}(x)$ can reconstruct $\mathcal{M}^1(y)$ arbitrarily high below $\lambda(x)$. This follows from 6.1.

Next, 5.1 generalizes as follows:

Lemma 6.7 *Under the hypotheses of theorem 6.5, we have that for x, y countable transitive with $x \subseteq y$,*

$$x \in OD(z \mapsto \mathcal{M}(z), y)^D \Leftrightarrow x \in \mathcal{M}^1(y).$$

The proof is just like that of 5.1, using 6.6 in place of 3.6. These two lemmas imply that if the $\mathcal{M}^1$ operator is in D, then it has Wadge rank θ_1 there. The proof is the same as that which shows the $\mathcal{M}^0$ operator sits at θ_0.

To see that the $\mathcal{M}^1$ operator is in D, we relativise the argument of 6.1. Let δ be the least Woodin cardinal above the least $< \lambda$-strong cardinal of $\mathcal{M}$, and let

$$i \colon \mathcal{M} \to \mathcal{N}$$

come from a genericity iteration below δ and above that strong cardinal, with $\mathcal{M} \in \mathcal{N}[g]$, where g is $\mathrm{Col}(\omega, i(\delta))$-generic. We will show in the next section that

$$\mathcal{N}[g] = \mathcal{M}^1[\langle \mathcal{N} | i(\delta), g \rangle].$$

We can therefore apply the argument of 6.1, doing a further genericity iteration $j \colon \mathcal{N} \to \mathcal{P}$ in the window between the first $< i(\lambda)$-strong cardinal of $\mathcal{N}[g]$ and its next Woodin, and add a universally Baire code of the $\mathcal{M}^1$ operator to some $\mathcal{P}[g][h]$. By derived model invariance, we may assume the I in our theorem started out going to $\mathcal{N}$ and then $\mathcal{P}$, so we are done. $\square$

One can avoid using the identity $\mathcal{N}[g] = \mathcal{M}^1[\langle \mathcal{N} | i(\delta), g \rangle]$, borrowed from the next section, in the proof above. It is enough that $\mathcal{N}[g]$ can reconstruct an iterate $\mathcal{P}$ of $\mathcal{M}^1[\langle \mathcal{N} | i(\delta), g \rangle]$ via a construction which provides full background extenders for the total extenders on the $\mathcal{P}$-sequence. Note here that because of the background condition, every set in $V_{i(\lambda)} \cap \mathcal{N}[g]$ is generic over $\mathcal{P}$ by a forcing of size $< i(\lambda)$. This means that $\mathcal{P}$ is close enough to $\mathcal{N}[g]$ that the argument of 6.1 still works.

7. The ∗-transform

In this section, we describe a method for translating mice with extenders overlapping some δ into relativised mice having δ as a cutpoint.

Suppose that $\mathcal{M}$ is an $\omega_1 + 1$-iterable, sound x-mouse, and $\mathcal{M}$ projects to x. Let $\mathcal{T}$ be a normal iteration tree on $\mathcal{M}$ played according to its unique iteration strategy, and let $\mathcal{N}$ be the last model of $\mathcal{T}$. Suppose δ is a cardinal of $\mathcal{N}$, and not measurable in $\mathcal{N}$ or any $\mathrm{Ult}(\mathcal{N}, E)$ with $\mathrm{lh}(E) \geq \delta$. Let τ be the order type of

$$S_\delta^\mathcal{N} = \{\kappa < \delta \mid \exists E(E \text{ is on the } \mathcal{N}\text{-sequence})$$
$$\text{and } \mathrm{crit}(E) = \kappa \text{ and } \mathrm{lh}(E) \geq \delta\}.$$

For $\alpha < \tau$, let

$$\kappa_\alpha = \alpha^{\text{th}} \text{ member of } S_\delta^\mathcal{N},$$

and

$$\gamma_\alpha = \text{ least } \gamma \text{ such that } \kappa_\alpha < \nu(E_\gamma^\mathcal{T}),$$

where we assume that γ_α exists for all $\alpha < \tau$, and

$$\mathcal{P}_\alpha = \mathcal{M}_{\gamma_\alpha} | \xi_\alpha,$$

where ξ_α is the largest ξ such that $\mathcal{M}_{\gamma_\alpha} | \xi$ and $\mathcal{M}_{\gamma_\alpha} | \mathrm{lh}(E_{\gamma_\alpha}^\mathcal{T})$ have the same subsets of κ_α. Note that $\mathcal{P}_\alpha$ is the model to which an extender with critical point κ_α would be applied in any normal continuation of $\mathcal{T}$. We now define

$$\Phi_\delta(\mathcal{T}) = \langle \mathcal{P}_\alpha \mid \alpha < \tau \rangle.$$

$\Phi_\delta(\mathcal{T})$ is essentially the sub-phalanx of $\Phi(\mathcal{T})$ containing the models we might actually go back to in a normal continuation of $\mathcal{T}$ using extenders of length $\geq \delta$. For example, if $\mathcal{T}$ is the genericity iteration of $\mathcal{M}$ we used in the proof of 6.1, which lived in the window between the least $< \lambda$-strong of $\mathcal{M}$ and the next Woodin δ, and $i \colon \mathcal{M} \to \mathcal{N}$ is the iteration map of $\mathcal{T}$, then $\Phi_{i(\delta)}(\mathcal{T}) = \langle \mathcal{M} \rangle$.

Theorem 7.1 (Closson, Neeman, Steel) *If* $\mathcal{M}, \mathcal{T}, \mathcal{N}$, *and* δ *are as above, and* g *is* $\mathrm{Col}(\omega, \xi)$ *generic over* $\mathcal{N}$, *for some* $\xi \leq \delta$, *and* $\Phi_\delta(\mathcal{T})$ *is in the least admissible set over* $\langle \mathcal{N}|\delta, g \rangle$, *then*

$$\mathcal{N}[g] =^* \mathcal{R},$$

for some $\langle \mathcal{N}|\delta, g \rangle$-*mouse* $\mathcal{R}$.

Remark 7.2 The equality $\mathcal{N}[g] = \mathcal{R}$ cannot literally be true, since the two are structures for different languages. What we mean is that the two structures are fine-structurally equivalent, in that they have the same projecta, standard parameters, and Levy hierarchy past some point. (It can happen that their universes are different, although they are the same if $\mathcal{N}[g]$ is admissible.) We prefer not to spell out here the details of this notion of equivalence, which we call *intertranslatability*. We write $=^*$ for it.

Proof. We define the *-transform*, which associates to initial segments $\mathcal{Q}[g]$ of $\mathcal{N}[g]$ mice $\mathcal{Q}[g]^*$ over $\langle \mathcal{N} | \delta, g \rangle$, in such a way that $\mathcal{Q}[g]$ and $\mathcal{Q}[g]^*$ are intertranslatable. Let $\Phi_\delta(\mathcal{T}) = \langle \mathcal{P}_\alpha \mid \alpha < \tau \rangle$.

To begin with, letting $\alpha > \delta$ be least such that $\mathcal{N} | \alpha \models \mathsf{KP}$, there is clearly a unique mouse $\mathcal{R}$ over $\langle \mathcal{N} | \delta, g \rangle$ such that $\mathcal{N} | \alpha[g] =^* \mathcal{R}$, and we let $(\mathcal{N} | \alpha[g])^*$ be this unique $\mathcal{R}$.

Let U be the tree of all finite sequences $\langle E_0, ..., E_n \rangle$ such that each E_i is an extender with $\mathrm{crit}(E_i) < \delta$ and $\mathrm{lh}(E_i) \geq \delta$, and E_0 is on the $\mathcal{N}$-sequence, and E_{i+1} is on the sequence of $\mathrm{Ult}(\mathcal{M}_{\gamma_\alpha}, E_i)$ for α such that $\mathrm{crit}(E_i) = \kappa_\alpha$. For $\langle E_0, ..., E_n \rangle$ in U, we write $\mathcal{P}(\vec{E})$ for $\mathrm{Ult}(\mathcal{M}_{\gamma_\alpha}, E_n)$, where $\kappa_\alpha = \mathrm{crit}(E_n)$. Here we understand that $\mathcal{P}(\emptyset) = \mathcal{N}$. Because our initial $\mathcal{M}$ is iterable, U is wellfounded. We shall define $\mathcal{Q}[g]^*$ for all $\mathcal{Q} \trianglelefteq \mathcal{P}(\vec{E})$ where $\vec{E} \in U$ such that $o(\mathcal{Q})$ is at least the first admissible over $\mathcal{N} | \delta$. The definition is by induction on the U-rank of $\vec{E}$, with a subinduction on $o(\mathcal{Q})$.

The inductive clauses are as follows:

(a) if $o(\mathcal{Q}) = \omega\alpha + \omega$, then $\mathcal{Q}[g]^*$ is obtained from $\mathcal{Q} | \alpha[g]^*$ by taking one step in the J-hierarchy.

(b) if $o(\mathcal{Q})$ is a limit ordinal and $\mathcal{Q}$ is passive, then $\mathcal{Q}[g]^* = \bigcup \{ (\mathcal{Q} | \eta)[g]^* \mid \eta < o(\mathcal{Q}) \}$.

(c) if $\mathcal{Q}$ is active with last extender E, and $\mathrm{crit}(E) > \delta$, then letting $\mathcal{Q} = (\mathcal{R}, E)$, we set $\mathcal{Q}[g]^* = (\mathcal{R}[g]^*, E)$.

(d) if $\mathcal{Q}$ is active with last extender E, and $\mathrm{crit}(E) = \kappa_\alpha < \delta$, then $\mathcal{Q}[g]^* = \mathrm{Ult}_n(\mathcal{P}_\alpha, E)[g]^*$, where n is least such that $\rho_{n+1}(\mathcal{P}_\alpha) \leq \kappa_\alpha$.

Our non-measurability assumption on δ guarantees that these cases are exhaustive.

The detailed verification that $\mathcal{Q}[g]$ and $\mathcal{Q}[g]^*$ are intertranslatable takes some work. However, the basic idea is quite simple. Since $\Phi_\delta(\mathcal{T})$ is coded into $\langle \mathcal{Q}|\delta, g \rangle$, $\mathcal{Q}[g]$ can recover $\mathcal{Q}[g]^*$ by employing the inductive definition we just gave. Conversely, if we want to recover $\mathcal{Q}[g]$ from $\mathcal{Q}[g]^*$, all is trivial unless inductive clause (d) applies. Adopting the notation there, we simply note that $\mathcal{Q}^*$ can recover $\mathcal{P}_\alpha$ and E as an appropriate core of itself, and associated core-embedding extender. It is worth noting that in this direction, we do not need to use g.

(For example, let $\mathcal{Q}$ be the first level of $\mathcal{N}$ such that case (d) applies in defining $\mathcal{Q}[g]^*$. It is not hard to see that $\mathrm{crit}(E) = \kappa_0$ then. Let n be least such that $\rho_{n+1} \leq \kappa_0$. We have $\mathrm{Ult}_n(\mathcal{P}_0, E)[g]^* = \mathrm{Ult}_n(\mathcal{P}_0, E)[g]$ because E is the first extender overlapping δ. We can regard $\mathrm{Ult}_n(\mathcal{P}_0, E)[g]$ as a mouse $\mathcal{Q}[g]^*$ over $\langle \mathcal{N}|\delta, g \rangle$. The universes of $\mathcal{Q}[g]$ and $\mathcal{Q}[g]^*$ are different in this case, but one has that the $r\Sigma_{n+1}^{\mathcal{Q}[g]^*}$ subsets of δ are the same as the $r\Sigma_1^{\mathcal{Q}[g]}$ subsets of δ. Moreover, $\rho_{n+1}(\mathcal{Q}[g]^*) \leq \delta$, and $\mathcal{Q}[g]^*$ is $n+1$-sound as a mouse over $\langle \mathcal{N}|\delta, g \rangle$. Similarly, $\rho_1(\mathcal{Q}) \leq \delta$, and $\mathcal{Q}$ is 1-sound. So $\mathcal{Q}[g]$ and $\mathcal{Q}[g]^*$ are intertranslatable.) $\qquad\square$

We can use the $*$-transform to compute the length of certain Solovay sequences.

Theorem 7.3 *(a) For $\mathcal{M} = M_{\mathrm{adr}}^\sharp$ and λ the sup of its Woodins, $D(\mathcal{M}, \lambda) \models \theta = \theta_\omega$.*

(b) For $\mathcal{M} = M_{\mathrm{dc}}^\sharp$ or $\mathcal{M} = M_{\mathrm{inlim}}^\sharp$, and λ the sup of its Woodins in either case, we have $D(\mathcal{M}, \lambda) \models \theta = \theta_{\omega_1}$.

Proof. We prove (b), the proof of (a) being similar. Let I be $\mathcal{I}(\mathcal{M}, \lambda)$-generic, and $A \in \mathrm{Hom}_I^*$. By 6.5, it suffices to show that A is projective in some operator $\mathcal{M}^\tau$, where $\tau < \omega_1$.

Let A have a $< \lambda_\infty$-UB code in $M_\infty[g]$, where g is generic on $\mathrm{Col}(\omega, \mu_0)$. Let n be large enough that $\mu = \mathrm{crit}(i_{n,\infty}) > \mu_0$. Let $\mathcal{T}$ be the iteration tree giving rise to $i_{0,n}$. Let ξ be the least Woodin of M_n^I above μ. We can do a genericity iteration of M_n^I in the window (μ, ξ), producing a normal tree $\mathcal{U}$ and associated embedding $k \colon M_n^I \to \mathcal{P}$ so that $\Phi_\mu(\mathcal{T}) \in \mathcal{P}[g][h]$ for some h which is $\mathcal{P}$- generic over $\mathrm{Col}(\omega, k(\xi))$.

Let γ be the least Woodin of $\mathcal{P}$ strictly above $k(\xi)$, and let Σ be the iteration strategy for $\mathcal{M}$. By the proof of 3.0.1, A is projective in the window-based fragment $\Gamma = \Sigma|(\mathcal{T}^\frown \mathcal{U}, [k(\xi), \gamma])$, and so it suffices to show this fragment is projective in some $\mathcal{M}^\tau$. Let τ be the order type of the cardinals of $\mathcal{P}$ which are strong past γ. Note that because ξ and γ were chosen to be *least*, the cardinals of $\mathcal{P}$ which are strong past γ are all $< \mu$, and moreover

$$\Phi_\gamma(\mathcal{T}^\frown \mathcal{U}) = \Phi_\mu(\mathcal{T}).$$

We can compute Γ from $\mathcal{M}^\tau$ using $\mathcal{Q}$-structures. Letting $\mathcal{W}$ be played according to Γ, and $b = \Gamma(\mathcal{W})$, we have that b is the unique cofinal branch c of $\mathcal{W}$ such that $\mathcal{Q}(c, \mathcal{W})$ is iterable when backed up by the phalanx $\Phi_{\delta(\mathcal{W})}(\mathcal{T}^\frown \mathcal{U}^\frown \mathcal{W}) = \Phi_\mu(\mathcal{T})$. (We use here that $\mathcal{W}$ does not drop.) The branch oracle $\mathcal{Q}(b, \mathcal{W})$ may involve extenders overlapping $\delta(\mathcal{W})$, but these can be transformed away, and we get b is the unique cofinal branch c of $\mathcal{W}$ such that $\mathcal{Q}(\mathcal{W})[g][h]^* \trianglelefteq \mathcal{M}^\tau(\langle \mathcal{M}(\mathcal{W}), (g, h) \rangle)$. The key here is that $\mathcal{Q}(\mathcal{W})[g][h]^*$ reaches no further than this in the hierarchy of mice over $\langle \mathcal{M}(\mathcal{W}), (g, h) \rangle)$. That can be proved by looking a little more closely at the definition of the $*$-transform. $\qquad \square$

Remark 7.4 The author first discovered a verion of the way $\mathcal{Q}[g]^*$ recovers $\mathcal{Q}[g]$ which works without the g coding $\Phi_\delta(\mathcal{T})$, and used this to prove 7.3. Itay Neeman then observed that the process simplified considerably in the presence of a g coding $\Phi_\delta(\mathcal{T})$, and would likely lead in that case to a level-by-level intertranslation. Erik Closson worked out the intertranslation in full detail.

8. A long Solovay sequence

We shall show in this section that $\theta_{\omega_1} < \theta$ in the derived model associated to $M^\sharp_{\text{wlim}}$, and then give some indication as to how to prove that in fact $\theta = \theta_\theta$ holds there.

It is easy to say what the mouse operator sitting at θ_{ω_1} is. Let us fix $\mathcal{M} = M^\sharp_{\text{wlim}}$ throughout this section, and let the $\mathcal{M}^\alpha$-operator be obtained from $\mathcal{M}$ as in 6.4.

Definition 8.1 *For any countable transitive set* x,

$$\mathcal{M}^{\omega_1}(x) = \mathcal{M}^{\omega_1^x}(x),$$

where ω_1^x *is the height of the least admissible set to which* x *belongs.*

Clearly $\mathcal{M}^\alpha$ is projective in $\mathcal{M}^{\omega_1}$, for all $\alpha < \omega_1$. It is enough then to show that, letting λ be the Woodin limit of Woodins in $\mathcal{M}$, and $D = D(\mathcal{M}^I_\infty, \lambda^I_\infty)$ where $I \in \mathcal{I}(\mathcal{M}, \lambda)$-generic, that $\mathcal{M}^{\omega_1} \in D$.

Claim 1. If $\mathcal{M}^{\omega_1} \upharpoonright V_{\lambda_\infty}^{M_\infty[g]} \in M_\infty[g]$, for some g generic over M_∞ for a poset of size $< \lambda_\infty$, then $\mathcal{M}^{\omega_1} \in D$.

Proof. The same proof that worked for $\mathcal{M}^0$ in 6.1 works here. Since $\mathcal{M}^{\omega_1}$ condenses to itself, and determines itself on small generic extensions, we get club many generically correct hulls of $M_\infty[g]$, and hence a UB code in $M_\infty[g]$ for $\mathcal{M}^{\omega_1}$. □

Now let

$$\mathcal{M} \models \kappa \text{ is } \lambda + \omega\text{-reflecting in } \lambda.$$

That there is such a κ follows from the Woodinness of λ, and this is all of the Woodin property we shall need to show that $\mathcal{M}^{\omega_1} \in D$. Let ξ be the least Woodin of $\mathcal{M}$ above κ, and let

$$k \colon \mathcal{M} \to \mathcal{P}$$

come from a genericity iteration $\mathcal{T}$ on $\mathcal{M}|\xi$ with all critical points $> \kappa$, such that

$$\mathcal{M} \in \mathcal{P}[g], \text{ where } g \text{ is } \mathrm{Col}(\omega, k(\xi))\text{-generic over } \mathcal{P}.$$

We may as well assume that $\mathcal{T}$ is the first tree used in I, and therefore it is enough to show

Claim 2. $\mathcal{M}^{\omega_1} | V_{k(\lambda)}^{\mathcal{P}[g]} \in \mathcal{P}[g]$.

Proof. We show that $\mathcal{P}[g]$ can compute $\mathcal{M}^{\omega_1}(x)$ by using the $*$-transform. To this end, let η be successor cardinal of $\mathcal{P}[g]$ such that $k(\xi) < \eta < k(\lambda)$. Let F_η be the first extender F on the $\mathcal{P}$-sequence such that $\kappa < \mathrm{crit}(F) \le \eta$ and $\mathrm{lh}(F) \ge \eta$, if there is one, and let F_η be a principal ultrafilter otherwise. We have that $\mathrm{crit}(F_\eta) < \eta$ and F_η is total on $\mathcal{P}$. Note that $\mathrm{crit}(F_\eta) > k(\xi)$. Set

$$\mathcal{P}_\eta = \mathrm{Ult}(\mathcal{P}, F_\eta).$$

The choice of F_η guarantees that there are no extenders G on the sequence of $\mathcal{P}_\eta$ such that $\kappa < \mathrm{crit}(G) \le \eta$ and $\mathrm{lh}(G) \ge \eta$. Thus

$$\Phi_\eta(\mathcal{T}^\frown \langle F_\eta \rangle) = \langle \mathcal{M} \rangle,$$

and since $\mathcal{M}$ is coded into $\langle \mathcal{P}|k(\xi), g \rangle$, we can define the $*$-transform of $\mathcal{P}_\eta[g]$ at η as in the proof of 7.1. We write

$$\mathcal{Q}[g]^{*,\eta}$$

for the $\langle \mathcal{P}|\eta, g\rangle$-mouse we get by applying the transform at η to an appropriate $\mathcal{Q}[g]$.

We now show by induction on $\alpha < k(\lambda)$:

Subclaim. If η is a successor cardinal of $\mathcal{P}$ such that $k(\xi) < \eta < k(\lambda)$, and $\alpha \leq \omega_1^{\langle \mathcal{P}|\eta, g\rangle}$, then $\mathcal{M}^\alpha(\langle \mathcal{P}|\eta, g\rangle) \trianglelefteq \mathcal{Q}[g]^{*,\eta}$, for some proper initial segment $\mathcal{Q}$ of $\mathcal{P}_\eta$.

Proof. We have done the case $\alpha = 0$ in the proof of 6.1. Let α be a limit ordinal, and $\alpha \leq \omega_1^{\langle \mathcal{P}|\eta, g\rangle}$. By induction, we get that the function $f(\beta) = \mathcal{M}^\beta(\langle \mathcal{P}|\eta, g\rangle)$, defined on all $\beta < \alpha$, is in $\mathcal{P}_\eta[g]$. But then $f \in \mathcal{P}_\eta[g]^{*,\eta}$, so $f \in \mathcal{Q}[g]^{*,\eta}$ for some $\mathcal{Q} \lhd \mathcal{P}_\eta$. Clearly then $\mathcal{M}^\alpha(\langle \mathcal{P}|\eta, g\rangle) \trianglelefteq \mathcal{Q}[g]^{*,\eta}$.

Now suppose the subclaim holds for α. Fix η such that $k(\xi) < \eta < \lambda$, η is a successor cardinal of $\mathcal{P}[g]$, and $\alpha < \omega_1^{\langle \mathcal{P}|\eta, g\rangle}$. We want to show $\mathcal{M}^{\alpha+1}(\langle \mathcal{P}|\eta, g\rangle)$ is a proper initial segment of $\mathcal{P}_\eta[g]^{*,\eta}$. Note that κ is $k(\lambda) + \omega$ reflecting in $k(\lambda)$ in $\mathcal{P}_\eta$, as it is not moved by our embedding from $\mathcal{M}$ to $\mathcal{P}_\eta$. Let A be the theory in $\mathcal{P}_\eta|(k(\lambda) + \omega)$ of parameters in $\mathcal{P}_\eta|k(\lambda)$, and let E be an extender on the $\mathcal{P}_\eta$-sequence so that $\mathrm{crit}(E) = \kappa$, and

$$i_E(A) \cap \eta^+ = A \cap \eta^+$$

holds in $\mathcal{P}_\eta$. Let

$$\mathcal{Q} = \mathcal{P}_\eta|\mathrm{lh}(E).$$

It is enough to show $\mathcal{M}^{\alpha+1}(\langle \mathcal{P}|\eta, g\rangle) \trianglelefteq \mathcal{Q}[g]^{*,\eta}$. But $\mathcal{Q}[g]^{*,\eta} = \mathrm{Ult}(M, E)[g]^{*,\eta}$, which reaches the Woodin-limit-of-Woodins hypothesis, so it is enough to show that $\mathrm{Ult}(M, E)[g]^{*,\eta}$ is $\mathcal{M}^\alpha$-closed.

Now by our induction hypothesis, $\mathcal{P}[g] \models$ " for all $\beta \leq \alpha$, for all successor cardinals ν such that $k(\xi) < \nu < k(\lambda)$ and $\beta \leq \omega_1^{\langle \mathcal{P}|\nu, g\rangle}$, $\mathcal{P}_\nu[g]^{*,\nu}$ has a proper initial segment satisfying "I am $\mathcal{M}^\beta(\langle \mathcal{P}|\nu, g\rangle)$"". Let us call the sentence in quotes $\psi(k(\xi), \mathcal{M}, g, k(\lambda), \alpha)$, where we have displayed the parameters about which it speaks. We then get $\mathcal{P}_\eta[g] \models \psi[k(\xi), \mathcal{M}, g, k(\lambda), i_{F_\eta}(\alpha)]$, using that $k(\xi), k(\lambda), \mathcal{M}$, and g are fixed by i_{F_η}. But $i_{F_\eta}(\alpha) \geq \alpha$, so inspecting ψ, we see $\mathcal{P}_\eta[g] \models \psi[k(\xi), \mathcal{M}, g, k(\lambda), \alpha]$. Letting τ be a term such that $\mathcal{M} = \tau^g$, we can fix $p \in \mathrm{Col}(\omega, k(\xi))$ such that

$$\mathcal{P}_\eta \models p \Vdash \psi(k(\xi), \tau, \dot{g}, k(\lambda), \alpha).$$

Because $i_E(A) \cap \eta^+ = A \cap \eta^+$ holds in $\mathcal{P}_\eta$,

$$\mathrm{Ult}(\mathcal{P}_\eta, E) \models p \Vdash \psi(k(\xi), \tau, \dot{g}, k(\lambda), \alpha).$$

But $\mathcal{M}$ embeds into $\mathcal{P}_\eta$ with critical point $> \kappa$, and hence $\mathrm{Ult}(M, E)$ embeds into $\mathrm{Ult}(\mathcal{P}_\eta, E)$ with critical point $> \eta^{+,\mathcal{P}_\eta}$. Thus

$$\mathrm{Ult}(\mathcal{M}, E) \models p \Vdash \psi(k(\xi), \tau, \dot{g}, k(\lambda), \alpha).$$

It is easy to see that this implies that $\mathrm{Ult}(\mathcal{M}, E)[g]$ is $\mathcal{M}^\alpha$-closed below $i_E(\lambda)$.   □

The subclaim completes the proof of Claim 2.   □

Now we may assume that the first iteration tree used in I is the tree giving rise to k. In that case, we get the hypothesis of Claim 1 from Claim 2, and so we have $\mathcal{M}^{\omega_1} \in D$.

Now suppose $\gamma < \theta^D$. We want to show $\theta_\gamma < \theta$ holds in D. Assume first that there is a prewellorder $\leq^*$ of $\mathbb{R}$ of order type γ such that $\leq^*$ has a UB code in $\mathcal{M}$, or more precisely, is captured by $(\emptyset, \emptyset)$ in the sense of the proof of 3.6. (This is true if $\leq^*$ is projective, for example.) For $x \in R$ and $n < \omega$, we define a mouse operator $\mathcal{M}^{x,n}$ which, we shall show, has Wadge rank at least $\theta_{\omega|x|+n}$ in D. (Here and below, $|x|$ is the rank of x in $\leq^*$.)

We shall define mouse jump operators $\mathcal{N}^{x,n}$, and then let $\mathcal{M}^{x,n}$ be the HC-extension of $(\mathcal{N}^{x,n})$. The operator $\mathcal{N}^{x,n}$ will be defined on the Turing cone above x. The definition of $\mathcal{N}^{x,n}$ proceeds by induction on the lexicographic order on $\mathbb{R} \times \omega$ determined by $\leq^*$. If $|x| = 0$, then we set $\mathcal{N}^{x,0} = \mathcal{M} \upharpoonright \mathbb{R}$. If $|x| > 0$, then for $z \in \mathbb{R}$,

$$\mathcal{N}^{x,0}(z) = \bigcup \{\mathcal{N}^{y,n}(z) \mid y \leq_T x \wedge y <^* x \wedge n < \omega\}.$$

It is important here that $\mathcal{N}^{x,0}$ is a mouse-jump operator itself, and for that one needs to use the definability of $\leq^* \upharpoonright \{y \mid y \leq_T z\}$ over $\mathcal{M}(z)$. This is where we use the assumption that $\leq^*$ has a UB code in $\mathcal{M}$.

Finally,

$$\mathcal{N}^{x,n+1}(z) = \text{ minimal active } (\mathcal{N}^{x,n})^+\text{-closed } z\text{-mouse } \mathcal{R}$$

$$\text{such that } \mathcal{R} \models \text{ there is a Woodin limit of Woodins.}$$

One can show that the $\mathcal{M}^{x,n}$ are in D by an argument like that given in the proof of claim 2 above. It is also not hard to show $\mathcal{M}^{x,n+1}$ is not ordinal definable from $\mathcal{M}^{x,n}$ and a real over D, and is essentially Wadge least with this property. (However, we have no reason to believe that $\mathcal{N}^{x,0}$ is even approximately Wadge minimal among upper bounds for the set of all $\mathcal{N}^{y,n}$ such that $y <^* x$.

In order to remove the assumption that $\leq^*$ has a UB code in $\mathcal{M}$, we replace our Woodin limit mouse operator $\mathcal{M}$ with a *hybrid mouse operator*

$\mathcal{M}^\Sigma$, where $\Sigma \in D$ is an iteration strategy with condensation such that $\leq^*$ is Wadge reducible to Σ. Here $\mathcal{M}^\Sigma(z)$ is the minimal active Σ-mouse with a Woodin limit of Woodins. The invariance of the derived model proof (see 3.6) extends so as to show that $\mathcal{I}$-generic iterations of $\mathcal{M}^\Sigma(z)$ yield the derived model D as well. This can be used as above to show that the nestings of $\mathcal{M}^\Sigma$ guided by $\leq^*$ are in D.

Remark 8.2   (a) Erik Closson has identified the least tractable $(\mathcal{M}, \lambda)$ such that $\theta_{\omega_1} < \theta$ holds in the derived model $D(\mathcal{M}, \lambda)$. The precise description is somewhat technical, but the existence of a $\kappa < \lambda$ which is S-strong in λ, where $S = \{\mu < \lambda \mid \mu \text{ is } < \lambda\text{-strong }\}$, is more than enough.

(b) The theory $\mathsf{AD}^+ + \theta_{\omega_1} < \theta$ should be equiconsistent with the theory satisfied by the least mouse whose derived model satisfies $\theta_{\omega_1} < \theta$. This is open at the moment, however.

(c) We have no large-cardinal characterization of the least tractable $(\mathcal{M}, \lambda)$ such that $\theta_{\omega_2} < \theta$ holds in $D(\mathcal{M}, \lambda)$.

9. The mouse set conjectures: Framework of the induction

We now shift gears, and head toward some partial results on the mouse set conjectures MSC. We shall be assuming AD^+ for pretty much the rest of these notes, perhaps tacitly on occasion. We shall rely heavily on the concepts and results involved in Woodin's proof of the mouse set conjectures below iteration strategies for nontame mice, as presented in [23]. (We give Woodin's proof at the end of section 11.)

For our arguments, it is important to prove a local form of 1.0.2. Let us say that Γ is a *boldface pointclass* just in case Γ is a collection of sets of reals closed downward under Wadge reducibility. By results of Wadge and Martin, the inclusion order on boldface pointclasses closed under complements is a prewellorder. A *projectively closed pointclass* is a boldface pointclass closed under complements and real quantification.

Definition 9.1 *(AD)* P_α *is the* α^{th} *projectively closed pointclass, in the inclusion order. We associate a structure to* P_α *by setting*

$$\tilde{P}_\alpha = (HC \cup P_\alpha, \in).$$

So P_0 is the class of projective sets, and $P_\lambda = \bigcup_{\alpha < \lambda} P_\alpha$ for λ limit.

Definition 9.2 *Let $\varphi(v)$ be a Σ_n formula in the language of set theory expanded by a unary predicate symbol $\dot{A}$, and let y be countable and transitive. We call a pair $(\mathcal{M}, \Sigma)$ a $\langle \varphi, y \rangle$-witness just in case*

(a) $\mathcal{M}$ is an ms-mouse over y,

(b) Σ is a good ω_1-iteration strategy for $\mathcal{M}$, and

(c) for some λ,

$$\mathcal{M} \models \text{there are } n + 5 \text{ Woodin cardinals} < \lambda,$$

and

$$\mathcal{M} \models \exists A \in \text{Hom}_{<\lambda}((HC, \in, A) \models \varphi[y]).$$

Although finer local versions of the mouse set conjectures could be stated and proved (in the initial segment of the Wadge hierarchy for which we have any proofs at the moment), we shall be content with the following.

Definition 9.3 S_λ *is the assertion: for any real y, any $A \in P_\lambda$, and formula $\varphi(v)$, if $(HC, \in, A) \models \varphi[y]$, then there is a $\langle \varphi, y \rangle$-witness $(\mathcal{M}, \Sigma)$ such that $\Sigma \in P_\lambda$.*

S_λ is a local version of MSC2. It easily implies a corresponding local version of MSC1.

Definition 9.4 *We write $x \in OD^\alpha(y)$ just in case x is definable from y and ordinal parameters over $\tilde{P}_\alpha$. We write $x \in OD^{<\lambda}(y)$ just in case $x \in OD^\alpha(y)$ for some $\alpha < \lambda$.*

Definition 9.5 C_λ *is the assertion: whenever x and y are countable transitive sets, with $x \subseteq y$, and $x \in OD^{<\lambda}(y)$, then there is an ms-mouse $\mathcal{M}$ over y such that $x \in \mathcal{M}$, and $\mathcal{M}$ has an ω_1-iteration strategy in P_λ.*

Lemma 9.6 *Let λ be a limit ordinal, and suppose S_λ holds; then C_λ holds.*

The lemma is an easy consequence of the fact that "there is a γ-th real in $OD^{<\lambda}(y)$" can be expressed by "$\tilde{P}_\lambda \models \psi[y]$", for some Σ_1 formula ψ. We omit further proof.

In the remainder of this paper, we shall outline the proof of a special case of:

Theorem 9.7 (Neeman, Steel) *Let λ be a limit ordinal which is small enough that no ω_1-iteration strategy for a mouse as in the hypothesis of 1.2 belongs to P_λ; then S_λ holds.*

The proof goes by induction on λ. The important stages to consider are of course those at which new Σ_1^2 facts are verified.

Definition 9.8 *We write $\tilde{P}_\alpha \prec_1^{\mathbb{R}} \tilde{P}_\beta$ iff $\alpha \leq \beta$, and whenever $\phi(v)$ is Σ_1, $x \in \mathbb{R}$, and $\tilde{P}_\beta \models \phi[x]$, then $\tilde{P}_\alpha \models \phi[x]$.*

Definition 9.9 *Let $\alpha \leq \beta \leq \theta$; then we call $[\alpha, \beta]$ a Σ_1^2-gap iff*

(1) $\tilde{P}_\alpha \prec_1^{\mathbb{R}} \tilde{P}_\beta$,

(2) there is no $\gamma < \alpha$ such that $\tilde{P}_\gamma \prec_1^{\mathbb{R}} \tilde{P}_\alpha$, and

(3) there is no $\gamma > \beta$ such that $\gamma \leq \theta$ and $\tilde{P}_\beta \prec_1^{\mathbb{R}} \tilde{P}_\gamma$.

Woodin's proofs that Σ_1^2 has the scale property and that every Σ_1^2 collection of sets of reals has a Δ_1^2 member localize, giving

Theorem 9.10 (Woodin) *Assume AD^+; then for any α*

(a) the lightface pointclass $\Sigma_1^{\tilde{P}_\alpha}$ has the scale property, and

(b) if $x \in \mathbb{R}$, ϕ is Σ_1, and there is an $A \in P_\alpha$ such that $\tilde{P}_\alpha \models \phi[A, x]$, then there is a $\Delta_1^{\tilde{P}_\alpha}(x)$ set $A \in P_\alpha$ such that $\tilde{P}_\alpha \models \phi[A, x]$.

It follows that there is a last Σ_1^2 gap, namely $[\delta_1^2, \theta]$, where δ_1^2 is the sup of the lengths of the boldface Δ_1^2 prewellorders of the reals, or equivalently, the length of any regular Σ_1^2-norm on a complete Σ_1^2 set.

According to 9.10, the appearance of a new Σ_1^2 truth about a real generates new scales. Indeed, suppose β ends a Σ_1^2 gap, that is, it is not the case that $\tilde{P}_\beta \prec_1^{\mathbb{R}} \tilde{P}_{\beta+1}$. We get from 9.10(b) a $\Delta_1^{\tilde{P}_{\beta+1}}(x)$ set $A \subseteq \mathbb{R}$ such that $A \in P_{\beta+1} \setminus P_\beta$. It follows that every set in $P_{\beta+1}$ is in the scaled boldface pointclass $\Sigma_1^{\tilde{P}_{\beta+1}}(\mathbb{R})$. So the new Σ_1^2 truth witnessed by a set of reals in $P_{\beta+1}$ has quickly generated scales on all sets of reals in $P_{\beta+1}$.

In $L(\mathbb{R})$ the converse is true, in that no new scales appear in a Σ_1^2 gap. This is not true in general, however. If $\theta_0 < \theta$, then there is a scale on Π_1^2 which appears inside the gap $[\delta_1^2, \theta]$; in fact the least Wadge rank of such a scale is θ_0, and we get a pointclass with the scale property at that point. (These are consequences of AD^+ due to Woodin; see 9.18 below.)

There can be local versions of this phenomenon in gaps below $[\delta_1^2, \theta]$. One way to understand the picture is to assume that our AD^+ universe is an initial segment of $K(\mathbb{R})$, as it probably must be unless we have reached ω_1-iteration strategies for mice with superstrongs.[2] As shown in [24], in initial segments of $K(\mathbb{R})$, new scales correspond precisely to the verification of new Σ_1 formulae about reals, in exactly the same pattern as held in $L(\mathbb{R})$. The rub is that these formulae are allowed a predicate for the extender sequence $\vec{E}$ from which $K(\mathbb{R})$ is constructed, and our initial segment of $K(\mathbb{R})$ may not have the iteration strategies needed to define $\vec{E}$. New scales appear in lock-step with the verification of new Σ_1 facts about $\vec{E}$, but these only generate new Σ_1^2 facts when ω_1-iteration strategies for countable $M \prec K(\mathbb{R})|\alpha$ not formerly seen to be ω_1-iterable are constructed.

We begin now with our inductive proof of S_λ. If λ is least limit ordinal such that S_λ fails, then $\lambda = \gamma + \omega$ for some limit ordinal γ. Clearly, there is some least β such that $\gamma \leq \beta < \lambda$ and it is not the case that $\tilde{P}_\beta \prec_1^{\mathbb{R}} \tilde{P}_{\beta+1}$; that is, there is a least $\beta \in [\gamma, \lambda)$ which ends a Σ_1^2 gap. Let β^* be this β.

Let α^* begin the Σ_1^2 gap which ends at β^*. Let

$$\bar{\Gamma} = \Sigma_1^{\tilde{P}_{\alpha^*}}$$

be the scaled pointclass at the beginning of our gap. From this point through the end of section 14, we have fixed α^*, β^*, and $\bar{\Gamma}$, and we are trying to prove $\mathsf{S}_{\beta^*+\omega}$.

Lemma 9.11 *If $\bar{\Gamma}$ is not closed under real quantification, then $\mathsf{S}_{\beta^*+\omega}$ holds.*

Proof sketch. In this case we must have $\alpha^* = \beta^*$, and since we chose β^* least, we must then have that β^* is a limit ordinal. We therefore have S_{β^*} by induction. We can now use arguments like those in the "no gap" case of the core model induction ([27]) to get a mouse operator $\mathcal{M}$ such that $\mathcal{M}(z)$ knows $\Sigma_1^{\tilde{P}_{\beta^*+1}}(z)$ truth, and has an ω_1 iteration strategy in P_{β^*+1}, for all z. Now we can use the argument of [23] to show that $\mathsf{S}_{\beta^*+\omega}$ fails, then there is an $\mathcal{M}$-closed mouse $\mathcal{N}$ with infinitely many Woodin cardinals and a measurable above such that $\mathcal{N}$ has an iteration strategy Σ in $P_{\beta^*+\omega}$. This is impossible, as all sets in $L(P_{\beta^*+1})$ are projective in any such Σ. $\square$

We therefore assume that $\bar{\Gamma}$ is closed under real quantification.

The rest of what we can prove has a vague similarity to the proof of 9.11, in that we show that if the gap between $\bar{\Gamma}$, the pointclass we have already captured by mice, and $P_{\beta^*+\omega}$, the next pointclass to be captured, is not too large, then we can capture $P_{\beta^*+\omega}$ as desired (that is, $\mathsf{S}_{\beta^*+\omega}$ holds). The markers in our trek across this Σ_1^2 gap are the new scaled pointclasses. We now state some results of Martin and Woodin (see [8] and [3]) which characterize the next scaled pointclass after one closed under real quantifiers abstractly, in terms of reflection properties.

Definition 9.12 *We call a pointclass good if it is ω-parametrized, and closed under recursive substitution, number quantification, and existential real quantification. A good scaled pointclass is a good pointclass which has the Scale Property. An inductive-like pointclass is a good scaled pointclass closed under universal real quantification.*

The scale property for $\bar{\Gamma}$ follows from 9.10, so we are assuming henceforth that $\bar{\Gamma}$ is inductive-like.

Definition 9.13 *Let Γ be a good scaled pointclass, and $A \subseteq \mathbb{R}$. We say that A is countably captured over Γ just in case there is a real x such that for all countable $\sigma \subseteq \mathbb{R}$ with $x \in \sigma$, $A \cap \sigma \in C_\Gamma(\sigma \cup \{\sigma\})$. Such a real x we call a Γ-good parameter for A. We call x a Γ-good parameter for the sequence $\vec{A}$ just in case it is a Γ-good parameter for each A_i.*

Here $C_\Gamma(b)$ is the largest countable $\Gamma(b)$ collection of subsets of b, for b countable transitive. $C_\Gamma(b) = P(b) \cap L[T, b]$ for T the tree of a Γ-scale on a universal Γ set. (Harrington and Kechris, [2].) It is clear that if x is a Γ-good parameter for A, then so is any $y \geq_T x$. So if x_i is Γ-good for A_i, then any real coding all the x_i is Γ-good for $\vec{A}$.

Definition 9.14 *For any good pointclass Γ, the envelope of Γ is given by:*
$\mathrm{Env}(\Gamma) = \{A \mid A \text{ is countably captured over } \Gamma\}$.

It is easy to see from the definition that $\mathrm{Env}(\Gamma)$ is a boldface pointclass closed under complements. Martin's argument from [8, §4] shows

Theorem 9.15 (Martin [8]) *If Γ is closed under real quantifiers, then*

(a) $\mathrm{Env}(\Gamma)$ *is closed under real quantifiers, and hence a projectively closed boldface pointclass, and*

(b) there is no scale on a universal Γ-dual set whose sequence of associated prewellorders is in $\mathrm{Env}(\Gamma)$.

We include a proof of (b), because it is short: if (b) fails, then the relation $x \notin C_\Gamma(y)$ would be uniformized by a function $f \in \mathrm{Env}(\Gamma)$. The relation $R(y, n, m) \Leftrightarrow f(y)(n) = m$ would then be countably captured over Γ, and letting y_0 be a real which witnesses this, we would have $f(y_0) \in C_\Gamma(y_0)$, a contradiction.

The next scaled pointclass after one closed under real quantifiers lies just beyond its envelope, and is given by a *self-justifying system*.

Definition 9.16 *A self-justifying system is a countable set* $\mathcal{A} \subseteq P(\mathbb{R})$ *which is closed under complements (in $\mathbb{R}$), real quantification, and such that every $A \in \mathcal{A}$ admits a scale $\vec{\psi}$ such that $\leq_{\psi_i} \in \mathcal{A}$ for all i. We say a sequence $\vec{A} = \langle A_i \mid i < \omega \rangle$ is self-justifying iff $\{A_i \mid i < \omega\}$ is self-justifying. An sjs is a self-justifying system or sequence.*

Definition 9.17 *Let Γ be a good scaled pointclass pointclass, and let $\mathcal{A}$ be an sjs. We say that $\mathcal{A}$ seals Γ just in case there is a universal Γ set in $\mathcal{A}$, and $\mathcal{A} \subseteq \mathrm{Env}(\Gamma)$.*

The following is the basic result on the existence of self justifying systems which seal a given pointclass. It was proved by Woodin in the mid 90's.[3] The proof uses the method of obtaining Suslin representations from direct limits of mice. The last section of [26] constructs an sjs which seals some Γ by very much the same method.

Theorem 9.18 (Woodin) *Assume* AD^+. *Let α begin a Σ_1^2-gap, and S_α holds. Let $\Gamma = (\Sigma_1)^{\check{P}_\alpha}$. Suppose Γ is closed under real quantification, and suppose there are scaled sets which are not in Γ; then for any $B \in \Gamma$, there is a self-justifying system $\mathcal{A}$ such that $B \in \mathcal{A}$, and $\mathcal{A}$ seals Γ.*

It is not hard to see that if $\mathcal{A}$ seals Γ, then letting Γ^+ be the collection of sets of reals which are Wadge reducible to a countable union of sets in $\mathcal{A}$, we have that Γ^+ is the least boldface pointclass with the scale property and closed under countable unions properly including Γ.

---

[3]Steve Jackson's article ([3, Lemma 3.18]) exposits some older results of Martin which approximate the result. In the notation of [3], $\mathrm{Env}(\Gamma)$ is $\Lambda(\Gamma, \kappa)$ for κ the prewellordering ordinal of Γ. It is introduced by a somewhat different definition, ostensibly stronger than countable capturing. One can use 9.15 to show the two definitions equivalent. Jackson also found a direct proof of the equivalence.

Our goal in these notes will be to show that $S_{\beta^*+\omega}$ holds as long as there is no local θ_0 between α^* and β^*:

Assumption (†): There is no ξ such that $\alpha^* < \xi \leq \beta^*$, $P_\xi = P(\mathbb{R}) \cap L(P_\xi)$, and $L(P_\xi) \models AD^+ + \theta_0 < \theta$.

One can formulate this in terms of mice as: $S_{\beta^*+\omega}$ holds so long as there is no iteration strategy in $P_{\beta^*+\omega}$ for a mouse $\mathcal{M}$ having a limit of Woodin cardinals λ, some cardinal $\kappa < \lambda$ which is $< \lambda$ strong, and such that $\mathcal{M}$ is closed under lower part mice built over its initial segments below λ which have iteration strategies in $\bar{\Gamma}$. We shall not need the equivalence of (†) with this statement, however.

Recall that $\bar{\Gamma} = (\Sigma_1^2)^{P_{\alpha^*}} = (\Sigma_1^2)^{P_{\beta^*}}$. By 9.10 every $\bar{\Gamma}$-dual set admits a scale (in P_{β^*+1}), and thus by 9.18, there is a sjs $\vec{A}$ such that $\vec{A}$ seals $\bar{\Gamma}$. Assumption (†) implies at once that

$$P_{\beta^*+\omega} \subseteq L(\vec{A}, \mathbb{R}),$$

for any such sjs $\vec{A}$. For otherwise, $P(\mathbb{R}) \cap L(\vec{A}, \mathbb{R}) \subseteq P_{\beta^*}$ by Wadge. But then $\bar{\Gamma} = (\Sigma_1^2)^{L(\vec{A}, \mathbb{R})}$, and hence in $L(\vec{A}, \mathbb{R})$, $\vec{A}$ yields a scale on a universal Π_1^2 set. Thus $L(\vec{A}, \mathbb{R}) \models \theta_0 < \theta$, and letting ξ be such that $P_\xi = P(\mathbb{R}) \cap L(\vec{A}, \mathbb{R})$, we have a counterexample to (†).

With regard to our trek upward along scales from $\bar{\Gamma}$ to $P_{\beta^*+\omega}$, assumption (†) says that we have one potentially significant leap, from $\bar{\Gamma}$ to some $\vec{A}$ which seals it, followed by steps within $L(\vec{A}, \mathbb{R})$. The latter steps are easily understood, in the same way that steps within $L(\mathbb{R})$ are understood.

We shall first prove what might seem like a weak approximation to $C_{\beta^*+\omega}$:

Lemma 9.19 *If assumption* (†) *holds, then for a Turing cone of reals x, there is a ms-mouse $\mathcal{M}_x$ over x such that*

(a) $\mathcal{M}_x$ projects to ω,

(b) $\mathcal{M}_x$ has an ω_1-iteration strategy in $P_{\beta^*+\omega}$, but

(c) $\mathcal{M}_x$ has no ω_1-iteration strategy in P_{α^*}.

We shall show in section 14 that in fact the conclusion of Lemma 9.19 implies $S_{\beta^*+\omega}$, in general, without the restrictive assumption (†). (But assuming AD^+, of course.) Our proof of 9.19 will take up sections 10-13, In these sections we assume AD^+, and that α^*, β^*, and $\bar{\Gamma}$ are as above.

10. The background universe N_x^*

As in [23], we will produce the ms-mice verifying 9.19 by using a full background extender construction, done in a sufficiently strong background universe.

Definition 10.1 *Let M be a countable transitive model of* ZFC, *let Σ be an (ω, ω_1)-iteration strategy for M, and let $A \subseteq \mathbb{R}$. Then we say (M, Σ) captures A (at δ, via (T, U)) just in case*

(a) $M \models$ "δ is Woodin, and (T, U) are absolutely complementing for $\mathrm{Col}(\omega, \delta)$", and

(b) for all reals x, $x \in A$ iff there is an iteration map $i \colon M \to P$, coming from an iteration according to Σ, such that $x \in p[i(T)]$.

If A is captured by some (M, Σ) such that Σ has the Dodd-Jensen property, then A is Suslin and co-Suslin. (The Dodd-Jensen property lets us define a direct limit for the system of all countable iterates of M via Σ. Letting (T_∞, U_∞) be the image of (T, U) in this direct limit, we can use genericity iterations to see that (T_∞, U_∞) witnesses that A is Suslin and co-Suslin.) Conversely, it is a basic result of Woodin that assuming AD$^+$, every Suslin and co-Suslin set A is captured by some (M, Σ) such that Σ has the Dodd-Jensen property. In fact

Theorem 10.2 (Woodin, see [5] and [23].) *Assume* AD$^+$, *let Γ be a good scaled pointclass not closed under $\forall^{\mathbb{R}}$, and let $A \in \Gamma$. Then for any real x, there is an (M, Σ) such that*

1. $x \in M$, and (M, Σ) captures A,

2. Σ has condensation, and hence the Dodd-Jensen property, and

3. Σ is projective in Γ.

See lemma 3.12 of [23] for a proof of 10.2. That proof relies heavily on [5]. MSC implies a fine-structural strengthening of 10.2; see theorem 16.6 below.

We shall be dealing with relativised, hybrid mice M of various types below. They are relativised in that they may be built over some countable transitive x, which is usually self-wellordered, so that the resulting model satisfies AC. Our convention is that the language of a mouse built over x has a name for x (but not names for elements of x). Our mice may be hybrids, in that at successor steps we may be closing under something stronger than first order definability (while occasionally adding extenders at limit steps). The function we close under at successor steps must have condensation properties which yield a good theory of the resulting hybrid mice.

In this situation we shall write $M|\eta$ for the η-th level of M.

We shall also deal with *hybrid mouse operators*, that is, functions $x \mapsto M_x$, where M_x is a hybrid mouse over x, defined on an HC-cone of countable transitive x.

We need a strengthening of 10.2, one in which the capturing mouse knows something about its own iteration strategy. This comes from combining 10.2 with Woodin's analysis of HOD. (See [5], [19, §8], [22].)

Theorem 10.3 (Woodin) *Assume* AD^+, *let* Γ *be a good scaled pointclass not closed under* $\forall^{\mathbb{R}}$, *and let* $A \in \Gamma$. *Then there is a function* $F(x) = (M_x, N_x^*, \Sigma_x, \delta_x)$, *defined on a cone of countable transitive* x, *such that* $F \in L_2(\Gamma, \mathbb{R})$ *(when coded naturally as a set of reals), and*

(a) (M_x, Σ_x) *captures* A *at* δ_x *via some* (T, U), *and*

(b) $x \in M_x$, $\Sigma_x \upharpoonright V_{\delta_x}^{M_x} \in M_x$,

(c) $N_x^* = L(M_x, \Lambda)$, *where* Λ *is the restriction of* Σ_x *to finite stacks* $\vec{T}$ *of normal iteration trees based on* $V_{\delta_x}^{M_x}$ *such that* $\vec{T} \in M_x$,

(d) $V_{\delta_x}^{M_x} = V_{\delta_x}^{N_x^*}$, $M_x|\delta_x = N_x^*|\delta_x$, *and*

(e) $N_x^* \models \mathsf{ZFC} + \delta_x$ *is Woodin.*

Proof sketch. For those to whom it might make sense, here is a very rough sketch. Let $\langle \Gamma_i \rangle$ be a strictly increasing sequence of good scaled pointclasses, each containing A, and projective in Γ (i.e., in $iL_1(\Gamma, \mathbb{R})$). Let J_i be an iteration strategy with condensation for coarse Γ_i Woodin mouse Q_i. (See lemma 3.12 of [23] for the construction of such J_i and Q_i.) Letting $J = \langle J_i \mid i < \omega \rangle$ and $Q = \langle Q_i \mid i < \omega \rangle$, we can, for any real x, build $J - hybrid$ mice over $\langle Q, x \rangle$. These are mice over $\langle Q, x \rangle$ constructed from extender sequences as usual, except that the mice are told the action of the iteration strategies in J on trees which they construct. The condensation properties of J guarantee that if this is done properly, the resulting mice have a fine structure, and behave like ordinary ms-mice. The desired M_x in 10.3 is then

$\mathcal{M}_1^J(x)|\kappa_x$, the minimal iterable 1-J-Woodin hybrid mouse over $\langle Q, x \rangle$, cut off at its least inaccessible cardinal κ_x above its Woodin cardinal. Such an iterable hybrid J-Woodin mouse exists, and has an iteration strategy in $P_{\beta^* + \omega}$, because we can simply repeat the construction in lemma 3.12 of [23] in the hybrid setting. Let δ_x be the Woodin cardinal of M_x, and let Σ_x be its canonical iteration strategy. Because M_x^- knows J, and Σ_x moves J properly, we get that $(M_x, \Sigma_x, \delta_x)$ captures A.

Finally, we set $N_x^* = L(M_x, \Lambda)$, where Λ is the restriction of Σ_x to iteration trees $\mathcal{T}$ based on $V_{\delta_x}^{M_x}$ such that $\mathcal{T} \in M_x$. The key is that for a cone of x, adding Σ_x to M_x neither adds bounded subsets to δ_x, nor kills the Woodiness of δ_x. This is is a direct limit system argument involving ideas in the computation of $\mathrm{HOD}^{L[x, G]}$, which Woodin used to prove the corresponding theorem about adding an appropriate fragment of the canonical iteration strategy for $M_1(x)$ to $M_1(x)$.[4]

This end our sketch. $\qquad\square$

Remark 10.4 It should be possible to improve 10.3 by requiring that F be projective in Γ, rather than one step away from that. It is because we don't have this stronger result that we need the room we get by only trying to prove S_λ for limit λ. Presumably there is a way to use more care here, and get S_η for all η. At any rate, the argument of section 15 shows $S_{\beta^* + \omega}$ implies $S_{\beta^* + n}$ for all n.

Given $A \in P_{\beta^* + \omega}$, 10.3 gives us for a cone of x a background universe N_x^* having x in it which

(i) has a Woodin cardinal δ_x,

(ii) knows how to iterate itself for finite stacks $\vec{\mathcal{T}}$ of normal trees based on $N_x^* | \delta_x$ such that $\vec{\mathcal{T}} \in L_{\kappa_x}(N_x^* | \delta_x)$, where $L_{\kappa_x}(N_x^* | \delta_x) \models \mathsf{ZFC}$, and

(iii) knows A.

---

[4]The reason we have resorted to a sequence of $J's$ and $Q's$ is that the computation of $\mathrm{HOD}^{L^J[x, G]}$, as an iterate of $M_1^J(x)$ together with a fragment of the canonical strategy for this iterate, seems to require that generic extensions of iterates of $M_1^J(x)$ know the action of J. Since the $Q's$ are not fine-structural, the usual Boolean-valued comparison argument will not work. However, J_{i+1} lets us compute the $C_{\Gamma_{i+1}}$-closure operator of $M_1^J(x)$, which determines easily the $C_{\Gamma_{i+1}}$-closure operator of $M_1^J(x)[g]$, which then determines J_i on $M_1^J(x)[g]$ via Q-structures.

The notation N_x^* is somewhat misleading, in that 10.3 gives us different functions $x \mapsto N_x^*$, depending on which $A \in P_{\beta^*+\omega}$ we are trying to capture. We shall make the dependence on A explicit when we need to by

Definition 10.5 *Let* $x \mapsto (M_x, N_x^*, \Sigma_x, \delta_x)$ *be as in 10.3 with respect to* A; *then we call the function* $x \mapsto (M_x, N_x^*, \Sigma_x, \delta_x)$ *a coarse mouse operator capturing* A.

11. The $L[\vec{E}]$-model N_x

Let $x \mapsto N_x^*$, etc., come from a coarse mouse operator capturing some A. We associate an ordinary mouse operator $x \mapsto N_x$ by letting N_x be the output of the full background extender K^c-construction over x, done inside N_x^* up to δ_x. Since N_x^* is ω_1-iterable in V (where AD^+ holds), the full background extender construction does indeed converge to an x-premouse N_x of height δ_x. Note that N_x is ω_1-iterable in V, since trees on it can be lifted to trees on N_x^*. The lifting process itself will be definable in N_x^*, so long as N_x^* knows how to iterate itself for the lifted trees. This it does know, so long as the lifted trees are in M_x. So N_x^* knows how to iterate N_x, for trees in M_x.

The following simple lemma implies that N_x is a universal weasel in the sense of N_x^*. Its proof is an adaptation of the proof by Mitchell and Schindler in [11] that K^c is universal.

Lemma 11.1 *Assume* ZFC, *and let* δ *be Woodin. Let* N *be the output of a maximal full background extender construction of length* δ, *done over some* x, *and assume this construction does not break down, so that* $o(N) = \delta$. *Suppose no initial segment of* N *satisfies "there is a superstrong cardinal". Let* W *be a premouse over* x *of height* $\leq \delta$, *and suppose* $\mathcal{P}, \mathcal{Q}$ *are the final models above* W, N *respectively in a successful coiteration. Then* $\mathcal{P} \trianglelefteq \mathcal{Q}$.

Proof. Assume instead that W iterates past N; that is, that $\mathcal{Q}$ is a proper initial segment of $\mathcal{P}$. Since δ is inaccessible, we get by standard arguments that $\mathcal{P} = \mathcal{M}_\delta^{\mathcal{T}}$ and $\mathcal{Q} = \mathcal{M}_\delta^{\mathcal{U}}$, where $(\mathcal{T}, \mathcal{U})$ are the coiteration trees. We also have that $\delta = o(\mathcal{Q})$, while

$$\delta = i_{\alpha,\delta}^{\mathcal{T}}(\mu)$$

for some α and $\mu < \delta$, which we now fix. Letting

$$\kappa_\beta = i_{\alpha,\beta}^{\mathcal{T}}(\mu),$$

we get a club $C \subseteq [0,\delta]_T \cap [0,\delta]_U$ such that

$$\beta \in C \Rightarrow \beta = \kappa_\beta \wedge i^{\mathcal{U}}_{0,\delta} \text{``}\beta \subseteq \beta.$$

Let us define

$$f(\gamma) = \text{ least } \beta > \gamma \text{ such that } \beta \in C.$$

Using the Woodiness of δ, we can find a limit point κ of $\mathcal{U}$, and a $j \colon V \to M$ such that

$$V_{j(f)(\kappa)+\omega} \subseteq M \text{ and } j(\dot{E}^N) \restriction j(f)(\kappa) = \dot{E}^N \restriction j(f)(\kappa).$$

Now our backgrounded model N was the output of a maximal construction, so letting E^* be the extender of j restricted to $j(f)(\kappa)$, we have that $E = E^* \cap N \in N$. We claim that some initial segment of E is of superstrong type in N.

For that, it is enough to show that for all $g \in N$ such that $g \colon \kappa \to \kappa$, $i_E(g)(\kappa) < j(f)(\kappa)$. For then, $E \restriction \sup\{i_E(g)(\kappa) \mid g \in N\}$ is the desired superstrong initial segment of E. So fix $g \in N$. We shall show that in fact, $g(\gamma) < f(\gamma)$ for all sufficiently large $\gamma < \kappa$.

Note that $i_{0,\kappa}(g) \colon \kappa \to \kappa$, and since W iterated past N, we can then find $\eta < \kappa$ such that

$$i^{\mathcal{U}}_{0,\kappa}(g) = i^{\mathcal{T}}_{\eta,\kappa}(h),$$

for $h \colon \kappa_\eta \to \kappa_\eta$. Now let $\kappa_\eta < \gamma < \kappa$. We then have

$$\begin{aligned}
g(\gamma) &\leq i^{\mathcal{U}}_{0,\kappa}(g(\gamma)) = i^{\mathcal{U}}_{0,\kappa}(g)(i^{\mathcal{U}}_{0,\kappa}(\gamma)) \\
&= i^{\mathcal{T}}_{\eta,\kappa}(h)(i^{\mathcal{U}}_{0,\kappa}(\gamma)) < f(\gamma),
\end{aligned}$$

as desired. The last inequality holds because if $\beta = f(\gamma)$, then $\beta \in C$, so $i^{\mathcal{U}}_{0,\kappa}(\gamma) < \beta$ and $i^{\mathcal{T}}_{\eta,\kappa}(h) \restriction \beta = i^{\mathcal{T}}_{\eta,\beta}(h) \colon \beta \to \beta$. $\square$

We can apply 11.1 inside N^*_x, and we get that N_x is universal for premice of height $\leq \delta$ there. This is useful, because we have a reasonable fragment of the iteration strategy for N_x in N^*_x. In particular, we get

Lemma 11.2 *Let $x \mapsto (M_x, N^*_x, \Sigma_x, \delta_x)$ be a coarse mouse operator which captures some A such that the complete $\bar{\Gamma}$ set of reals is Wadge reducible to A. Let N_x be the full bacground-certified extender model of $N^*_x|\delta_x$. Then for a cone of x, N_x is lower part full, in the sense that whenever η is a cardinal of N_x, and M is an η-sound mouse over $N_x|\eta$ which projects to η and has an ω_1-iteration strategy in $\bar{\Gamma}$, then $M \trianglelefteq N_x$.*

Proof. Because N_x^* captures A, we get both that $M \in N_x^*$, and that N_x^* has in it the resriction of the canonical iteration strategy for M to iteration trees in $L(N_x^*|\delta_x)$. But then N_x^* knows enough of their respective iteration strategies to successfully coiterate N_x with M. By 11.1, the N_x side comes out longer, and this easily implies $M \trianglelefteq N_x$. □

This is already enough for Woodin's proof that MSC holds in the initial segment of the Wadge hierarchy below iteration strategies for nontame mice, which is properly contained in the minimal model of $\mathsf{AD}^+ + \neg(\dagger)$. . We give here the proof of Lemma 9.19 under this more restictive assumption; section 14 will show that 9.19 implies the full MSC.

Theorem 11.3 (Woodin) *Assume* AD^+, *and let* Γ *be a good scaled point-class closed under real quantifiers. Suppose* $C_\Gamma(x)$ *is captured by mice for all* x; *that is, suppose* C_λ *holds, where* λ *is the Wadge ordinal of* Γ. *Suppose every* Γ-*dual set admits a scale, and let* $P_\beta = \mathrm{Env}(\Gamma)$ *be the envelope of* Γ; *then either*

(a) for a Turing cone of reals x, *there is ms-mouse* $\mathcal{M}$ *over* x *such that* $\mathcal{M}$ *has an* ω_1 *iteration strategy in* $P_{\beta+2}$, *but no* ω_1 *strategy in* Γ, *or*

(b) there is an ω_1 *iteration strategy for a nontame ms-mouse in* $P_{\beta+2}$[5].

Proof. Let A be a universal Γ set, and let $x \mapsto (M_x, N_x^*, \Sigma_x, \delta_x)$ be a coarse mouse operator capturing A. Our construction guarantees that each N_x^* has an ω_1 iteration strategy in $P_{\beta+2}$. Suppose that (a) fails on a cone, and let x be in this cone. It is enough to get a nontame mouse over some set, since then we can rebuild from it a nontame mouse over $\emptyset$.

Claim. For club many $\eta < \delta_x$, $N_x^*|\eta$ is Woodin with respect to all $f: \eta \to \eta$ such that $f \in C_\Gamma(N_x^*|\eta)$.

Proof. Work in N_x^*, and let (T, U) be our δ_x^+-absolutely complementing pair capturing A. Let S be a transitive model of ZFC^- with $(T, U, N_x^*|\delta_x) \in S$, and let $\pi: H \to S$ where H is transitive, $(T, U, N_x^*|\delta_x) \in \mathrm{ran}(\pi)$, and $\pi(\eta) = \delta_x$ for $\eta = \mathrm{crit}(\pi)$. Using the collapses of T and U, and the fact that C_Γ-fulllness is Γ-dual, we get that $C_\Gamma(N_x^*|\eta) \subseteq H$, The elementarity of π implies that $N_x^*|\eta$ is Woodin with respect to all $f \in H$, so we have the desired conclusion at η. Clearly, there are club many such η. □

---

[5] One could probably get much better bounds on the iteration strategies given in (a) and (b); $\Sigma_4^1(\vec{A})$ for any sjs containing U should suffice.

We may assume that N_x has boundedly many Woodin cardinals below δ_x, as otherwise δ_x is a Woodin limit of Woodin cardinals in $L(N_x)$, and we have a nontame mouse. (To see that δ_x is Woodin in $L(N_x)$, just note that it is Woodin in $L(N_x^*|\delta_x)$, and $L(N_x)$ is the output of a maximal full background extender construction in $L(N_x^*)$.) Now take η in the club of our claim, and such that η is a cardinal of N_x, but not Woodin in N_x.

Let Q be the largest initial segment of N_x satisfying "η is Woodin", so that Q projects to η, and in fact defines a new $f: \eta \to \eta$ witnessing $N_x|\eta$ is not Woodin. If η is a cutpoint of Q, then we can regard Q as a ms-mouse over $N_x|\eta$, and since we are in the cone where (a) fails, we get that f is in $C_\Gamma(N_x|\eta)$, and hence f is in $C_\Gamma(N_x^*|\eta)$. By the argument of [12][§11], there is a $g \in C_\Gamma(N_x^*|\eta)$ witnessing that $N_x^*|\eta$ is not Woodin. (g is obtained from f and $N_x|\eta$ in a simple way. Since η is a cardinal of N_x, $N_x|\eta$ is the η-th model of the construction of N_x, and thus $N_x|\eta$ is definable over $N_x^*|\eta$.) This contradicts η being in our club.

Thus η is not a cutpoint of Q, and hence Q is nontame.   □

Corollary 11.4 (Woodin) *Assume* AD^+ *and* $\theta_0 < \theta$; *then there is an* ω_1-*iterable nontame mouse.*

Proof. We apply the theorem with $\Gamma = \Sigma_1^2$. Since $\theta_0 < \theta$, every Γ-dual set admits a scale. Clearly, alternative (a) cannot hold, as every iterable mouse has an iteration strategy in Γ by the basis theorem. So (b) holds, and we are done.   □

12. Two hybrid mouse operators at θ_0

Let $\vec{A}$ be an sjs which seals $\bar{\Gamma}$. We can associate two hybrid mouse operators to $\vec{A}$ as follows.

Definition 12.1 *For any countable transitive* b, b^+ *is the b-mouse obtained by stacking collapsing mice with* ω_1-*iteration strategies in* $\bar{\Gamma}$, *through* ω *cardinals.*

Descriptive-set-theoretically, the sets in b^+ are just those in $L[T, b]$ of rank less than the ω-th cardinal above $|b|$ of $L[T, b]$, where T is the tree of a $\bar{\Gamma}$-scale on a universal $\bar{\Gamma}$ set. Our induction hypothesis tells us these are precisely the sets captured by mice with $\bar{\Gamma}$ strategies. Most of the time, b will be selfwellordered, and in fact, mostly it will be a premouse over some real x. We write $\mu_i(b^+)$ for the i-th cardinal above $|b|$ in the sense of b^+.

From [23] we have that for any $A \in \text{Env}(\bar{\Gamma})$, there is a real x such that whenever $x \in b$, and $i < \omega$, there is a term $\tau_i^A(b)$ capturing A over b^+ at $\mu_i(b^+)$. That is, whenever g is generic over b^+ for $\text{Col}(\omega, \mu_i(b^+))$, then $\tau_i^A(b)_g = A \cap b^+[g]$. In fact, this is true whenever x is a $\bar{\Gamma}$-good parameter for A. (See 9.13.)

Definition 12.2 *Let b be countable and transitive, and let $\vec{A}$ be an sjs which seals $\bar{\Gamma}$, and suppose there is some $x \in b$ which is a $\bar{\Gamma}$-good parameter for $\vec{A}$; then*

$$b \oplus \vec{A} = (b^+, T),$$

where

$$T(i, \tau) \Leftrightarrow \tau = \tau_i^{A(i)_0}(b).$$

Thus $b \oplus \vec{A}$ is a b-mouse with an extra amenable predicate T identifying the sequence of term relations capturing the A_i's at all the $\mu_j(b^+)$. Condensation for term relations (see [23, lemma 3.7]) gives

Theorem 12.3 (Woodin) *Let $\pi \colon P \to b \oplus \vec{A}$ be Σ_1 elementary; then $P = c \oplus \vec{A}$, where $\pi(c) = b$.*

In particular, $b \oplus \vec{A}$ is Σ_1 sound, in the sense that every element is Σ_1 definable from parameters in $b \cup \{b\}$.

Let us fix a real $x_{\vec{A}}$ such that whenever b is countable transitive and $x_{\vec{A}} \in b$, then $b \oplus \vec{A}$ exists. We write

$$\mathcal{H}_0^{\vec{A}}(b) = b \oplus \vec{A}.$$

So 12.3 says that our hybrid mouse operator $\mathcal{H}_0^{\vec{A}}$ condenses well.

Let $x \mapsto (M_x, N_x^*, \Sigma_x, \delta_x)$ be a coarse mouse operator in $P_{\beta^*+\omega}$ which captures $\vec{A}$. Let N_x be the full background certified extender model of $N_x^*|\delta_x$. We also suppose that there is a (J, Q) as in the proof sketch for 10.3, so that $M_x = M_1^J(x)|\kappa_x$ for all x, and J_i is an iteration strategy with condensation for Q_i, for all $i < \omega$.

Definition 12.4 $\kappa_\alpha(x)$ *is the α-th strong cardinal of N_x.*

Definition 12.5 *For x such that M_x, etc., are defined, we set $\mathcal{H}_1^{\vec{A}}(x) = \text{Hull}_1^{N_x|\kappa_0(x)\oplus\vec{A}}(x).$*

Thus $\mathcal{H}_1^{\vec{A}}(x)$ is the collapse of the set of points a such that for some i, a is first order definable over $N_x|\kappa_0(x))^+|\mu_i(N_x|\kappa_0(x))^+$ from some $\tau_j^{A_k}$'s.

Each of $\mathcal{H}_0^{\vec{A}}(x)$ and $\mathcal{H}_1^{\vec{A}}(x)$ take one step out of the pure $L[\vec{E}]$ hierarchy and into the $\vec{A}$-hybrid hierarchy, but the transition occurs much later in $\mathcal{H}_1^{\vec{A}}(x)$ than it does in $\mathcal{H}_0^{\vec{A}}(x)$. Both structures are Σ_1 sound and have Σ_1 projectum x, but they are nevertheless not equal, because in effect they represent different indexing conventions.

Let $P \in N_x^*|\delta_x$ be transitve, with $x \in P$. Let R be the output of the full background extender construction over P, done inside $N_x^*|\delta_x$. We shall write

$$N_x^* \models W \text{ is an } \vec{A}\text{-iterable strong weasel over } P \ ,$$

if in N_x^*, there is an iteration map $j\colon R \to S$ arising from a finite stack of normal iteration trees played according to the strategy induced by $\Sigma_x{}^6$, and an elementary $i\colon W \to S$ is such that $i(W|\kappa \oplus \vec{A}) = S|i(\kappa) \oplus \vec{A}$, where κ is the least strong cardinal of W. This implies W is truly iterable in V, via a strategy which moves the term relations for $\vec{A}$ at the least strong cardinal correctly. (It would be more natural to require the term relations at all cardinals be moved correctly, but we do not need that notion.)

In contrast to $\mathcal{H}_0^{\vec{A}}(x)$, $\mathcal{H}_1^{\vec{A}}(x)$ depends not just on $\vec{A}$ and x, but on N_x^*. The next lemma reduces this dependence to some extent.

Lemma 12.6 *Let $P \in N_x^*|\delta_x$ be transitive, with $x \in P$. Suppose that*

$$N_x^* \models W \text{ is a strong weasel over } P \ ,$$

then

$$\mathcal{H}_1^{\vec{A}}(P) = \text{Hull}_1^{W|\kappa \oplus \vec{A}}(P),$$

where κ is the least strong cardinal of W.

Proof. Let $j\colon R \to S$ and $i\colon W \to S$ witness that W is an $\vec{A}$-iterable strong weasel in N_x^*. Then

$$\text{Hull}_1^{W|\kappa \oplus \vec{A}}(P) = \text{Hull}_1^{S|i(\kappa) \oplus \vec{A}}(P)$$

$$= \text{Hull}_1^{R|\mu \oplus \vec{A}}(P),$$

where $\mu = j^{-1}(i(\kappa))$ is the least strong cardinal of R.

---

[6]It is not hard to generalize 11.1 to this situation, and show that if $o(S) = \delta$, then S is universal in N_x^*, in that no weasel of height δ iterates past S inside N_x^*.

In N_x^*, we re-build a one-J-Woodin hybrid mouse over P, using full background extenders from the N_x^*-sequence with critical points $> o(P)$. We get a proper class model

$$T \models \text{``I am } M_1^J(P)\text{''}$$

which is truly iterable (in a way which preserves the J-operation). Let U be the output of the maximal full background extender construction over P, as done inside T. We can compare U with R inside N_x^*, and the proof of 11.1 easily shows that no weasel can iterate past U in N_x^* (two layers of backgrounding yield as much universality as one), so U vand R have a common iterate via strategies which move the $\vec{A}$-term relations correctly. Thus

$$\text{Hull}_1^{R|\mu \oplus \vec{A}}(P) = \text{Hull}_1^{U|\nu \oplus \vec{A}}(P),$$

where ν is the least strong cardinal of U.

Now we compare T with the true $M_1^J(P)$, which was the background universe used to define N_P, and thence $\mathcal{H}_1^{\vec{A}}(P)$. Since we have two J-iterable models of "I am $M_1^J(P)$", we get $k(T) = l(M_1^J(P))$, and thence $k(U) = l(N_P)$. The iteration maps k, l move the $\vec{A}$-term relations correctly, and thus

$$\text{Hull}_1^{U|\nu \oplus \vec{A}}(P) = \mathcal{H}_1^{\vec{A}}(P),$$

completing the proof. $\qquad \square$

Remark 12.7 Using lemma 12.6, we can characterize the structure $\mathcal{H}_1^{\vec{A}}(N_x|\kappa_0(x))$ as follows: let $i \colon N_x^* \to M$ be the ultrapower embedding coming from the background extender for the order zero measure on $\kappa_0(x)$ of N_x. The structure $i(N_x)$ can be regarded as a mouse over $N_x|\kappa_0(x)$. (A mouse over $N_x|\kappa_0(x)$ can have no extenders on its sequence with critical point $\kappa_0(x)$. Although $i(N_x)$ has no *total* measures on its sequence with critical point $\kappa_0(x)$, it does have partial measures with this critical point. However, we can inductively "translate" partial measures μ to $\text{Ult}_n(Q, \mu)$, where Q is the longest initial segment of N_x, and n is as large as possible, so that this ultrapower makes sense. Much as with the $*$-transform, this translation gives a level-by-level equivalence between $i(N_x)$ and a mouse P over $N_x|\kappa_0(x)$. To save notation, we also write $i(N_x)$ for P.) Then so regarded, we have

$$\mathcal{H}_1^{\vec{A}}(N_x|\kappa_0(x)) = \text{Hull}_1^{i(N_x|\kappa_0(x)) \oplus \vec{A}}(N_x|\kappa_0(x)).$$

To see this, just apply lemma 12.6, with $W = i(N_x)$ and $P = N_x|\kappa_0(x)$.

We now prove a condensation property for the $\mathcal{H}_1^{\vec{A}}$ operator parallel to that proved in 12.3 for the $\mathcal{H}_0^{\vec{A}}$ operator. We shall obtain this property by showing that $\mathcal{H}_0^{\vec{A}}(P)$ is easily interdefinable with $\mathcal{H}_1^{\vec{A}}(P)$, for a cone of P. We obtain this equivalence on a cone by using the jump operator comparison techniques of [28]. The proof of this equivalence uses heavily the assumption that the conclusion of Lemma 9.19 fails. (We have not used that assumption yet in this section.) So for the rest of this section, we assume that for a cone of x, every ms-mouse over x which has an ω_1 strategy in $P_{\beta^*+\omega}$ and projects to x has such a strategy in P_{α^*}.

Let $x_0 \in \mathbb{R}$ be such that N_x^*, etc., are defined for all reals $x \geq_T x_0$. Let $x \geq_T x_0$, and let $k < \omega$. We set

$$I_k(x) = \mathrm{Th}^{(x \oplus \vec{A} | \mu_k(x))}(\{\tau_i^{A(i)_0}(x) \mid i < k\}),$$

and

$$J_k(x) = \mathrm{Th}^{P_x \oplus \vec{A} | \mu_k(P_x))}(\{\tau_i^{A(i)_0}(P_x) \mid i < k\}),$$

where P_x is such that $\mathcal{H}_1(x) = P_x \oplus \vec{A}$. (In other words, P_x is $\mathcal{H}_1(x)$ cut off at its largest limit cardinal.) It is easy to see that for all $k < \omega$

(1) I_k and J_k are uniformly Turing invariant functions, and hence jump operators in the sense of [28],

(2) for all $x \geq_T x_0$, $I_k(x) \leq_T J_k(x)$,

(3) for any $x \geq_T x_0$, $J_k(x) \in \mathcal{H}_1^{\vec{A}}(x)$, and hence $J_k(x) \in C_{\bar{\Gamma}}(x)$, and

(4) for any $x \geq_T x_0$ and any $y \in C_{\bar{\Gamma}}(x)$, there is a n such that $y \leq_T I_n(x)$.

Item (4) is a simple consequence of the soundness of $\mathcal{H}_0^{\vec{A}}(x)$.

Using (3),(4), and the comparability of jump operators, we get for each $k < \omega$ integers e_k, n_k and a real z_k such that

$$J_k(x) = \{e_k\}^{\langle I_{n_k}(x), z_k \rangle},$$

for all $x \geq_T z_k$. That is, the jump operator J_k is below the jump operator I_{n_k} in the Martin order, with e_k and z_k giving the reduction. Let z be a real coding x_0, $\langle (e_k, n_k) \mid k < \omega \rangle$, and $\langle z_k \mid k < \omega \rangle$. For $x \geq_T z$, $\mathcal{H}_0^{\vec{A}}(x)$ is simply coded by the recursive join of the $I_k(x)$ for $k < \omega$, while $\mathcal{H}_1^{\vec{A}}(x)$ is simply coded by the recursive join of the $J_k(x)$ for $k < \omega$, moreover, these two recursive joins are uniformly (granted z) Turing equivalent to one another.

Henceforth, we write z_0 for the real parameter z of the last paragraph. Notice now that the definitions of $I_k(x)$ and $J_k(x)$ make sense for x an arbitrary countable transitive set with $z_0 \in x$. (In fact, $x_0 \in x$ is enough.) Let us write $\mathcal{H}_{0,k}^{\vec{A}}(x)$ and $\mathcal{H}_{1,k}^{\vec{A}}(x)$ for the transitive, pointwise definable structures whose theories are $I_k(x)$ and $J_k(x)$ respectively. Thus for $i = 0, 1$, $\mathcal{H}_{i,k}^{\vec{A}}(x) \in x^+$ for all k, and $\mathcal{H}_i^{\vec{A}}(x)$ is the direct limit of the $\mathcal{H}_{i,k}^{\vec{A}}(x)$, for $k < \omega$, under the natural maps. Using z_0 together with condensation for $\mathcal{H}_0$, we get a condensation result for $\mathcal{H}_1$.

Theorem 12.8 *Assume that for a cone of x, every ms-mouse over x which has an ω_1 strategy in $P_{\beta^*+\omega}$ and projects to x has such a strategy in P_{α^*}. Let P and Q be countable and transitive, with $z_0 \in P$, and let $\pi \colon P \to Q$. The following are equivalent:*

(1) π extends to a Σ_1 elementary map $\pi^* \colon \mathcal{H}_0^{\vec{A}}(P) \to \mathcal{H}_0^{\vec{A}}(Q)$,

(2) π extends to a Σ_1 elementary map $\pi^* \colon \mathcal{H}_1^{\vec{A}}(P) \to \mathcal{H}_1^{\vec{A}}(Q)$.

Moreover, the map π^* is unique in each case.

The trick here is just to apply the reductions coded into z_0 to reals coming from generic enumerations of P and Q. We omit further detail.

13. New mice modulo (†)

In this section we complete the proof of Lemma 9.19. Fix a sjs $\vec{A}$ which seals $\bar{\Gamma}$. Let y_0 be a real and φ a Σ_n formula in the language of set theory expanded by $\dot{A}$ such that for some $A \in P_{\beta^*+1}$, we have $(\mathrm{HC}, \in, A) \models \varphi[y_0]$, but there is no such A in P_{β^*}.[7]

By (†), we have $P_{\beta^*+\omega} \subseteq L(\vec{A}, \mathbb{R})$. By the results of [23] relativised to $\vec{A}$, we have that Σ_1^2 facts true in $L_\gamma(\vec{A}, \mathbb{R})$ are witnessed by hybrid $\vec{A}$-mice with iteration strategies in $L_\gamma(\vec{A}, \mathbb{R})$.[8] This gives us a function $r \mapsto M(r)$

---

[7]It is easy to prove, using Skolem functions given by scales, that φ can be taken to be Σ_3.

[8]An $\vec{A}$-premouse $\mathcal{M}$ is just like an ordinary ms-premouse, except that if $\mathcal{M}|\xi = (\mathcal{M}|\gamma)^+$, then we must have $(\mathcal{M}|\gamma) \oplus \vec{A}$, or in other words $(\mathcal{M}|\xi, T)$ for T the term-relation predicate for $\vec{A}$, as the next level of $\mathcal{M}$. We can construct such mice in an appropriate background universe by adding extenders to the sequence as usual, coring down as usual, and adding term-relation predicates when we reach levels of the form $(\mathcal{M}|\gamma)^+$. We demand of iteration strategies that they move the $\oplus\vec{A}$ operation correctly. We can obtain this kind of iterability, and use it to develop a fine structure theory analogous to that of [12], because the $\oplus\vec{A}$ operation condenses well, in the sense of 12.3.

which is in P_{β^*+3}, and is defined at all countable transitive r such that $y_0 \in r$, such that for each countable transitive r in its domain

(a) $M(r)$ is an $\vec{A}$-mouse over r,

(b) $M(r)$ has a good ω_1-iteration strategy in P_{β^*+3}, and

(c) $o(M(r)) = \gamma + \omega$ for some γ, and for some λ,

$$M(r)|\gamma \models \lambda \text{ is a cardinal,}$$

$$M(r)|\gamma \models \text{there are } n+5 \text{ Woodin cardinals} < \lambda,$$

and

$$M(r)|\gamma \models \exists A \in \mathrm{Hom}_{<\lambda}((\mathrm{HC}, \in, A) \models \varphi[y_0]).$$

We assume $M(r)$ is chosen to be the minimal such $\vec{A}$-mouse over r, so that $M(r)|\gamma$ projects to r, where γ is as in (c).

Let $\tilde{M}$ be a set of reals coding the function $r \mapsto M(r)$ in some natural way. Now let $\mathcal{N} = (x \mapsto (M_x, N_x^*, \Sigma_x, \delta_x))$ a coarse mouse operator in P_{β^*+3} which captures $\vec{A}$ and $\tilde{M}$. From $\vec{A}$ and $\mathcal{N}$ we get the operators $\mathcal{H}_0^{\vec{A}}$ and $\mathcal{H}_1^{\vec{A}}$ of the last section. Let us assume toward contradiction that for the cone of $x \geq_T w_0$, every ms-mouse over x which has an ω_1 strategy in $P_{\beta^*+\omega}$ and projects to x has such a strategy in P_{α^*}. This gives us a real z_0 as in the preamble to 12.8. Fix $x \geq_T w_0, z_0$.

Claim 1. There is an $\eta < \delta_x$ such that

(a) there is a fully elementary $\pi : M(N_x^*|\eta) \to M(N_x^*|\delta_x)$ such that $\eta = \mathrm{crit}(\pi)$ and $\pi(\eta) = \delta_x$, and

(b) there is some $B \subseteq \eta$ such that $B \in N_x$, but $B \notin M(N_x^*|\eta)$.

Proof. Work in N_x^*, where we have $M(N_x^*|\delta_x)$ because we have a tree which projects to $\tilde{M}$ in all size δ_x extensions. Using this tree, it is easy to see there are club many $\eta < \delta_x$ satisfying (a).[9]

Since $M(N_x^*|\delta_x)$ has cardinality δ_x, we have a function $f : \delta_x \to \delta_x$ which eventually dominates each $g \in M(N_x^*|\delta_x)$. We can assume $f(\gamma)$ is inaccessible, for all γ. Now let η be in the club from (a), and such that $f \restriction \eta$

---

[9]One could avoid using the tree by taking $M(r)$ to be φ-minimal for the appropriate φ, thereby guaranteeing condensation for the M-operator.

dominates each $g \colon \eta \to \eta$ in $M(N_x^* | \eta)$, and such that there is a $j \colon V \to M$ with critical point η with $V_{j(f)(\eta)} \subseteq M$ and $j(\vec{E}) \restriction j(f)(\eta) = \vec{E} \restriction j(f)(\eta)$, where $\vec{E}$ is the extender sequence of N_x. We get such a j because δ_x is Woodin. As before, if $f \restriction \eta$ eventually dominates all $g \colon \eta \to \eta$ in N_x, then η would be Shelah in N_x, a contradiction. Thus there is a $g \colon \eta \to \eta$ which is in N_x, but not in $M(N_x^* | \eta)$. □

Fix η as in the claim. We may assume $\kappa < \eta$, where $\kappa = \kappa_0(x)$ is the first strong cardinal of N_x. Let γ be least such that there is a subset of η definable over $N_x | \gamma$ which is not in $M(N_x^* | \eta)$, and set $\mathcal{R} = N_x | \gamma$. Our plan is to show that $M(N_x^* | \eta)$ is intertranslatable with some initial segment of $\mathcal{R}[N_x^* | \eta]$, and thus that some initial segment of $\mathcal{R}$ is a $\langle \varphi, y_0 \rangle$-witness. The translation is a variant of the $*$ transform.

Note that $\mathcal{R} \models \eta$ is Woodin. Moreover, the total extenders from the $\mathcal{R} | \eta$ sequence whose support is a cardinal of $\mathcal{R}$ extend to extenders over N_x^* with strength $\geq$ this support. It follows that $N_x^* | \eta$ is generic over $\mathcal{R}$ for the η-generator extender algebra of $\mathcal{R}$ at η. Thus $\mathcal{R}[N_x^* | \eta]$ makes sense. It is not clear that we can regard it as a mouse over $N_x^* | \eta$, however, because η may not be a cutpoint of $\mathcal{R}$. Our smallness assumption (†) does give

Claim 2. Let E be an extender on the $\mathcal{R}$ sequence such that $\mathrm{crit}(E) \leq \eta \leq \mathrm{lh}(E)$; then $\mathrm{crit}(E) = \kappa$.

Proof. Since κ is the least strong, it is a limit of cutpoints on the N_x-sequence, so $\kappa \leq \mathrm{crit}(E)$. Since η is Woodin in $\mathcal{R}$ and by (†) there is no iterable mouse with a measurable Woodin, $\mathrm{crit}(E) < \eta$. Also, η is Woodin in $\mathcal{R}$, and hence Woodin in $\mathrm{Ult}(\mathcal{R}, E)$, and thus $\mathrm{crit}(E)$ is a limit of Woodins in $\mathcal{R}$. If $\kappa < \mathrm{crit}(E)$, then κ is $< \mathrm{crit}(E)$-strong in $\mathcal{R}$. Thus the minimal active mouse satisfying "there is a λ which is a limit of Woodins and such that some $\kappa < \lambda$ is $< \lambda$-strong" is an initial segment of $\mathcal{R}$, and hence has an ω_1 iteration strategy. This contradicts (†). □

Let us write $g = N_x^* | \eta$. We now define an analog of the $*$-*transform*. Our analog associates, to initial segments $\mathcal{Q}[g]$ of $\mathcal{R}[g]$, $\vec{A}$-mice $\mathcal{Q}[g]^a$ over g, in such a way that $\mathcal{Q}[g]$ and $\mathcal{Q}[g]^a$ are intertranslatable.

To begin with, letting $\alpha > \eta$ be least such that $\mathcal{R} | \alpha \models \mathsf{KP}$, there is clearly a unique mouse $\mathcal{S}$ over g such that $\mathcal{R} | \alpha[g] =^* \mathcal{S}$, and we let $(\mathcal{R} | \alpha[g])^a$ be this unique $\mathcal{S}$.

Let U be the tree of all finite sequences $\langle E_0, ..., E_n \rangle$ such that each E_i is an extender with $\mathrm{crit}(E_i) = \kappa$ and $\mathrm{lh}(E_i) \geq \eta$, and E_0 is on the $\mathcal{R}$-sequence, and E_{i+1} is on the sequence of $\mathrm{Ult}(N_x, E_i)$ for all $i < n$. For $\vec{E} = \langle E_0, ..., E_n \rangle$ in U, we write $\mathcal{P}(\vec{E})$ for $\mathrm{Ult}(N_x, E_n)$. Here we understand that $\mathcal{P}(\emptyset) = \mathcal{R}$. Because N_x is iterable, U is wellfounded. We shall define $\mathcal{Q}[g]^a$ for all $\mathcal{Q} \trianglelefteq \mathcal{P}(\vec{E})$ where $\vec{E} \in U$ such that $o(\mathcal{Q})$ is at least the first admissible over $\mathcal{R}|\eta$. The definition is by induction on the U-rank of $\vec{E}$, with a subinduction on $o(\mathcal{Q})$.

The inductive clauses are as follows:

(a) if $o(\mathcal{Q}) = \omega\alpha + \omega$, then $\mathcal{Q}[g]^a$ is obtained from $\mathcal{Q}|\alpha[g]^a$ by taking one step in the J-hierarchy.

(b) if $o(\mathcal{Q})$ is a limit ordinal and $\mathcal{Q}$ is passive, then $\mathcal{Q}[g]^a = \bigcup\{(\mathcal{Q}|\eta)[g]^a \mid \eta < o(\mathcal{Q})\}$.

(c) if $\mathcal{Q}$ is active with last extender E, and $\mathrm{crit}(E) > \eta$, then letting $\mathcal{Q} = (\mathcal{S}, E)$, we set $\mathcal{Q}[g]^a = (\mathcal{S}[g]^a, E)$.

(d) if $\mathcal{Q}$ is active with last extender E, and $\mathrm{crit}(E) = \kappa$, then $\mathcal{Q}[g]^a = (\mathrm{Ult}(N_x, E)|\gamma)[g]^a \oplus \vec{A}$, where γ is the least cardinal cutpoint ξ of $\mathrm{Ult}(N_x, E)$ such that $\nu(E) \leq \xi$.

Note that the ultrapower in case (d) makes sense, and that the least ξ as in (d) is $< i_E(\kappa)$. Note also that in case (d), we have that $\mathrm{Ult}(N_x, E)|\gamma^+ \trianglelefteq \mathrm{Ult}(N_x, E)$, so we could also write $\mathcal{Q}[g]^a = (\mathrm{Ult}(N_x, E)|\gamma^+[g]^a, T)$, where T is the term relation predicate for $\vec{A}$.

Again, the detailed verification that $\mathcal{Q}[g]$ and $\mathcal{Q}[g]^a$ are intertranslatable takes some work. The basic idea is as follows. Since $N_x|\kappa \oplus \vec{A} \in g$, $\mathcal{Q}[g]$ can recover $\mathcal{Q}[g]^a$ by employing the inductive definition we just gave. (N_x has the same subsets of κ as does $N_x|\kappa \oplus \vec{A}$, because κ is a cutpoint of N_x, and we are in the cone where 9.19 fails. The embedding associated to $\mathrm{Ult}(N_x|\kappa \oplus \vec{A}, E)$ moves the term relations for $\vec{A}$ correctly, and these image term relations can be used to compute the $\oplus \vec{A}$ part of $(\mathrm{Ult}(N_x|\kappa \oplus \vec{A}, E)|\gamma)[g]^a \oplus \vec{A}$.) Conversely, if we want to recover $\mathcal{Q}[g]$ from $\mathcal{Q}[g]^a$, all is trivial unless inductive clause (d) applies.

We sketch how to do the recovery in case (d). Adopting the notation there, we wish to recover E from $\mathrm{Ult}(N_x, E)|\gamma[g]^a \oplus \vec{A}$. We can recover $\mathrm{Ult}(N_x, E)|\gamma$ by induction. Since $z_0 \leq_T x$, we can then use $\mathrm{Ult}(N_x, E)|\gamma \oplus \vec{A}$ to obtain $\mathcal{H}_1^{\vec{A}}(\mathrm{Ult}(N_x, E)|\gamma)$. We can regard $\mathrm{Ult}(N_x, E)$ as a weasel over

$\mathrm{Ult}(N_x, E)|\gamma$, and we have then that

$$N_x^* \models \mathrm{Ult}(N_x, E) \text{ is an } \vec{A}\text{-iterable strong weasel over } \mathrm{Ult}(N_x, E)|\gamma.$$

So by Lemma 12.6, there is a Σ_1-elementary

$$\pi \colon \mathcal{H}_1^{\vec{A}}(\mathrm{Ult}(N_x, E)|\gamma) \to i_E(N_x|\kappa \oplus \vec{A}),$$

such that $\pi \upharpoonright \gamma = $ identity. (Here we regard $i_E(N_x|\kappa \oplus \vec{A})$ as a mouse over $\mathrm{Ult}(N_x, E)|\gamma$.) We have $i_E``(N_x|\kappa \oplus \vec{A}) \subseteq \mathrm{ran}(\pi)$, because $(N_x|\kappa \oplus \vec{A})$ is the Σ_1 hull of κ. Thus setting

$$j = \pi^{-1} \circ i_E \upharpoonright (N_x|\kappa \oplus \vec{A}),$$

we have

$$j \colon N_x|\kappa \oplus \vec{A} \to \mathcal{H}_1^{\vec{A}}(\mathrm{Ult}(N_x, E)|\gamma \oplus \vec{A}).$$

Now we can recover j from $N_x|\kappa \oplus \vec{A}$ and $\mathcal{H}_1^{\vec{A}}(\mathrm{Ult}(N_x, E)|\gamma \oplus \vec{A}))$, which we have available, as the uncollapse embedding associated to the Σ_1 hull of $N_x|\kappa$. Since $\pi \upharpoonright \gamma = $ identity, the extender of j agrees with that of i_E out to γ. Since $\gamma \geq \nu(E)$, we can thus recover E from j.

This ends our sketch of the intertranslatability of $\mathcal{Q}[g]$ and $\mathcal{Q}[g]^a$, for $\mathcal{Q} \trianglelefteq \mathcal{R}$.

Remark 13.1 There is a fine point here. In order to say that $\mathcal{Q}[g]^a$ is always a level of an $\vec{A}$-*mouse*, we must understand "level" in such a way that $\xi + 1$-st level of an $\vec{A}$-mouse $\mathcal{M}$ is not $\mathcal{M}|\xi \oplus \vec{A}$, but rather the first level of $(\mathcal{M}|\xi)^+$. The $\vec{A}$-active levels of $\mathcal{R}[g]^a$ come exactly from translating extenders with critical point κ, as in case (d). The following easily proved fact tells us how often this happens: let $\vec{E} \in U$, let $N = \mathcal{P}(\vec{E})$, and let μ, ν be indices of extenders on the N-sequence with critical point κ, with ν the least such index $> \mu$. Then $N|\nu = (N|\mu)^+|\nu$, and ν is the least cardinal $> \mu$ of $(N|\mu)^+$.

Now $\mathcal{R}$ defines a subset of η not in $M(N_x^*|\eta)$, and thus $\mathcal{R}[g]^a$ cannot be a proper initial segment of $M(N_x^*|\eta)$. It follows that, setting $\gamma + \omega = o(M(N_x^*|\eta))$, $M(N_x^*|\eta)|\gamma$ is a proper initial segment of $\mathcal{R}[g]^a$. Let λ be such that for $r = N_x^*|\eta$,

$$M(r)|\gamma \models \lambda \text{ is a cardinal,}$$

$$M(r)|\gamma \models \text{there are } n + 5 \text{ Woodin cardinals} < \lambda,$$

and

$$M(r)|\gamma \models \exists A \in \mathrm{Hom}_{<\lambda}((\mathrm{HC}, \in, A) \models \varphi[y_0]).$$

Inspecting our translation procedure, one sees that because λ is a limit of $\vec{A}$-active levels of $M(r)$, $M(r)|\lambda = (\mathcal{R}|\lambda)[g]^a$, and $(\mathcal{R}|\lambda)[g]$ has the same universe as $(\mathcal{R}|\lambda)[g]^a$. Thus $\mathcal{R}|(\lambda)[g]$ has an $n+5$-th Woodin cardinal $\delta > \eta$, and

$$(\mathcal{R}|\lambda)[g] \models \exists A \in \mathrm{Hom}_{\delta^+}((\mathrm{HC}, \in, A) \models \varphi[y_0]).$$

A result of Woodin (cf. [25, Theorem 5.1]) shows that then

$$\mathcal{R}|\lambda \models \exists A \in \mathrm{Hom}_{\delta^+}((\mathrm{HC}, \in, A) \models \varphi[y_0]).$$

Thus $\mathcal{R}|\lambda$ is a $\langle \varphi, y_0 \rangle$-witness. The first initial segment of $\mathcal{R}$ which is such a witness is easily seen to project to ω, and have no ω_1-iteration strategy in P_{β^*}. This is a contradiction, completing the proof of Lemma 9.19. $\qquad \square$

14. Proof of $S_{\beta^*+\omega}$

In this section, we prove

Theorem 14.1 *Assume* AD^+*, and let* $[\alpha, \beta]$ *be a* Σ_1^2 *gap, with* $\alpha < \beta$. *Suppose* S_α *holds. Suppose also there is a function* J *defined on a Turing cone of reals* y*, with* J *coded by a set of reals in* $P_{\beta+\omega}$*, and such that for* $y \in \mathrm{dom}(J)$*, we have* $J(y) = (\mathcal{S}_y, \Omega_y)$ *where*

(a) $\mathcal{S}_y$ *is a ms-mouse over* y *which projects to* ω*,*

(b) Ω_y *is an* ω_1*-iteration strategy for* $\mathcal{S}_y$*, and*

(c) $\mathcal{S}_y$ *has no* ω_1*-iteration strategy in* P_β*.*

Then $S_{\beta+\omega}$ *holds.*

By 9.19, we have that (†) implies there is a J as in the hypotheses of the theorem above when $\alpha = \alpha^*$ and $\beta = \beta^*$. So a proof of 14.1 will complete our proof of $S_{\beta^*+\omega}$ under the smallness assumption (†). It is worth emphasizing, however, that there is no smallness assumption in 14.1.

Proof of 14.1. We first derive a strengthened form of the hypotheses of 14.1.

Lemma 14.2 *Assume the hypotheses of 14.1, and let Γ be a good scaled pointclass such that $\Gamma \subseteq P_{\beta+\omega}$. Then for a Turing cone of reals y, there is an ms-mouse $\mathcal{M}$ over y such that*

(a) $\mathcal{M}_y$ *projects to* ω,

(b) $\mathcal{M}_y$ *has an* ω_1*-iteration strategy in* $P_{\beta+\omega}$, *but*

(c) $\mathcal{M}_y$ *is not coded by a real in* $C_\Gamma(y)$.

Proof. Let J be as in the hypothesis of 14.1. We may assume that J is coded by a set of reals in Γ. Let $x \mapsto (M_x, N_x^*, \Sigma_x, \delta_x)$ be a coarse mouse operator capturing U, where U is a universal Γ set. We can assume that the iteration strategies Σ_x (and for that matter, the whole operator) are in $P_{\beta+\omega}$. Fix any real x in the domain of this operator (which is a Turing cone), and in the cone where J is defined. It is enough to find a real $y \geq_T x$ and a mouse $\mathcal{M}_y$ over y such that the conclusion of 14.2 holds.

Claim 1. $N_x \models$ there are finitely many Woodin cardinals.

Proof. Suppose not, and let λ be a limit of Woodin cardinals in N_x. Because we are in the cone where J is defined, $\mathcal{S}_x$ is a mouse over x whose unique ω_1-iteration strategy Ω_x is in Γ, but not in P_β. Now N_x^* is sufficiently correct that

$$N_x^* \models \mathcal{S}_x \text{ is } \delta_x + 1 \text{ iterable above } \delta.$$

By 11.2, we have that $\mathcal{S}_x \trianglelefteq N_x$.

Going further in that direction, let $\gamma < \delta_x$, and let W be the result of a full background construction over x done in N_x with critical points above γ. It is not hard to modify the proof of 11.1 to show that $N_x^* \models W$ is universal, and thus that $\mathcal{S}_x \trianglelefteq W$.

By the last paragraph, we can compute $\Omega_x \restriction N_x$ inside N_x by reducing iterations of $\mathcal{S}_x$ to iterations of N_x above sufficiently large γ, using UBH in N_x to compute the correct branches for trees of size $< \gamma$. (See [21] for an expansion of this point.) This yields definable class trees T and T^* which project to a set $\tilde{\Omega}$ of reals coding Ω_x, and to $\mathbb{R} \setminus \tilde{\Omega}$, in all size $< \delta_x$ extensions of N_x. Moreover, if $j\colon N_x \to M$ comes from an iteration following the strategy induced by Σ_x, $j(T)$ and $j(T^*)$ still project to $\tilde{\Omega}$ and its complement in set-generic extensions of M.

Now let I be an $\mathbb{R}$-genericity iteration of N_x below λ, using the strategy induced by Σ_x, and let D be the resulting derived model. We have just shown $\Omega \in D$, and hence $P_{\beta+\omega} \subseteq D$. We may assume that x was chosen large enough that there is a real z such that $z \in \mathrm{OD}^D(x)$, but $z \notin \mathrm{OD}^{P_{\beta+\omega}}(x)$. (For a cone of x there is such a z, because the pointclass $(\mathbf{\Sigma}_1^2)^\mathbf{D}$ properly includes $P_{\beta+\omega}$, because our gap ended, and thus we can uniformize $z \notin \mathrm{OD}^{\beta+\omega}(x)$ by a $(\mathbf{\Sigma}_1^2)^\mathbf{D}$ function.) But since $z \in \mathrm{OD}^D(x)$, we get $z \in N_x$, and since N_x has an iteration strategy in $P_{\beta+\omega}$, this implies $z \in \mathrm{OD}^{P_{\beta+\omega}}(x)$, a contradiction. $\qquad\square$

Working in N_x^*, we can find club many $\eta < \delta_x$ such that

(a) η is a cardinal, and $N_x|\eta$ is the η-th model of the N_x-construction, and

(b) η is Γ-Woodin cardinal, in that whenever $f : \eta \to \eta$ and $f \in C_\Gamma(N_x^*|\eta)$, then there is an extender in $N_x^*|\eta$ witnessing the Woodin property with respect to f.

(Proof: Let T, T^* be trees witnessing that U is captured vis-a-vis the collapse of δ_x. Let $M \models \mathsf{ZFC}^-$ with $T, T^* \in M$, and let $\pi : H \to M$ with $T, T^* \in \mathrm{ran}(\pi)$, and $\pi(\eta) = \delta_x$, for $\eta = \mathrm{crit}(\pi)$. By arguments we have given already, η is Γ-Woodin.)

By claim 1, we can fix an η having properties (a) and (b) such that η is not Woodin in N_x. Let $\mathcal{Q} = N_x|\gamma$, where γ is least such that there is a $f : \eta \to \eta$ such that no extender from $N_x|\eta$ witnesses the Woodin property with respect to f, with f definable over $N_x|\gamma$. Note that there is no extender E on the $\mathcal{Q}$ sequence such that $\mathrm{crit}(E) \leq \eta \leq \mathrm{lh}(E)$, for otherwise, $\mathrm{crit}(E)$ would be a cardinal in N_x because η is, and then $\mathrm{crit}(E)$ would be a limit of Woodin cardinals in N_x, contrary to claim 1. Thus we can consider $\mathcal{Q}$ as a mouse over $N_x|\eta$. By (a), $N_x|\eta$ is first order definable over $N_x^*|\eta$, and by (b) then, $\mathcal{Q} \notin C_\Gamma(N_x|\eta)$. Let g be $\mathrm{Col}(\omega, \eta)$-generic over $L[T, N_x|\eta]$, where T is the tree of a Γ-scale on U, and let y be a real naturally coding $\langle N_x|\eta, g \rangle$. Then $\mathcal{Q}[g]$ can be regarded as a mouse $\mathcal{M}_y$ over y, and it is easy to see that the conclusions of 14.2 hold for $\mathcal{M}_y$. $\qquad\square$

The comparability of inner model operators gives

Corollary 14.3 *Let $\Gamma \subseteq P_{\beta+\omega}$ be a good scaled pointclass; then there is a mouse operator J_Γ coded by a set of reals in $P_{\beta+\omega}$, and such that for all $a \in \mathrm{dom}(J_\Gamma)$*

1. $J_\Gamma(a) = (\mathcal{M}_a, \Omega_a)$, where $\mathcal{M}_a$ is an ms-mouse over a which projects to a, and has (unique) ω_1 iteration strategy Ω_a, and

2. $C_\Gamma(a) \subseteq \mathcal{M}_a$.

Proof. By 14.2 we get a real z and a mouse jump operator J defined on all reals $a \geq_T z$ having property (1), and (2) weakened to "$\mathbb{R} \cap \mathcal{M}_a \not\subseteq C_\Gamma(a)$". ($J$ just picks the least mouse over a not in $C_\Gamma(a)$. Clearly J is uniformly Turing invariant.) Using [28], we see that J satisfies the full (2) on the cone of all $a \geq_T x$, for some $x \geq_T z$. Finally, we can extend J to the desired J_Γ acting on all countable transitive a such that $x \in a$ by Lemma 4.5. $\qquad\Box$

We now complete the proof of $\mathsf{S}_{\beta+\omega}$. Let y be a real and $A \in P_{\beta+\omega}$, and suppose

$$(\mathrm{HC}, \in, A) \models \varphi[y].$$

We seek a $\langle \varphi, y \rangle$-witness $(\mathcal{M}, \Sigma)$ such that $\Sigma \in P_{\beta+\omega}$. Suppose that φ is Σ_n.

Let $k = n + 10$, and let $\langle \Gamma_i \mid i \leq k \rangle$ be a strictly increasing sequence of good scaled pointclasses, with $A \in \Gamma_0$, and $\Gamma_k \subseteq P_{\beta+\omega}$. Let $J = J_{\Gamma_0}$ be as in 14.3 for Γ_0, with its domain being the HC-cone with base x_0. We assume Γ_1 is chosen large enough that there is a set of reals in Γ_1 which codes J. Let U_i be a universal Γ_i set, and let $x \mapsto (M_x, N_x^*, \Sigma_x, \delta_x)$ be a coarse mouse operator in $P_{\beta+\omega}$ which captures U_k. Fix a real x in the domain of this operator, and such that $y, x_0 \leq_T x$.

Working in N_x^*, let $\langle \mathcal{N}_\xi \mid \xi < \delta_x \rangle$ be the stages of the full background extender construction done *over* y through δ_x stages. We shall show that some $\mathcal{N}_\xi$ is a $\langle \varphi, y \rangle$-witness.

For $i \leq k$, we have trees T_i, T_i^* such that whenever $j \colon N_x^* \to S$ comes from an iteration via Σ_x, and g is S-generic over $\mathrm{Col}(\omega, \delta_x)$, then

$$U_i \cap S[g] = p[j(T_i)] \cap S[g]$$

and

$$(\mathbb{R} \setminus U_i) \cap S[g] = p[j(T_i^*)] \cap S[g].$$

Working in N_x^*, let us call η a Γ_i-Woodin cardinal iff by using (T_i, T_i^*) to compute $C_\Gamma(N_x^*|\eta)$, we see that η is Woodin with respect to all functions in $C_\Gamma(N_x^*|\eta)$. Let us put

$$\eta_k = \text{ the least } \Gamma_k\text{-Woodin cardinal,}$$

and for $1 \leq i \leq k$,

$$\eta_{i-1} = \text{ the least } \Gamma_{i-1}\text{-Woodin cardinal} > \eta_i.$$

So $\eta_k < \eta_{k-1} < \ldots < \eta_0$, and η_i is Γ_i-Woodin. Let $W = \mathcal{N}_{\eta_0}$. We shall show that W is a $\langle \varphi, y \rangle$-witness.

Claim 1. Let $\eta = \eta_i$, where $1 \leq i \leq k$; then $\mathcal{N}_\eta = W|\eta$, η is a cutpoint of W, and η is Woodin in W.

Proof. Because η is the first Γ_i-Woodin above some point, η is not measurable in N_x^*, and there is in N_x^* no $j\colon V \to M$ with $\mathrm{crit}(j) < \eta$ and $N_x^*|\eta \subseteq M$. (Proof: let m be largest such that $\mathrm{crit}(j) \leq \eta_m$. Then η_m is Γ_m-Woodin in N_x^* as certified by T_m, T_m^*, so η_m is Γ_m-Woodin in M as cerified by $j(T_m), j(T_m^*)$, by a standard argument. We can then pull back to get that $\mathrm{crit}(j)$ is a limit of Γ_m-Woodins, as certified by T_m, T_m^*. But $\mathrm{crit}(j) > \eta_{m+1}$ if $m < k$, so in any case, we have a contradiction.))

Now let $\mathcal{R} = \mathcal{N}_\eta$. It will be enough to show that if $\eta < \xi < \eta_0$, then $\rho_\omega(\mathcal{N}_\xi) \geq \eta$ and for every $f\colon \eta \to \eta$ definable over the core of $\mathcal{N}_\xi$, some extender from the $\mathcal{R}$-sequence witnesses the Woodin property with respect to f. So suppose not, and let $\mathcal{Q} = \mathcal{N}_\xi$ where ξ is the least counterexample. By the remarks of the last paragraph, η is a cutpoint of $\mathcal{Q}$.

Clearly $\rho_\omega(\mathcal{Q}) \leq \eta$. Let $\mathcal{S}$ be the η-core of $\mathcal{Q}$, that is, the transitive collapse of the Σ_n-hull of η in $\mathcal{Q}$, where n is least such that $\rho_n(\mathcal{Q}) \leq \eta$. We can regard $\mathcal{S}$ as a sound mouse over $\mathcal{R} = \mathcal{S}|\eta$. $\mathcal{S}$ has an iteration strategy easily computed from τ, where τ is the restriction of Σ_x to trees based on $N_x^*|\eta_0$ which are above $\eta = \eta_i$. But at worst we have Γ_{i-1}-Woodins in this interval, so Ω is coded by a set of reals in Γ_i, and in fact, $\mathcal{S}$ is coded by a subset of η in $C_{\Gamma_i}(\mathcal{R})$.

Because η is Γ_i-Woodin, $\mathcal{S}$ cannot define a bad $f\colon \eta \to \eta$. So let $\rho < \eta$ and $B \subseteq \rho$, with B definable over $\mathcal{S}$ but not in $\mathcal{R}$. Let T be the tree of a Γ_i scale on U_i. We work in $L(T, N_x^*|\eta)$, which satisfies that η is Woodin. Note $\mathcal{S} \in L(T, N_x^*|\eta)$. Let $M = L_\gamma(T, N_x^*|\eta) \models \mathsf{ZFC}^-$, and let

$$\pi\colon H \to M$$

with $\rho < \bar{\eta} = \mathrm{crit}(\pi) < \eta$, and

$$\pi((\bar{\mathcal{S}}, \bar{\eta})) = (\mathcal{S}, \eta).$$

We have that $\bar{\eta}$ is a cutpoint of $\bar{\mathcal{S}}$, $\bar{\mathcal{S}}|\bar{\eta} = \mathcal{R}|\bar{\eta}$, and B is definable over $\bar{\mathcal{S}}$.

Working now in V, let us compare $\bar{\mathcal{S}}$ with $\mathcal{R}$. We do this as in the proof of 12.6, moving the model on the $\mathcal{R}$ side of the coiteration by the background extender embeddings. This gives us iteration trees $\mathcal{T}$ on $\bar{\mathcal{S}}$ with last model $\mathcal{P}$, and $\mathcal{U}$ on N_x^* with embedding $j\colon N_x^* \to Z$, such that $j(\mathcal{R}) \trianglelefteq \mathcal{P}$. It is not hard to see that $\bar{\mathcal{S}} \in j(N_x^*|\eta)$. Let us now work in $L([j(T), j(N_x^*|\eta)]$. Since $\mathcal{S}$ has an iteration strategy in Γ_i for trees above η, so does $\bar{\mathcal{S}}$ for trees above $\bar{\eta}$. The strategy is unique, and using $j(T)$ we can compute it. Thus if we compare $\bar{S}$ with $j(\mathcal{R})$ in $L([j(T), j(N_x^*|\eta)], \bar{S}$ iterates via $\mathcal{T}$ to $\mathcal{P}$, while $j(\mathcal{R})$ does not move. It is easy to use the proof of our lemma on universality at Woodin cardinals (11.1) in $L([j(T), j(N_x^*|\eta)]$, where $j(\eta)$ is Woodin, to get a contradiction. $\square$

By our claim, x is generic over W for an extender algebra at η_k. Moreover, we can regard $W[x]$ as a mouse over $\langle W|\eta_k, x \rangle$.

Claim 2. $C_{\Gamma_0}(W[x]|\eta_1) \subseteq W[x]$.

Proof. Consider $J(W[x]|\eta_1) = (\mathcal{M}, \Omega)$. It will be enough to show $\mathcal{M} \trianglelefteq W[x]$, because $a = W[x]|\eta_0$ has x in it, so that (2) of 14.3 applies. So suppose not. Working in V, we compare $\mathcal{M}$ with $W[x]$. We do this as in [17], moving the model on the $W[x]$ side of the coiteration by the background extender embeddings. This gives us iteration trees $\mathcal{T}$ on $\mathcal{M}$ by Ω with last model $\mathcal{P}$, and $\mathcal{U}$ on N_x^* with embedding $j: N_x^* \to Z$, such that $j(W[x]) \trianglelefteq \mathcal{P}$. Let us work now in the universe $L[j(T), j(N_x^*|\eta_0)]$, where T is the tree of a Γ_0-scale on U_0. We have $\mathcal{M}$ and the restriction of Ω to trees in $L[j(T), j(N_x^*|\eta_0)]$ which are countable in V, as this can be computed from $j(T)$. It follows that the tree $\mathcal{T}$ leading from $\mathcal{M}$ to $\mathcal{P}$ is in $L[j(T), j(N_x^*|\eta_0)]$, and in this universe, it represents $\mathcal{M}$ iterating past $j(W[x])$ without the latter moving. However, $j(W)$ is the output of a maximal K^c-construction in $L[j(T), j(N_x^*|\eta_0)]$, and the proof of universality (11.1) applied shows that therefore $\mathcal{M}$ cannot iterate past $j(W)[x] = j(W[x])$ in $L[j(T), j(N_x^*|\eta_0)]$. This contradiction proves Claim 2. $\square$

Again, let T be the tree of a Γ_0-scale on U_0. In $L[T, N_x^*|\eta_0]$ there are trees S, S^* on $\omega \times \eta_0$ which project to A and its complement in all size $< \eta_0$ generic extensions, and whose images under iterations according to Σ_x also have this property. (Take S and S^* to be trees coming from Γ_0-scales coded into U_0, and note that Σ_x moves its tree for U_0 correctly.) By Claim 2, $S, S^* \in W[x]$. In $W[x]$, S and S^* are $< \eta_0$ absolute complements, and hence $p[S]$ is Hom_{η_2}. Now $W[x]$ has enough Woodin cardinals that its iteration strategy moving S and S^* consistently with A guarantees

$$W[x] \models \text{there are } n + 5 \text{ Woodin cardinals} < \eta_2,$$

and

$$W[x] \models \exists A \in \mathrm{Hom}_{\eta_2}((\mathrm{HC}, \in, A) \models \varphi[y]).$$

Using the Woodin cardinals of W between η_k and η_2 and Woodin's $(\Sigma_1^2)^{\mathrm{Hom}_\infty}$ generic absoluteness theorem ([25, §5]), we get

$$W \models \text{there are } n+5 \text{ Woodin cardinals} < \eta_2,$$

and

$$W \models \exists A \in \mathrm{Hom}_{\eta_0}((\mathrm{HC}, \in, A) \models \varphi[y]).$$

Thus W is the desired $\langle \varphi, y \rangle$-witness. This completes the proof of 14.1.

$\square$

Combining 9.19 with 14.1, we have a proof of $S_{\beta^*+\omega}$ under the smallness assumption (†). That is,

Corollary 14.4 *Assume* AD^+ *and* (†); *then for all limit* λ, S_λ *holds, and hence* MSC *holds.*

15. The consistency strength of $\mathsf{AD}^+ + \theta_0 < \theta$

The author's proof that $\mathsf{AD}_\mathbb{R}$ is equiconsistent with the $\mathsf{AD}_\mathbb{R}$-hypothesis also shows that $\mathsf{AD}^+ + \theta_0 < \theta$ is equiconsistent with ZFC plus the existence of one cardinal which is strong to the sup of ω Woodin cardinals. This latter result is due to Woodin. Both results are tightly connected to MSC at the level in question. In this section, we shall give a proof of Woodin's equiconsistency result for $\mathsf{AD}^+ + \theta_0 < \theta$ that uses the method by which we have obtained MSC at this level, and generalizes to yield the $\mathsf{AD}_\mathbb{R}$ equiconsistency.

Theorem 15.1 (Woodin) *The following theories are equiconsistent:*

(a) $\mathsf{ZF} + \mathsf{AD}^+ + \theta_0 < \theta$,

(b) $\mathsf{ZFC} + \exists \kappa \exists \lambda (\lambda$ *is a limit of Woodin cardinals, and* κ *is* $< \lambda$-*strong).*

Proof sketch. Given M satisfying the theory in (b), its derived model at λ satisfies the theory in (a), by theorem 2.11.

For the other direction, let us assume $\mathsf{ZF} + \mathsf{AD}^+ + \theta_0 < \theta$. We may also assume that we are in the minimal model of $\mathsf{ZF} + \mathsf{AD}^+ + \theta_0 < \theta$ containing all reals and ordinals, that is, that there is no $\xi < \theta$ such that $L(P_\xi) \cap P(\mathbb{R}) = P_\xi$, and $L(P_\xi) \models \theta_0 < \theta$. Thus by the preceding sections, we have MSC, in fact, we have S_λ for all limit λ. We shall construct a class model of the theory in (b).

Fix an sjs $\vec{A}$ which seals Σ_1^2, and a real x_0 such that each A_i is $OD(x_0)$. Woodin's proof that AD yields inner models with ω Woodin cardinals shows that for each countable transitive set c, there is a proper class model $M_\omega^{\vec{A}}(c)$ over c with ω Woodins, with the property that all its countable elementary submodels are ω_1-iterable $\vec{A}$-mice. (Hence they are iterable in a way that moves the $\vec{A}$-term relations correctly.)

We need to recall some features of the way we translated certain generic extensions $\mathcal{Q}[g]$ of ordinary ms-mice $\mathcal{Q}$ into $\vec{A}$-mice $\mathcal{Q}[g]^a$, replacing extenders with applications of the $\oplus \vec{A}$ operation. (See section 13.) This relied on two hybrid mouse operators $\mathcal{H}_0^{\vec{A}}$ and $\mathcal{H}_1^{\vec{A}}$, and on being above a real which enables one to compute $\mathcal{H}_1(x)$ from $\mathcal{H}_0(x)$. In the present context, we have

$$\mathcal{H}_0^{\vec{A}}(b) = b \oplus \vec{A},$$

for b in the HC-cone above x_0, as before. The $\mathcal{H}_1$ operator of section 12 depended on a coarse mouse operator $x \mapsto (M_x, N_x^*, \Sigma_x, \delta_x)$ capturing $\vec{A}$ and perhaps more. We do not have such a coarse mouse operator before us now, and in some sense, we have to consider all possible ones. But now for each coarse mouse operator we could choose, the associated $\mathcal{H}_1$ operator will be intercomputable with $\mathcal{H}_0$ on an HC cone via some effective process coded into some real z. (This process is described at the end of section 12, where the real in question is called z_0.) Let us call a real z a *potential code* iff

(i) z determines an operator $\mathcal{H}_1^{\vec{A},z}$ defined on the HC cone above x_0, z, via the procedure for computing $\mathcal{H}_1^{\vec{A}}(b)$ from $\mathcal{H}_0^{\vec{A},z}(b)$, and vice-versa, described at the end of section 12, and

(ii) the resulting operator $\mathcal{H}_1^{\vec{A},z}$ behaves sufficiently like the $\mathcal{H}_1^{\vec{A}}$ operator of section 12.

We shall leave it to the reader to determine what "sufficiently like" means, from the proof to follow. The main thing is that the set of potential codes is projective in $\vec{A}$, and if z is actually the z_0 of section 12 associated to some coarse mouse operator capturing $\vec{A}$, then z is a potential code.

Let us call c *adequate* iff c is a countable transitive structure with the relevant first order properties of one of the background universes $N_x^*|\delta_x$ as in section 10. We leave it to the reader to fully abstract these properties from the argument to follow. Among them are ZFC, that there is a distinguished real $x = x(c)$, and that the $L[\vec{E}, x]$-construction succeeds in producing a model N with a strong cardinal. If c has these properties, we write $N = N^c$ and $\kappa = \kappa^c$.

Let us call $\langle c, z \rangle$ *nice* iff z is a potential code, c is adequate, and $x_0, z \leq_T x(c)$.

Finally, let us call $\langle c, z \rangle$ *good* iff

(1) $\langle c, z \rangle$ is nice,

(2) $M_\omega^{\vec{A}}(c) \models o(c)$ is Woodin, and for all $\alpha < o(c)$, $P(\alpha) \cap M_\omega^{\vec{A}}(c) \subseteq c$, and

(3) there is a proper class premouse $\mathcal{Q}$ such that $N^c \trianglelefteq \mathcal{Q}$, κ^c is the unique cardinal strong past $o(c)$ in $\mathcal{Q}$, and it is strong in $\mathcal{Q}$ to the sup of the Woodin cardinals in $M_\omega^{\vec{A}}(c)$, and using $\mathcal{H}_0^{\vec{A}}$ to translate away extenders from $\mathcal{Q}$ overlapping $o(c)$, we get

$$\mathcal{Q}[c]^a = M_\omega^{\vec{A}}(c),$$

moreover, starting with $M_\omega^{\vec{A}}(c)$ and using $\mathcal{H}_1^{\vec{A}, z}$ to invert this translation, we get back $\mathcal{Q}$.

The translation from $\mathcal{Q}[c]$ to $\mathcal{Q}[c]^a$ is defined more carefully in section 13. We shall show that there is a good c. It is easy then to see that the associated $\mathcal{Q}$ is a model of the theory in (b). The Woodin cardinals of $M_\omega^{\vec{A}}(c)$ are Woodin in $\mathcal{Q}$, and κ^c is strong in $\mathcal{Q}$ to their sup.

Claim. There is a good $\langle c, z \rangle$.

Proof. Assume not.

For $\langle c, z \rangle \in \mathrm{HC}$, put

$$F(c, z) = 0 \Leftrightarrow \langle c, z \rangle \text{ is not nice .}$$

Suppose now $\langle c, z \rangle$ is nice. The key is that the translation process is step-by-step invertible, so that clauses (2) and (3) can be expressed in the form

$$M_\omega^{\vec{A}}(c) \models \varphi[c, z],$$

where φ is a Π_1 formula in the language of $\vec{A}$-mice. Since one of (2) and (3) fails, we can set

$$F(c, z) = \text{ minimal } \omega_1\text{-iterable } \vec{A}\text{-mouse } \mathcal{P} \text{ over } c$$
$$\text{such that } \mathcal{P} \models \neg\varphi[c, z] .$$

Remark 15.2 The reader may wonder how one can express with a Π_1 assertion over $M_\omega^{\vec{A}}(c)$ that κ^c is strong to the sup of the Woodins in the structure $\mathcal{Q}$ obtained by inverting the translation procedure. The reason is just that our Π_1 formula φ can just say that *every* level of $M_\omega^{\vec{A}}(c)$ that is of the form $P \oplus \vec{A}$ yields an extender with critical point κ^c through the inversion process given by $\mathcal{H}_1^{\vec{A},z}$. There is a one-one correspondence between $\vec{A}$-active levels of $M_\omega^{\vec{A}}(c)$ and extenders with critical point κ^c which overlap $o(c)$, and occur finitely many ultrapowers, by such overlapping extenders, away from $\mathcal{Q}$.

F is a total function on HC × HC, and it is clear by inspection that F is $\Delta_1^2(\vec{A}, x_0)$. This gives us a good scaled pointclass containing F, and thus a way to "beat F" with some c. More precisely, let $f \colon \mathbb{R} \to \mathbb{R}$ be such that whenever $u \in \mathbb{R}$ codes a $\langle c, z \rangle$, then $f(u)$ codes $F(c, z)$, and f is $\Delta_1^2(\vec{A}, x_0)$. Let $x \mapsto (M_x, N_x^*, \Sigma_x, \delta_x)$ be a coarse mouse operator that captures $\vec{A}$ and $\{(u, n, m) \mid f(u)(n) = m\}$. From this coarse operator we get an operator $\mathcal{H}_1^{\vec{A}}$ and a code z with the properties of z_0 at the end of section 12. z is a potential code, and we have $\mathcal{H}_1^{\vec{A}} = \mathcal{H}_1^{\vec{A},z}$. Fix x such that $x_0, z \leq_T x$, and set $c = N_x^* | \delta_x$.

Clearly, $\langle c, z \rangle$ is nice. Let $\mathcal{P} = F(c, z)$, so that $\mathcal{P}$ is an ω_1-iterable $\vec{A}$-mouse over c, and

$$\mathcal{P} \models \neg\varphi[c, z].$$

Also, $\mathcal{P} \in N_x^*$, because our coarse operator captured what it did. Let us now work in N_x^*, where $o(c) = \delta_x$ is Woodin, and form a Skolem hull embedding

$$\pi \colon S \to N_x^* | \tau,$$

for some large τ, with S transitive, and $\eta = \mathrm{crit}(\pi)$ an inaccessible cardinal of N_x^*, and

$$\pi(\eta) = \delta_x,$$

and everything relevant in $\mathrm{ran}(\pi)$. In particular, let

$$\pi(\bar{\mathcal{P}}) = \mathcal{P}.$$

We also arrange $\kappa^c < \eta$, and that η is not Woodin in N^c, which we can do because otherwise, some proper initial segment of N^c already satisfies the theory in (b) of our theorem.

Set

$$Q = N^c|\gamma,$$

where $\gamma > \eta$ is least such that over $N^c|\gamma$ one can define a function $g\colon \eta \to \eta$ witnessing that η is not Woodin. Set also $\bar{c} = N_x^*|\eta$, so that $\pi(\bar{c}) = c$. Now $\bar{c}$ is generic over Q for the extender algebra of Q at η, and moreover, we can form $Q[\bar{c}]^a$ by the construction of section 13. Both $Q[\bar{c}]^a$ and $\bar{\mathcal{P}}$ are ω_1-iterable sound $\vec{A}$-mice over $\bar{c}$ which project to $\bar{c}$, so one is an initial segment of the other. Since $Q[\bar{c}]^a$ can recover Q, it defines a failure of η to be Woodin, whereas $\bar{\mathcal{P}}$ defines no such failure. Thus $\bar{\mathcal{P}} \trianglelefteq Q[\bar{c}]^a$. But

$$Q[\bar{c}]^a \models \varphi[\bar{c}, z],$$

because $Q[\bar{c}]^a$ was in reality obtained by the translation procedure using $\mathcal{H}_0^{\vec{A}}$, and $\mathcal{H}_1^{\vec{A},z}$ to invert it. Since φ is Π_1, we have

$$\bar{\mathcal{P}} \models \varphi[\bar{c}, z].$$

This contradicts the elementarity of π, and proves the claim. □

The claim easily yields the theorem, as we described above. □

The author has extended this argument so as to prove that $\mathrm{Con}(\mathsf{AD}_{\mathbb{R}}) \Rightarrow \mathrm{Con}(\mathsf{AD}_{\mathbb{R}}\text{-hypothesis})$. Combining this with Woodin's 2.11(b), we have that $\mathsf{AD}_{\mathbb{R}}$ is equiconsistent with the $\mathsf{AD}_{\mathbb{R}}$-hypothesis.

16. Global MSC implies the local MSC

In this section, we prove

Theorem 16.1 *Assume* AD^+ *and* MSC2; *then for all* β, S_β *holds.*

The theorem is of no direct use in proving MSC, but it does suggest that trying to prove S_λ by induction on λ is a reasonable approach. The proof of the theorem may also indicate some features one should look for in an inductive proof of S_β.

We remark on a more general local form of MSC at the end of this section.

Lemma 16.2 *Assume* AD^+ *and* MSC. *Let* Γ *be a good scaled pointclass which is not closed under* $\exists^{\mathbb{R}}$, *with* $\Gamma \subseteq \boldsymbol{\Delta}_1^2$; *then there is a mouse operator* $\mathcal{M}$ *which is coded by a set of reals which is projective in some* Γ *set, such that for all* $a \in \mathrm{dom}(\mathcal{M})$, $C_\Gamma(a) = P(a) \cap \mathcal{M}(a)$.

Proof. Let $\Gamma^* = \exists^{\mathbb{R}} \forall^{\mathbb{R}} \Gamma$ and for any real r let $H(r)$ denote the universal $\Gamma^*(r)$ subset of ω. H is a jump operator in the sense of [28]. By 4.2 (2), we have a mouse jump operator $\mathcal{N}$ such that $\mathcal{N} \equiv_m H$. Let $\mathcal{M}$ be the HC-extension of $\mathcal{N}$ (cf. 4.5). It is easy to see that $C_\Gamma(a) \subseteq \mathcal{M}(a)$, for an HC-cone of a. We can get $P(a) \cap \mathcal{M}(a) \subseteq C_\Gamma(a)$ by taking $\mathcal{M}(a)$ minimal. $\qquad\square$

It is worth noting that nothing in Lemma 16.2 requires that the unique ω_1 strategies for the $\mathcal{M}(a)$ be found anywhere near Γ. Indeed, if Γ is properly within some Σ_1^2 gap, this will not be the case. Nonetheless, the $\mathcal{M}$ operator is itself near Γ. In the case Γ occurs just after a new Σ_1^2 fact, we will parlay $\mathcal{M}$ into a mouse witnessing the fact with a strategy near Γ.

The following little lemma will be useful at one point.

Lemma 16.3 *Assume* AD$^+$, *and let $\mathcal{M}$ be a countable a-premouse which is ω_1-iterable (i.e., iterable for normal trees); then $\mathcal{M}$ is (ω_1, ω_1)-iterable (i.e., iterable for stacks of normal trees).*

Proof. Let Σ be an ω_1-iteration strategy for $\mathcal{M}$, which by the Basis theorem we may assume is Suslin and co-Suslin. Let $x \mapsto (M_x, \Sigma_x, N_x^*, \delta_x)$ be a coarse mouse operator capturing (a set of reals coding) Σ. Let x be in the domain of this operator, with $\mathcal{M}$ coded by a real recursive in x. Let N be the output of the maximal full background extender construction of N_x^*, done over a. N_x^* knows how to iterate $\mathcal{M}$, and the proof of universality from 11.1 shows that $\mathcal{M}$ embeds into an Ω-iterate of N, where Ω is the strategy for N induced by Σ_x. (A little more work would show $\mathcal{M}$ embeds into some initial segment of N itself.) But Ω is an (ω_1, ω_1)-strategy, so by pulling back, we have such a strategy for $\mathcal{M}$. $\qquad\square$

Proof of Theorem 16.1. Assume AD$^+$ and MSC2. We shall prove S$_\lambda$ for limit λ, and leave the minor adjustment needed for the full proof to a remark. Suppose A, y, and Γ^- are such that

$$(\text{HC}, \in, A) \models \varphi[y],$$

with $y \in \mathbb{R}$ and $A \in \Gamma^-$, and Γ^- a good scaled pointclass not closed under $\forall^{\mathbb{R}}$. Suppose φ is a Σ_k formula, and let Γ be, in the inclusion order, the $k + 5$-th good scaled pointclass containing Γ^-. It will be enough to find a $\langle \varphi, y \rangle$-witness $\mathcal{P}$ such that $\mathcal{P}$ has an ω_1-iteration strategy which is in $L_2(\Gamma, \mathbb{R})$.

Let $\mathcal{M}$ and r_0 be as in Lemma 16.2 with respect to Γ. We may assume $y \leq_T r_0$.

Claim 2. There is a function J defined on all countable, transitive, self-wellordered a such that $r_0 \in a$, and such that for each such a:

(a) $J(a)$ is a sound, ω_1-iterable mouse over a, and

(b) either $\mathbb{R} \cap a \neq \mathbb{R} \cap J(a)$, or $J(a) \models \mathsf{ZFC}^- + |a|^+$ exists, and

    $J(a) \models$ there are trees T, U on $\omega \times \omega_1$ which are absolute complements
          for $Col(\omega, a)$ and such that $(\mathrm{HC}, \in, p[T]) \models \varphi[y]$.

    and

(c) $J(a)$ is the least a-mouse satisfying (a) and (b).

Moreover, J is coded by a set of reals which is in $L_2(\Gamma, \mathbb{R})$.

Proof. It is enough to define, over $L_1(\Gamma, \mathbb{R})$, a mouse $J_0(a)$ satisfying (a) and (b). We can then just let $J(a)$ be the first initial segment of $J_0(a)$ satisfying (a) and (b).

We define $J_0(a)$ by closing under the $\mathcal{M}$-operator. More precisely, set

$$\mathcal{P}_0 = \mathcal{M}(a).$$

Suppose now we have $\mathcal{P}_\xi$. If $\mathbb{R} \cap \mathcal{M}(\mathcal{P}_\xi) \not\subseteq a$, then we stop our induction, and let $J_0(a)$ be the first level of $\mathcal{M}(\mathcal{P}_\xi)$ which contains a real not in a. ($J_0(a)$ is the first such level re-arranged as an a-mouse; it will follow from the construction that this is possible.) If $\mathbb{R} \cap \mathcal{M}(\mathcal{P}_\xi) \subseteq a$, we set

$$P_{\xi+1} = \mathcal{M}(P_\xi).$$

If $\mathcal{P}_{\xi+1} \models \omega_1$ exists, then we stop the induction. Finally, for λ a countable limit ordinal,

$$\mathcal{P}_\lambda = \bigcup_{\alpha < \lambda} \mathcal{P}_\alpha.$$

It is easy to see that $\mathcal{P}_\eta$ is a sound mouse over a whenever it is defined, and $\mathcal{P}_\eta$ projects to a at all stages η except possibly the last, when we may have $\eta = \xi + 1$ and $\rho_\omega(\mathcal{P}_\eta) = \omega_1^{\mathcal{P}_\eta} = o(\mathcal{P}_\xi)$.[10]

Since there is no ω_1-sequence of distinct subsets of a, our induction must stop at some countable stage. If it stops because a new real has been

---

[10]We do not have any form of condensation for the $\mathcal{M}$-operator, and this blocks the obvious attempt to prove $\mathcal{P}_\eta$ projects to a in this case.

constructed, then we have already defined $J_0(a)$. If not, suppose it stops with $\mathcal{P}_{\xi+1} = \mathcal{M}(\mathcal{P}_\xi)$ for the second reason. Let T_Γ be the tree of a Γ-scale on a universal Γ set. Note that $o(\mathcal{P}_\xi)$ is the ω_1 of $L[T_\Gamma, \mathcal{P}_\xi]$, and in $L[T_\Gamma, \mathcal{P}_\xi]$ we have a pair (T_0, U_0) of trees which project to A and $\mathbb{R} \setminus A$ in V. Working in $L[T_\Gamma, \mathcal{P}_\xi]$ and taking a Skolem hull, we get trees (T, U) on $\omega \times o(\mathcal{P}_\xi)$ which project in any $\mathrm{Col}(\omega, a)$-generic extension of $L[T_\Gamma, \mathcal{P}_\xi]$ to the intersections of A and $\mathbb{R} \setminus A$ with that extension. But then $(T, U) \in \mathcal{P}_{\xi+1}$, so we can set $J_0(a) = \mathcal{P}_{\xi+1} | \nu$ (re-arranged as mouse over a), where $T, U \in \mathcal{P}_{\xi+1} | \nu$ and $\mathcal{P}_{\xi+1} | \nu \models \mathsf{ZFC}^-$. It is clear that $J_0(a)$ thinks T and U are $\mathrm{Col}(\omega, a)$-absolutely complementing, so to show that (b) holds, it is enough to see that $(\mathrm{HC}, \in, p[T])^{J_0(a)} \prec_{\Sigma_k} (\mathrm{HC}, \in, A)$. But $p[T]^{J_0(a)} = A \cap a$ by construction, and the Σ_k-elementarity holds because $\mathbb{R} \cap a = \mathbb{R} \cap J_0(a)$ is C_Γ-closed, and we took Γ to be k good scaled pointclasses above Γ_0, where $A \in \Gamma_0$.

This proves claim 2. $\qquad\qquad\square$

Fix J as in claim 2. $J(a)$ is the unique a-mouse satisfying a certain first order theory, which we call T^J. (This is a theory in the language of set theory, expanded by a name for a.) Therefore, we have condensation:

Claim 3. I $\pi : P \to J(a)$ is fully elementary, and r_0 is in the range of π then $P = J(\pi^{-1}(a))$.

Definition 16.4 *Let a be self-wellordered, with $r_0 \in a$.*

(a) *An a-premouse $\mathcal{N}$ is J-closed iff $J(\mathcal{N} | \xi)$ is an initial segment of $\mathcal{N}$, for every cardinal ξ of $\mathcal{N}$.*

(b) *$\mathcal{N}$ is T^J-closed iff for every cardinal ξ of $\mathcal{N}$, there is a $\gamma \leq o(\mathcal{N})$ such that $(\mathcal{N} | \gamma, \mathcal{N} | \xi) \models T^J$. We write $J^{\mathcal{N}}(\mathcal{N} | \xi) = \mathcal{N} | \gamma$ in this case.*

(c) *$\mathcal{N}$ is J-correct iff whenever $J^{\mathcal{N}}(\mathcal{N} | \xi) = \mathcal{N} | \gamma$, then $J(\mathcal{N} | \xi) = \mathcal{N} | \gamma$.*

(d) *By $M_n^J(a)$ we mean the least ω_1-iterable a-mouse $\mathcal{M}$ such that some ordinals $\delta_0 < \ldots < \delta_{n-1}$ are Woodin cardinals of $\mathcal{M}$, $\mathcal{M} | \delta_{n-1}$ is J-closed, and $\mathcal{M} = J(\mathcal{M} | \delta_{n-1})$.*

Clearly, an ω_1-iterable model is J-correct, and hence T^J-closed iff it is J-closed. So $M_n^J(a)$ is just the least iterable a-mouse satisfying a certain theory.

Claim 4. $M_n^J(a)$ exists, for every n and every selfwellordered a with $r_0 \in a$. Moreover, fixing n, the operator $a \mapsto M_n^J(a)$ is coded by a set of reals which is in $L_2(\Gamma, \mathbb{R})$.

Proof. Let Γ_0 be a good scaled pointclass containing Γ, contained in projective-in-Γ, and such that J is coded by a set of reals in Γ_0. Let Γ_i be the i-th good scaled pointclass including Γ_0. Fix $n \geq 0$, and let U be a universal Γ_{n+5} set. By 10.3, we can fix a coarse mouse operator $x \mapsto (M_x, \Sigma_x, N_x^*, \delta_x)$ which captures U, and which is in $L_2(\Gamma, \mathbb{R})$. It is enough to define the desired $a \mapsto M_n^J(a)$ in a projective way from this operator.

So fix a selfwellordered a with $r_0 \in a$. Let x_0 be a real coding a, and let $x \geq_T x_0$ be in the domain of our coarse mouse operator capturing U. Working in N_x^*, we shall do a variant of the full background extender construction over a. We shall show that all models in this variant construction are ω_1-iterable in V, and that the construction reaches a level which believes it is $M_n^J(a)$. By iterability, this level must really be $M_n^J(a)$. Since the variant construction is uniformly definable over N_x^* from a, we then have that $a \mapsto M_n^J(a)$ is projective in $x \mapsto (M_x, \Sigma_x, N_x^*, \delta_x)$, as desired.

There is no reason as yet to believe the usual construction reaches $M_n^J(a)$, for although it produces a model N which is universal in N_x^*, we cannot show $J(N|\xi) \trianglelefteq N$ without being able to iterate $J(N|\xi)$ above ξ inside N_x^*. We know that $J(N|\xi)$ is iterable above ξ in V, but we have no useful bound on the complexity of its iteration strategy. So we modify the usual construction by simply adding steps which close under J. Note here that because our coarse mouse operator captured U, $J \upharpoonright (N_x^* | \delta_x) \in N_x^*$.

More precisely, working in N_x^*, we construct an extender model over a via approximations $\mathcal{N}_\xi$, for $\xi < \delta_x$. If $\mathcal{N}_\xi = (J_\alpha^{\vec{E}}, \in, \vec{E}, \emptyset)$ and there is an extender F such that $(J_\alpha^{\vec{E}}, \in, \vec{E}, F)$ is a premouse and F has a full background extender, then we pick such an F subject to the usual conditions, and set $\mathcal{N}_{\xi+1} = (J_\alpha^{\vec{E}}, \in .\vec{E}, F)$. If there is no such F, or if $\mathcal{N}_\xi$ is already active, then we let $\mathcal{Q}$ be the first initial segment of $J(\mathcal{N}_\xi)$ such that $\rho_\omega(\mathcal{Q}) < o(\mathcal{N}_\xi)$ and $\mathcal{Q}$ is not ω-sound, and put $\mathcal{N}_{\xi+1} = \mathcal{C}_\omega(\mathcal{Q}) = $ the ω-th core of $\mathcal{Q}$.[11] If there is no such initial segment $\mathcal{Q}$ of $J(\mathcal{N}_\xi)$, then we just set $\mathcal{N}_{\xi+1} = J(\mathcal{N}_\xi)$. If $\lambda \leq \delta_x$ is a limit, then $\mathcal{N}_\lambda$ is the lim inf of the $\mathcal{N}_\alpha$, for $\alpha < \lambda$, as usual. Finally, we set $N = \mathcal{N}_{\delta_x}$.

---

[11] $\mathcal{Q}$ is ω-sound when regarded as a mouse over $\mathcal{N}_\xi$, but here we are regarding it as a mouse over a. $\mathcal{Q} = \mathcal{N}_\xi$ is possible.

Subclaim A.

1. If $\mathcal{N}$ is J-correct and $\pi\colon \mathcal{M} \to \mathcal{N}$ is elementary, then $\mathcal{M}$ is J-correct.

2. Every $\mathcal{N}_\xi$ is J-correct.

3. Let κ be a cutpoint of $\mathcal{N}_\xi$ and suppose $\rho_\omega(\mathcal{N}_\xi|\gamma) \leq \kappa$, and $J(\mathcal{N}_\xi|\gamma)$ is not a proper initial segment of $\mathcal{N}_\xi|\gamma$; then $\mathcal{N}_\xi|\gamma$ is ω_1-iterable above κ.

Proof. Part (1) follows at once from condensation for J. Part (2) is an easy induction on ξ, using part (1) to show that coring down preserves J-correctness.

Part (3) is a refinement of (2). Suppose $\mathcal{T}$ is a normal iteration tree on $\mathcal{N}_\xi|\gamma$ above κ. Let E be the first extender used in $\mathcal{T}$, let $\eta \leq \gamma$ be least such that $\mathrm{lh}(E) \leq \eta$ and $\rho_\omega(\mathcal{N}_\xi|\eta) = \kappa$, and let ν be the κ^+ of $\mathcal{N}_\xi|\eta$. Then $\mathcal{T}$ can be regarded as a tree on $\mathcal{N}_\xi|\eta$. Tracing the ancestry of $\mathcal{N}_\xi|\eta$ is our construction yields a $\sigma \leq \xi$ and an elementary $\pi\colon \mathcal{N}_\xi|\eta \to \mathcal{N}_\sigma$ such that, letting $\mu = \pi(\nu)$, we have that $\mathcal{N}_\sigma \trianglelefteq J(\mathcal{N}_\sigma|\mu)$. But then $\mathcal{N}_\sigma$ has an iteration strategy Σ above μ, and we can assume $\mathcal{T}$ is being played by the pullback Σ^π, and use Σ^π to continue iterating. □

Subclaim B.

1. If κ is a cardinal of $\mathcal{N}_\xi$ and $\kappa > \text{order-type}(<)$, where $<$ is a $\mathrm{rud}(a)$ wellorder of a, then $\mathcal{N}_\xi|\kappa$ is J-closed.

2. An extender F on the sequence of $\mathcal{N}_\xi$ has a background extender provided by the construction (perhaps via some resurrection of F) if and only if $\mathcal{N}_\xi|\mathrm{lh}(F)$ is J-closed.

Proof. Straightforward. □

Subclaim C. Each $\mathcal{N}_\xi$ is $(\omega, \omega_1, \omega_1)$-iterable.

Proof sketch. By Lemma 16.3, it suffices to consider normal iterations. We lift such iterations of $\mathcal{N}_\xi$ to iterations of N_x^* by resurrecting background extenders as in [12], when these exist. The problem here is that we may use an extender which is justified by an application of J, and hence has no background extender to be resurrected. Let us consider the first stage at which this happens.

We must be in the following situation. We have T on $\mathcal{N}_\xi$ played according to the strategy of lifting to N_x^* and using Σ_x to form a tree $\mathcal{U}$ there. T has a last model $\mathcal{P}$, and $\mathcal{U}$ has a last model R, with

$$i^{\mathcal{U}}\colon N_x^* \to R$$

the canonical embedding. There is an embedding

$$\pi\colon \mathcal{P} \to \mathcal{Q},$$

where $\mathcal{Q}$ is a model on $i^{\mathcal{U}}(\langle \mathcal{N}_\gamma \mid \gamma < \delta_x \rangle)$. We have that all models on $i^{\mathcal{U}}(\langle \mathcal{N}_\gamma \mid \gamma < \delta_x \rangle)$ are J-correct, because iterations by Σ_x move the absolutely complementing trees for U in N_x^* correctly. So $\mathcal{Q}$, and hence $\mathcal{P}$, are J-correct. We are about to use an extender F from the sequence of $\mathcal{P}$ to extend T. Since we cannot resurrect a background extender for $\pi(F)$, $\mathcal{Q} \mid \mathrm{lh}(\pi(F))$ is not J-closed, and thus $\mathcal{P} \mid \mathrm{lh}(F)$ is not J-closed. Let $\lambda < \mathrm{lh}(F)$ be largest such that

$$\alpha < \lambda \Rightarrow J(\mathcal{P}\mid\alpha) \unlhd \mathcal{P}\mid\lambda.$$

This implies that $\lambda < \mathrm{crit}(F)$, and that λ is a cutpoint of $\mathcal{P}$ and not the critical point of an extender on the $\mathcal{P}$-sequence. Since F was the first E_α^T such that $\mathcal{M}_\alpha^T \mid \mathrm{lh}(E_\alpha^T)$ is not J-closed, all extenders used in T before F have length $< \lambda$.

Let η be least such that $\mathrm{lh}(F) \le \eta$ and $\rho_\omega(\mathcal{P}\mid\eta) \le \lambda$. It follows that the rest of our normal T will have to be a tree on $\mathcal{P}\mid\eta$ which is above λ. There will be no going back to earlier models of T; instead, we have permanently dropped. So it is enough to see that $\mathcal{P}\mid\eta$ is ω_1-iterable above λ. For that, it is enough that $\mathcal{Q}\mid\pi(\eta)$ is ω_1-iterable above $\pi(\lambda)$. But this follows at once from the proof of part (3) of Subclaim A, applied to the models of $i^{\mathcal{U}}(\langle \mathcal{N}_\gamma \mid \gamma < \delta_x \rangle)$. The proof does apply because $i^{\mathcal{U}}$ moves the J-operator correctly. $\square$

By Subclaim C, the $\mathcal{N}_\xi$ are all ω-solid, so that the construction does not stop, and does indeed produce a model $N = \mathcal{N}_{\delta_x}$.

Working in N_x^*, let us call η a Γ_i-Woodin cardinal iff by using the appropriate absolutely complementing trees to compute $C_\Gamma(N_x^*\mid\eta)$, we see that η is Woodin with respect to all functions in $C_\Gamma(N_x^*\mid\eta)$. Let us put

$$\eta_n = \text{the least } \Gamma_n\text{-Woodin cardinal},$$

and for $1 \le i \le n$,

$$\eta_{i-1} = \text{the least } \Gamma_{i-1}\text{-Woodin cardinal} > \eta_i.$$

So $\eta_n < \eta_{n-1} < ... < \eta_0$, and η_i is Γ_i-Woodin. One can show that $\eta_n, ..., \eta_1$ are Woodin in $\mathcal{N}_{\eta_0}$, by an argument very similar to the proof of claim 1 of 14.1. But also, $J(\mathcal{N}_{\eta_0}|\eta_1) \trianglelefteq \mathcal{N}_{\eta_0}$. Thus our construction has reached $M_n^J(a)$, and the proof of Claim 4 is complete. $\qquad\square$

Recall that φ was Σ_k. Let us put

$$R(a) = M_{k+5}^J(a),$$

for a in $\text{dom}(M_{k+5}^J)$. Let $\mathcal{S}$ be the minimal $\langle \varphi, y \rangle$-witness, which exists by MSC. Since $\mathcal{S}$ projects to ω, it has a unique ω_1-strategy Σ. It is enough to show that Σ is projective in Γ. This we shall do by using the R-operator to compute Σ.

Let $x \mapsto (M_x, \Sigma_x, N_x^*, \delta_x)$ be a coarse mouse operator in $L_3(\Gamma, \mathbb{R})$ which captures a set of reals coding $a \mapsto R(a)$. Let x be in the domain of this operator, with r_0 and some real coding $\mathcal{S}$ recursive in x, and R defined on the HC-cone above x. Working in N_x^*, let N be the output of a maximal full background extender construction over y of height δ_x. Since N has an ω_1 iteration strategy which is $L_3(\Gamma, \mathbb{R})$, it will be enough to show that $\mathcal{S} \trianglelefteq N$. So suppose not.

Let us compare $\mathcal{S}$ with N. On the $\mathcal{S}$ side we use Σ to pick branches, and we fold in a genericity iteration which guarantees that if $\mathcal{T}$ is a tree of limit length on $\mathcal{S}$ which our process has produced, then

$$\langle \mathcal{T}, x \rangle \text{ is } \mathcal{B}\text{-generic over } \mathcal{M}(\mathcal{T}),$$

where $\mathcal{B}$ is the $\delta(\mathcal{T})$-generator version of the extender algebra determined by the extender sequence of $\mathcal{M}(\mathcal{T})$. It is not a problem to fold such steps into the usual coiteration process, as both $\mathcal{T}$ and $\dot{\mathcal{M}}(\mathcal{T})$ are being revealed initial-segment-wise as we proceed toward $\text{lh}(\mathcal{T})$. On the N side, we do the iteration at the level of the background universe N_x^*, as in 12.6 and [17].

This process must terminate, and when it does we must have

$$i \colon N_x^* \to N^{**},$$

and

$$\mathcal{T} \text{ on } \mathcal{S} \text{ according to } \Sigma, \text{ with last model } \mathcal{P},$$

so that

$$i(N) \trianglelefteq \mathcal{P}.$$

Now by the proof of 11.1, $i(N)$ is universal in N^{**}, in the sense that no y-premouse iterates past it. So we will have the desired contradiction when we show $T \in N^{**}$.

The R operator is moved correctly by i, so $R \upharpoonright (N^{**}|i(\delta_x) + 1) \in N^{**}$. We use this to recover T. Suppose that $\lambda \leq \delta_x$, and we have already recovered $T \upharpoonright \lambda$ in N^{**}. Let $\delta = \delta(T \upharpoonright \lambda)$, let $\mathcal{B}$ be the δ-generator extender algebra of $\mathcal{M}(T \upharpoonright \lambda)$, and let g be the $\mathcal{B}$-generic object determined by $\langle T \upharpoonright \lambda, x \rangle$. Let a be selfwellordered and constructibly equivalent to $\langle \mathcal{M}(T \upharpoonright \lambda), g \rangle$. Working in N^{**}, we can find these objects, and then find $R(a)$.

Let $b = \Sigma(T \upharpoonright \lambda)$, and $\mathcal{Q} = \mathcal{Q}(b, T \upharpoonright \lambda)$, which exists because $\mathcal{S}$ projects to ω. As δ remains Woodin in $\mathcal{Q}$, g is $\mathcal{B}$-generic over $\mathcal{Q}$. But g codes $T \upharpoonright \lambda$, so the $*$-transform $\mathcal{Q}[g]^*$ exists. It can be regarded as a mouse over a. So regarding it, we have

Claim 5. $\mathcal{Q}[g]^* \trianglelefteq R(a)$.

Proof. If not, then we have a proper initial segment $\mathcal{P}$ of $\mathcal{Q}$ such that $\mathcal{P}[g]^* = R(a)$. Let δ_i be the i-th Woodin cardinal of $R(a)$, and $\gamma = \delta_{k+5}$. As γ is a cardinal in $\mathcal{P}[g]$, $\mathcal{P}|\gamma[g]$ has the same universe as $R(a)|\gamma$. Let $\mathcal{N} = \mathcal{P}|\gamma$. It follows that

$$\mathcal{N}[g] \models \delta_1, ..., \delta_{k+4} \text{ are Woodin,}$$

and

$$\mathcal{N}[g] \models \exists A \in \mathrm{Hom}_{\delta_{k+3}}((\mathrm{HC}, \in, A \models \varphi[y]).$$

But then a standard absoluteness argument (due to Woodin) using stationary tower forcing shows that the two displayed statements are true in $\mathcal{N}$, not just in $\mathcal{N}[g]$. But this implies that $\mathcal{N}$ is a $\langle \varphi, y \rangle$-witness. Since $\mathcal{S}$ was the minimal such witness, and we have dropped in going from $\mathcal{S}$ to $\mathcal{N}$, this is a contradiction. $\qquad\square$

By claim 5, we can recover $\mathcal{Q}$ from $R(a)$ by inverting the $*$-transform. But $\mathcal{Q}$ determines b, and hence we can recover b as well. This shows $T \in N^{**}$, the desired contradiction. We conclude that in fact $\mathcal{S} \trianglelefteq N$, and hence $\mathcal{S}$ has an ω_1 iteration strategy which is in $L_3(\Gamma, \mathbb{R})$. This completes the proof of S_λ for limit λ.

One gets the full Theorem 16.1 by using 10.2 in the places where we used 10.3. This enables one to remain within the projective-in-Γ sets. In the end, our iteration strategy for $\mathcal{S}$ is projective in Γ, as required. $\qquad\square$

Remark 16.5 Even our local MSC is not as local as one might wish, because it says nothing about how sets of reals properly inside Σ^2_1 gaps are captured locally by mice. We believe that the proof of theorem 16.1 can be modified so as to show

Theorem 16.6 *Assume* $\mathsf{AD}^+ + \mathsf{MSC}$, *and let* $A \subseteq \mathbb{R}$ *be Suslin and co-Suslin. Then there is an ms-premouse* $\mathcal{M}$, *and some* g *generic over* $\mathcal{M}$ *for some* $\mathrm{Col}(\omega, \kappa)$, *and a Woodin cardinal* $\delta > \kappa$ *of* $\mathcal{M}$, *and an* ω_1 *iteration strategy* Σ *for* $\mathcal{M}$ *in the window* (κ, δ) *such that* $(\mathcal{M}[g], \Sigma)$ *captures* A.

The conclusion means that there are δ-absolutely complementing trees $T, U \in \mathcal{M}[g]$ such that for all reals x, $x \in A$ iff there is a Σ-iteration map i such that $x \in p[i(T)]$. Thus theorem 16.6 is a fine-structural refinement of theorem 10.2; the capturing mouse is a generic extension of an ms-premouse now. One needs the generic extension.

Another localization question one can ask is: suppose $(\mathcal{M}, \lambda)$ is tractable, and let Γ be a Wadge initial segment of Hom^*_I, for some $\mathbb{R}$-genericity iteration of $\mathcal{M}$. Must $L(\Gamma, \mathbb{R})$ be the derived model of some ms-mouse? We do not know the answer.

17. MSC implies capturing via $\mathbb{R}$-mice

In this section, we prove

Theorem 17.1 *Assume* $\mathsf{AD}^+ + \mathsf{MSC}$. *Let* α *begin a* Σ^2_1 *gap, and suppose* $\Gamma = (\Sigma^2_1)^{P_\alpha}$ *is closed under real quantification. Let* U *be a universal* Γ *set, and let* $A \in \mathrm{Env}(\Gamma)$; *then there is an ms-mouse* M *over* $\mathbb{R}$ *such that*

(a) M *is* $\mathbb{R}$-*sound, and* A *is definable over* M, *but not in* M,

(b) *for each* $k < \omega$, *the* Σ_k-*theory in* M *with parameters from* $\mathbb{R} \cup p_k(M)$ *is projective in* $\langle A, U \rangle$, *and*

(c) *every countable elementary submodel of* M *has an* ω_1 *iteration strategy coded by a set of reals in* P_α.

Proof. Assume Γ, U, A are a counterexample. By Woodin's basis theorem, we may assume A, U are $\boldsymbol{\Delta}^2_1$. We also assume that A is a Wadge-minimal counterexample with respect to Γ. Letting

$$\mathcal{B} = \{B \mid B <_w A\},$$

and

$$S = \text{ minimal } \mathbb{R}\text{-mouse } P \text{ such that } \mathcal{B} \subseteq P \ ,$$

we have then that every countable elementary submodel of S has an ω_1-iteration strategy which is in P_α. Let S_0 be a set of reals which is projective in $\langle A, U \rangle$, and codes S in some natural way. It is easy to see there is such a set.

Claim 1. $L[S] \cap P(\mathbb{R}) = \mathcal{B}$.

Proof. Otherwise, the first level of $L[S]$ past S itself which projects to $\mathbb{R}$ would be an $\mathbb{R}$-mouse M as required in the theorem. We show this now. Part (c) is clear. Let k be least such that $\rho_k(M) = \mathbb{R}$, and let C be the Σ_k theory of M with parameters from $\mathbb{R} \cup p_k(M)$. Wadge's lemma gives A is Wadge reducible to either C or $\mathbb{R} \setminus C$, so part (a) holds. For (b), note that C is a wellordered union of sets of reals in $\mathcal{B}$. Moreover, there is a pointclass Ω contained in the projective-in-A sets such that $A \in \Omega$, and Ω is closed under wellordered unions. It follows that C is projective in A, and from that we easily get (b). $\qquad\square$

Now let $x_0 \in \mathbb{R}$ be such that whenever σ is a countable set of reals with $x_0 \in \sigma$, then $A \cap \sigma \in C_\Gamma(\sigma)$. Let Ω be an inductive-like pointclass with the scale property containing A, U, and their complements, and let T be the tree of an Ω-scale on a universal Ω set. Put

$$N = L[T, x_0],$$

and

$$A^* = A \cap N, S_0^* = S_0 \cap N, \text{ and } U^* = U \cap N.$$

Using the Skolem functions given by T, we get

$$(\text{HC}^N, \in, A^*, U^*) \prec (\text{HC}, \in, A, U).$$

Now it is part of the first order theory of $(\text{HC}, \in, A, U)$ that there is no M satisfying (a)–(c) of the theorem. (Note $P_\alpha = \{B \mid B <_w U\}$.) It is also part of this theory that A is Wadge minimal having this property with respect to U. It follows that

$$N \models \neg\exists M (M \text{ is sound, projects to } \mathbb{R}, \text{ and}$$

A^* is definable over M, and the theory

of M with real parameters is projective in $\langle A^*, U^* \rangle$, and

every countable elementary submodel of M has an

ω_1 iteration strategy coded by a set of reals $<_w U^*$).

But $x_0 \in \mathbb{R}^N$, so $A^* \in C_\Gamma(\mathbb{R}^N)$, and thus by MSC, we can fix a mouse Q over $\mathbb{R}^N$ such that $A^* \in Q$, and Q has an ω_1 iteration strategy Σ in Γ. Since N has T in it, we get that $Q \in N$, and $\Sigma^* \in Q$, where Σ^* is the restriction of Σ to trees in N of size $< \omega_1^V$. (Note ω_1^V is measurable in N.) It follows that

$$N \models \text{ every countable elementary submodel of } Q \text{ has an}$$

ω_1 iteration strategy coded by a set of reals $<_w U^*$.

Working in N, let $\mathcal{B}^* = \{B \subseteq \mathbb{R} \mid B <_w A^*\}$, and let S^* be the mouse over $\mathbb{R}^N$ coded by S_0^*. Then $cB^* = P(\mathbb{R}^N) \cap S^*$, and S^* is countably iterable in N, and S^* is a union of mice projecting to $\mathbb{R}^N$. It follows that

$$S^* = Q|\xi,$$

for some ξ.

Claim 2. $L[S^*] \cap P(\mathbb{R}^N) = \mathcal{B}^*$.

Proof. (Let Ω be a nonselfdual pointclass with the prewellordering property, closed under $\forall^{\mathbb{R}}$, properly included in the projective-in-$\langle A, U \rangle$ sets, with $A, U,$ and S_0 in Ω. Let κ be the prewellordering ordinal of Ω, and let μ be the ω-club ultrafilter on κ. We have that $L[\mu, S] \cap P(\mathbb{R}) = \mathcal{B}$, because otherwise the desired M witnessing the theorem would just be the first level of $L[\mu, S]$ over which a set of reals not in $\mathcal{B}$ is definable, as in the proof of claim 1. It follows that $L[S]$ has a club-in-κ class of indiscernibles, in the language with names for each element of $S \cup \{S\}$. The type of this set of indiscernibles in projective-in-$\langle A, U \rangle$, by the Coding Lemma. This enables to express the fact that $L[S] \cap P(\mathbb{R}) = \mathcal{B}$ as a first order statement about $(HC, \in, A, U)$. Since $(HC^N, \in, A^*, U^*) \prec (HC, \in, A, U)$, we have proved Claim 2. $\square$

Claim 3. Let γ be least such that $P(\mathbb{R}^N) \cap Q|(\gamma+1) \not\subseteq \mathcal{B}^*$; then $\mathcal{Q}|(\gamma+1) \models$ AD.

*Proof.*Work in N. Let ξ be such that $S^* = Q|\xi$; then ξ is the θ of $Q|\gamma$. We have that $Q|\gamma \models$ AD. Assume $Q|(\gamma + 1) \not\models$ AD; then γ ends a proper Σ_1 gap in Q, and moreover this gap is weak. Let n be least such that $\rho_n(Q|\gamma) = \mathbb{R}^N$. By the analysis of scales in $K^*(\mathbb{R})$ at the end of a weak gap (see [26]), every boldface $\boldsymbol{\Sigma}_n^{Q|\gamma}$ set of reals D can be written

$$D = \bigcup_{n<\omega} D_n,$$

where each $D_n \in \mathcal{B}^*$. But then each such D is projective in A^*. Hence every set of reals in $Q|(\gamma + 1)$ is projective in A^*. As $(\mathrm{HC}^N, \in, A^*, U^*) \prec (\mathrm{HC}, \in, A, U)$, each such set is determined, contradiction. $\square$

We need now a slight refinement of Woodin's result that if G and H are models of AD^+ containing all the reals, and they diverge, in that each has a set of reals not in the other, then every game on $\mathbb{R}$ with payoff in $G \cap H$ has a winning strategy in $G \cap H$. We need this for G and H being merely projectively closed:

Theorem 17.2 (Woodin, unpublished) *Let* $B, C \subseteq \mathbb{R}$, *and let* G *and* H *be transitive, rudimentarily closed sets such that* $B \in G \setminus H$ *and* $C \in H \setminus G$. *Suppose* $G \models \mathrm{AD}^+$ *and* $H \models \mathrm{AD}^+$. *Let* $D \subseteq \mathbb{R}^\omega$ *be in* $G \cap H$; *then the real game with payoff* D *has a winning strategy in* $G \cap H$.

We apply Woodin's theorem in N, with $G = Q|(\gamma + 1)$, where γ is as in claim 3, and H the rudimentary closure of $\mathbb{R}^N \cup \{A^*\}$. Note here

Claim 4. $S^* \not\models \mathrm{AD}_{\mathbb{R}}$.

Proof. Otherwise, every set of reals in S^* has a scale in S^*. By the analysis of scales in iterable $\mathbb{R}$-mice ([24]), new Σ_1 statements about reals are verified cofinally in S^*. (These statements can refer to the extender sequence of S^*.) It follows that $\rho_1(S^*) = \mathbb{R}^N$. That contradicts claim 2. $\square$

By Woodin's theorem, claim 4, and the minimality of A^* and γ respectively, we get that the subsets of $\mathbb{R}^N$ in $Q|(\gamma + 1)$ are precisely those which (in N) are projective in A^*. Thus if $M = Q|(\gamma + 1)$, then

> $N \models M$ is sound, projects to $\mathbb{R}$, and
>> A^* is definable over M, and the theory
>> of M with real parameters is projective in $\langle A^*, U^* \rangle$, and
>> every countable elementary submodel of M has an
>> ω_1 iteration strategy coded by a set of reals $<_w U^*$).

This contradiction completes the proof of 17.1. □

Corollary 17.3 *Assume* AD^+ + MSC. *Let* U *be the universal* Σ_1^2 *set of reals, and* A *any* $OD(\mathbb{R})$ *set of reals; then there is a countably iterable* $\mathbb{R}$-*mouse* M *such that* A *is definable over* M, *and every set of reals definable over* M *is projective in* A *and* U.

Corollary 17.4 *Assume* AD^+ + MSC. *Let* A *be set of reals; then there is an* $\mathbb{R}$-*premouse* M *such that* A *is definable over* M, *and every set of reals definable over* M *is projective in* A *and* U.

There are extensions of theorem 17.1 and its corollaries to hybrid strategy-mice: capturing over countable sets of reals implies capturing over $\mathbb{R}$. The key to formulating these is to notice that if Σ is an ω_1-iteration strategy for some countable M, and Σ condenses well, then there is at most one extension of Σ to uncountable iteration trees. So one can define a Σ-mouse over $\mathbb{R}$ to be a hybrid mouse all of whose countable elementary submodels containing M are hybrid Σ-mice over their own (countable) set of reals. One can use this to make sense of $K^\Sigma(\mathbb{R})$, and to generalize 17.1.

The results of this section should be compared to those at the beginning of section 4.

References

[1] Q. Feng, M. Magidor, and W.H. Woodin, Universally Baire sets of reals, in *Set theory of the Continuum*, H. Judah, W. Just, and W.H. Woodin eds., MSRI publications vol 26 (1992), Springer-Verlag.

[2] L.A. Harrington and A.S. Kechris, On the determinacy of games on ordinals, *Annals of Math. Logic*, vol. 20 (1981), 109–154.

[3] S. Jackson, Structural consequences of AD, Handbook of set theory, M. Foreman, A. Kanamori, and M. Magidor eds., to appear.

[4] R. Ketchersid, Toward $\mathsf{AD}_\mathbb{R}$ from the Continuum Hypothesis and an ω_1-dense ideal, Ph.D. thesis, Berkeley, 2000.

[5] P. Koellner and W.H. Woodin, Large cardinals from determinacy, Handbook of Set Theory, M. Foreman, A. Kanamori, and M. Magidor eds., to appear.

[6] P. Larson, The stationary tower, Memoirs of the AMS.

[7] D.A. Martin, Measurable cardinals and analytic games, Bulletin of the AMS, 1968.

[8] D.A. Martin, The largest countable this, that, and the other, in *Cabal Seminar 79 81*, A.S. Kechris, D.A. Martin, and Y.N. Moschovakis eds., Lecture Notes in Math, vol. 1019 (1983), Springer-Verlag, Berlin, 97-106.

[9] D.A. Martin and R.M. Solovay, A basis theorem for Σ_3^1 sets of reals, *Annals of Mathematics* **89** (1969), 138–159.

[10] D.A. Martin and J.R. Steel, A Proof of Projective Determinacy, *Journal of the American Mathematical Society* bf 2 (1989), 71–125.

[11] W.J. Mitchell and R.D. Schindler, A universal extender model without large cardinals in V, J. Symb. Logic, vol. 69 (2004), pp. 371–386.

[12] W.J. Mitchell and J.R. Steel, Fine structure and iteration trees, Lecture Notes in Logic 3, Springer-Verlag, Berlin 1994.

[13] I. Neeman, Inner models in the region of a Woodin limit of Woodin cardinals, *Ann. of Pure and Applied Logic*, vol. 116 (2002) pp. 67–155.

[14] R.D. Schindler and J.R. Steel, The strength of AD, unpublished but available at http://www.math.uni-muenster.de/math/inst/logik/org/staff/rds.

[15] R.D. Schindler and J.R. Steel, The self-iterability of $L[\vec{E}]$, Journal of Symb. Logic, to apppear.

[16] R.D. Schindler and J.R.Steel, Problems in inner model theory, available at http://www.math.uni-muenster.de/math/inst/logik/org/staff/rds.

[17] F. Schlutzenberg and J.R. Steel, Comparing fine mice via coarse iteration, in preparation.

[18] T.A. Slaman and J.R. Steel, Definable functions on degrees, Cabal Seminar 81-85, A. Kechris, D. Martin, J. Steel eds., Springer Lecture Notes in Mathematics vol. 1333, pp. 37-55.

[19] J.R. Steel, An outline of inner model theory, Handbook of Set Theory, M. Foreman, A. Kanamori, and M. Magidor eds., to appear.

[20] J.R. Steel, The core model iterability problem , Lecture Notes in Logic 8, Springer-Verlag, Berlin 1996.

[21] J.R. Steel, Local K^c constructions, *Journal of Symb. Logic*, to appear.

[22] J.R. Steel, Woodin's analysis of $\mathrm{HOD}^{L(\mathbb{R})}$, unpublished, available at http://www.math.berkeley.edu/∼steel

[23] J. R. Steel, A theorem of Woodin on mouse sets, unpublished, available at http://www.math.berkeley.edu/∼steel

[24] J.R. Steel, *Scales in $K(\mathbb{R})$*, to appear in *The New Cabal*, available at http://www.math.berkeley.edu/∼steel

[25] J.R. Steel, The derived model theorem, unpublished, available at http://www.math.berkeley.edu/∼steel.

[26] J.R. Steel, Scales at the end of a weak gap, *Journal of Symb. Logic*, to appear.

[27] J.R. Steel, PFA implies $AD^{L(\mathbb{R})}$, *Journal of Symb. Logic* vol. 70 (2005) 1255-1296.

[28] J.R. Steel, A classification of jump operators, *Journal of Symb. Logic* vol. 47 (1982) 347-358.

[29] W.H. Woodin, unpublished lecture notes, Berkeley 1993-94.

[30] W.H. Woodin, Supercompact cardinals, sets of reals, and weakly homogeneous trees,*Proc. Nat. Acad. Sci. USA* **85**, 6587-6591.

[31] M. Zeman, Inner models and large cardinals, de Gruyter Series in Logic and Applications, vol. 5, Berlin 2002.

[32] A.S. Zoble, Stationary reflection and the determinacy of inductive games, Ph.D. thesis, U.C. Berkeley (2000).

TUTORIAL OUTLINE: SUITABLE EXTENDER SEQUENCES

W. Hugh Woodin

Department of Mathematics
University of California, Berkeley
Berkeley, CA, 94720 USA
E-mail: woodin@math.berkeley.edu

Contents

1. Introduction

This chapter serves two purposes. First it is an outline of a tutorial on Suitable Extender Sequences given as part of the program, *Computational Prospects of Infinity*, held in the summer of 2005 at the *Institute for Mathematical Sciences* (IMS) of the *National University of Singapore*. Second it also provides an outline of an expanded version, [7]. The expanded version contains proofs for essentially all of the theorems stated here and also contains a substantial amount of additional material which was either not covered in the tutorial or developed after the tutorial was given.

The subject of Suitable Extender Sequences is an attempt to understand the possibilities for an inner model theory for large cardinals beyond the level of *superstrong cardinals*. The original motivation was to seek clues for the large cardinal hypothesis which refutes the Ω Conjecture, if such a large cardinal hypothesis exists.

The surprising discovery is that entire problem seems to reduce to just the problem for exactly one supercompact cardinal. Further, subject to fairly general constraints, if this one problem can be solved then very likely no large cardinal hypothesis can refute the Ω Conjecture. But the implications of such a solution are more profound than this. The evidence suggests the possibility that if there is a supercompact cardinal then there is an ultimate version of L. This "L^∞" would be necessarily "close" to V and the inner model L^∞ would inherit essentially all known large cardinals from V. The relationship of L^∞ to V would be analogous to the relationship of L to V in the context that $0^\#$ does not exist.

The following two theorems illustrate these claims. Arguably part (3) of the first theorem would follow from any reasonable solution to the inner model problem for one supercompact cardinal. If E is an extender

$$\rho(E) = \sup \left\{ \alpha \mid V_\alpha \subset \mathrm{Ult}(V, E) \right\},$$

this is the *strength* of the extender E. The notion that a class of extenders witnesses δ is supercompact, is defined in Definition 2.5.

Theorem 1.1: (ZFC) *Suppose that there is an extendible cardinal. Then the following are equivalent.*

(1) *There is a proper class of regular cardinals which are not measurable in* HOD.
(2) *There is a cardinal δ such that* HOD $\vDash$ " *δ is a supercompact cardinal*" *and this is witnessed by the class of all $E \cap$ HOD such that E is an extender and $E \cap$ HOD $\in$ HOD.*

(3) *There exists a class $N \subseteq \mathrm{HOD}$ and there exists a cardinal δ such that*

$$N \vDash \mathrm{ZFC} \mid \text{`` } \delta \text{ is a supercompact cardinal''}$$

and this is witnessed by the class of all $E \cap N$ such that E is an extender, $\rho(E) = \mathrm{LTH}(E)$, and such that $E \cap N \in N$. □

The second theorem illustrates how close L^∞ would be to V, the conclusions (1)–(3) necessarily hold for any inner model N for there there is a cardinal δ such that

$$N \vDash \mathrm{ZFC} + \text{`` } \delta \text{ is a supercompact cardinal ''}$$

and this is witnessed by the class of all $E \cap N$ such that E is an extender, $\rho(E) = \mathrm{LTH}(E)$, and such that $E \cap N \in N$.

Theorem 1.2: (ZFC) *Suppose that δ is an extendible cardinal and that there is a regular cardinal above δ which is not measurable in HOD. Then the following hold.*

(1) *There exists an ordinal α such that for all cardinals $\gamma > \alpha$, if γ has countable cofinality then $\gamma^+ = (\gamma^+)^{\mathrm{HOD}}$.*
(2) *Suppose for each n there is a proper class of n-huge cardinals. Then for each n, $\mathrm{HOD} \vDash$ "There is a proper class of n-huge cardinals.".*
(3) *There exists an ordinal α such that for all $\gamma > \alpha$, if*

$$j : \mathrm{HOD} \cap V_{\gamma+1} \to \mathrm{HOD} \cap V_{j(\gamma)+1}$$

is an elementary embedding with critical point above α then $j \in \mathrm{HOD}$. □

Both theorems require a much weaker hypothesis than the existence of regular cardinals which are not measurable in HOD, see Section 5.3.

Before discussing closure properties of inner models with supercompact cardinals, we discuss iteration trees. The point of course is that iterability issues will be an important, if not central, part of any future inner model theory at the level of one supercompact cardinal.

2. Generalized iteration trees

2.1. *Long extenders*

Suppose that

$$j : V \to M$$

is an elementary embedding with critical point κ. Suppose that η is an ordinal, $\eta > \kappa$, and let $\hat{\eta}$ be the least ordinal such that $\eta \leq j(\hat{\eta})$. From

j one can define the *extender* of length η. This is, in essence, simply the function

$$F : \mathcal{P}(\hat{\eta}) \to V$$

given by: $F(A) = j(A) \cap \eta$.

The formal definition of the extender E specifies E as a family of ultra-filters. For each finite set $s \subset \eta$ let

$$E_s = \left\{ A \subset [\hat{\eta}]^{|s|} \mid A \in M \text{ and } s \in j(A) \right\}.$$

Thus E_s is an ultrafilter. The set

$$E = \left\{ (s, A) \mid s \in [\eta]^{<\omega} \text{ and } A \in E_s \right\}$$

is the extender of length η derived from j, it is also the (κ, η)-extender derived from j. If $\eta > j(\kappa)$ then the extender E is a *long extender*. Notice that if $\eta \leq j(\kappa)$ then $\hat{\eta} = \kappa$ and E is the usual (κ, η) extender derived from j.

Suppose that E is the (κ, η) extender derived from

$$j : V \to M.$$

The extender E gives an elementary embedding

$$j_E : V \to M_E$$

and there exists an elementary embedding (possibly the identity)

$$k : M_E \to M$$

such that $j = k \circ j_E$ and such that $k | \eta$ is the identity. Thus E is the extender of length η derived from

$$j_E : V \to M_E,$$

and so every extender contains all the information necessary to witness that E is an extender. Further, one can easily abstract out the relevant properties on E to ensure the construction of j_E yields an elementary embedding from which E can be derived.

The extender E is α-*strong* if $V_\alpha \subset M_E$. If $V_\alpha \subset M$ and $|V_\alpha|^M \leq \eta$ then E is α-strong.

Definition 2.1: Suppose that E is an extender of length η and that

$$j_E : V \to M_E$$

is the associated elementary embedding.

(1) An ordinal $\xi < \eta$, is a *generator* of E if

$$\xi \notin \left\{ j_E(f)(s) \mid s \in [\xi]^{<\omega}, f \in V_{\text{SPT}(E)+\omega} \right\}.$$

(2) If ξ is a generator of E and $\xi < j_E(\kappa)$, where κ is the critical point of j_E, then ξ is a *short generator* of E, otherwise ξ is a *long generator* of E. □

Notice that the extender E is uniquely specified by

$$E|X = \left\{ (s, A) \mid s \in [X]^{<\omega} \text{ and } A \in E_s \right\}$$

where X is the set of generators of E.

Given the definition above, a long extender E may have no long generators. In this case, E is a *degenerate* long extender. However, since we shall primarily be concerned with (coarse) extender models, this distinction is not really an issue and so we shall ignore it.

Remark 2.2: When extracting a long extender of length η from j we shall require at least that $\mathcal{P}(\eta) \subset M$. Otherwise the extender may encode too much.

If there exist a proper class of measurable cardinals then it is consistent that in all generic extensions of V, for any set a, there exist η and an elementary embedding

$$j : V \to M$$

such that $a \in L[E]$ where E is the (long) extender of length η derived from j.

This cannot happen in that case that E is not a long extender for then

$$L[E] = L[E_{\{\kappa\}}]$$

where κ is the critical point of j. Note

$$E_{\{\kappa\}} = \left\{ A \subseteq \kappa \mid \kappa \in j(A) \right\},$$

which is simply the normal measure on κ derived from j. In particular, the inner model $L[E]$ just depends on $\text{CRT}(E)$ and the family of inner models, $L[F]$, where F is a short extender, is wellordered under reverse containment, moreover $L[F_2] \subset L[F_1]$ if and only if $\text{CRT}(F_1) < \text{CRT}(F_2)$. □

We fix some notation.

Definition 2.3: Suppose that E is an extender.

(1) CRT(E) is the critical point of the elementary embedding

$$j_E : V \to M_E$$

given by E.

(2) LTH(E) is the length of the extender E.

(3) For each $\alpha <$ LTH(E) let SPT($E; \alpha$) be the least ordinal β such that $j_E(\beta) > \alpha$ and let SPT(E) = sup $\{$SPT($E; \alpha$) $\mid \alpha <$ LTH(E)$\}$.

(4) $\rho(E) =$ sup $\{\eta \mid V_\eta \subset M_E\}$.   □

Suppose that E is an extender. CRT(E) is the completeness of the ultrafilters associated to the extender, E, and LTH(E) specifies the domain of the extender, E. Note that SPT(E) is the least ordinal β such that $j_E(\beta) \geq$ LTH(E) and so SPT(E) is a minor variation of Steel's concept of the *space* of an extender, [4]. The extender, E, is ω-*huge* if $j_E(\rho(E)) = \rho(E)$ in which case $\rho(E)$ is the least ordinal ξ such that $j_E(\xi) = \xi$ and CRT(E) $< \xi$.

The following lemma from [7] gives a useful reformulation of supercompactness.

Lemma 2.4: *Suppose δ is a cardinal. Then the following are equivalent.*

(1) *δ is supercompact.*

(2) *For each $\lambda > \delta$ there exists an extender E such that SPT(E) $< \delta$, $\rho(E) > \lambda$, and such that j_E(CRT(E)) $= \delta$ where $j_E : V \to M_E$ is the elementary embedding given by E.*   □

Definition 2.5: A class, $\mathcal{E}$, of extenders *witnesses* that δ is supercompact if for each $\lambda > \delta$ there exists an extender $E \in \mathcal{E}$ such that

(1) SPT(E) $< \delta$ and $\rho(E) > \lambda$,

(2) j_E(CRT(E)) $= \delta$,

where $j_E : V \to M_E$ is the elementary embedding given by E.   □

2.2. Iteration trees

Following the definitions of Martin and Steel [1], a *premouse* is a pair (M, δ) such that M is transitive, $\delta \in M$, and

(1) $M \vDash$ ZC $+ \Sigma_2$-*Replacement*,

(2) Suppose that $F : M_\delta \to M \cap$ Ord is definable from parameters in M, then F is bounded in M.

(3) δ is strongly inaccessible in M.

The only difference between the definition here and the definition given in [1], is in the requirement (3). Here the requirement is that δ be strongly inaccessible.

A central notion in inner model theory is that of an *iteration tree* which specifies how models should be iterated. The notion of an iteration tree is due to Steel and we generalize the definition to the case of long extenders. There are many possible generalizations however finding a suitable one is a little subtle since the most natural generalization leads to a failure of iterability. For iteration trees of length ω the definition we adopt is due to Steel, [4].

Suppose that M is a transitive model of ZFC, $E \in M$, and

$$M \vDash \text{``}E \text{ is an extender''}.$$

we will use the notation $\mathrm{CRT}(E), \mathrm{LTH}(E), \mathrm{SPT}(E)$ and $\rho(E)$, to indicate the corresponding ordinals as computed in M. In fact it is not difficult to check that it is only $\rho(E)$ which could depend on M.

Suppose that (N, δ) is a premouse. Then $\mathrm{Ult}(N, E)$ *is defined* if $\eta < \delta$ and

$$N \cap V_{\eta+1} = M \cap V_{\eta+1}$$

where $\eta = \mathrm{SPT}(E)$. In general this much more agreement between N and M than is required, for example in the case that $\rho(E) = \mathrm{LTH}(E) = j_E(\mathrm{SPT}(E))$ one really only needs

$$N \cap V_\eta \subseteq M \cap V_\eta$$

where $\eta = \mathrm{SPT}(E)$. Finally if $\mathrm{Ult}(N, E)$ is defined and if

$$j_E : N \to \mathrm{Ult}(N, E)$$

is the associated embedding then j_E is an elementary embedding. This is the point of the restriction, $\mathrm{SPT}(E) < \delta$, and the reason for condition (2) in the definition of a premouse.

Definition 2.6: Suppose that (M, δ) is a premouse. An iteration tree , $\mathcal{T}$, on (M, δ) of length η is a tree order $<_\mathcal{T}$ on η with minimum element 0 and which is a suborder of the standard order, together with a sequence

$$\langle M_\alpha, E_\beta, j_{\gamma,\alpha} : \alpha < \eta, \beta + 1 < \eta, \gamma <_\mathcal{T} \alpha \rangle$$

such that the following hold.

(1) $M_0 = M$,

(2) $j_{\gamma,\alpha} : M_\gamma \to M_\alpha$ for all $\gamma <_T \alpha < \eta$

(3) Suppose that $\alpha + 1 < \eta$. Then $\alpha + 1$ has an immediate predecessor, α^*, in the tree order $<_T$ and:

(a) $E_\alpha \in j_{0,\alpha}(M \cap V_\delta)$ and $M_\alpha \vDash$ "E_α is an extender which is not ω-huge";

(b) If $\alpha^* < \alpha$ then $\mathrm{SPT}(E_\alpha) + 1 \leq \min \{\rho(E_\beta) \mid \alpha^* \leq \beta < \alpha\}$;

(c) $M_{\alpha+1} = \mathrm{Ult}(M_{\alpha^*}, E_\alpha)$ and

$$j_{\alpha^*,\alpha+1} : M_{\alpha^*} \to M_{\alpha+1}$$

is the associated embedding.

(4) If $0 < \beta < \eta$ is a limit ordinal then the set of α such that $\alpha <_T \beta$ is cofinal in β and M_β is the limit of the M_α where $\alpha <_T \beta$ relative to the embeddings; $j_{\alpha,\beta}$.   □

Remark 2.7:

(1) This definition, in case that none the extenders, E_α, are long extenders, coincides with the definition of an iteration tree in the sense of [1].

(2) The fundamental nature of the generalization of iteration tree in the sense of [1], that is provided by Definition 2.6, is *not* that long extenders are allowed, it lies in the cancellation rule; (3b).

(3) It is easy to prove by induction that the ultrapower, $\mathrm{Ult}(M_{\alpha^*}, E_\alpha)$, indicated in (3c) is defined; i.e. there is sufficient *agreement* between M_{α^*} and M_α (and this *certified* by the iteration tree).

(4) Suppose E is an extender. For each $s \in \mathrm{dom}(E)$ let

$$\kappa_s = \min \{|X| \mid E(s)(X) = 1\}$$

and define

$$\mathrm{sp}(E) = \sup \{\kappa_s \mid s \in \mathrm{dom}(E)\}.$$

Then $\mathrm{sp}(E)$ coincides with Steel's notion of the *space of an extender*, [4]. Now define

$$\rho^*(E) = \sup \{\eta \mid \mathcal{P}(\eta) \in \mathrm{Ult}(V, E)\}.$$

One can modify the definition of an iteration tree, replacing ρ by ρ^* and revising (3) to:

If $\alpha^* < \alpha$ then $\mathrm{sp}(E_\alpha) < \min \{\rho^*(E_\beta) \mid \alpha^* \leq \beta < \alpha\}$,

this is the definition of [4] for iteration trees of length ω. We have chosen the definition based on the parameters $(\mathrm{SPT}(E), \rho(E))$ simply in order to conform to the definitions of [1]. Restricting to extenders for which $\rho(E) = |V_{\rho(E)}|$, the two definitions coincide. For purposes of back-grounding an extender sequence construction, increasing the strength of the background extenders is irrelevant.

(5) Even using the parameters of $\text{SPT}(E)$ and $\rho(E)$, Definition 2.6 is not the most general definition for an iteration tree one might consider. Here we allow α^* to be the immediate predecessor of $\alpha + 1$ if

$$(V_{\eta+1})^{M_\alpha} = (V_{\eta+1})^{M_{\alpha^*}}$$

where $\eta = \text{SPT}(E_\alpha)$. But one might consider weakening this to the requirement that for each $\gamma < \text{LTH}(E_\alpha)$,

$$(V_{\eta+1})^{M_\alpha} = (V_{\eta+1})^{M_{\alpha^*}}$$

where $\eta = \text{SPT}(E_\alpha; \gamma)$ (again with this agreement certified by the iteration tree). A priori this seems a reasonable candidate for an iteration tree in the case of long extenders. However, by a result of Neeman, this definition leads to a failure of iterability even if one requires for each $\gamma < \text{LTH}(E_\alpha)$,

$$(V_{\eta+\omega})^{M_\alpha} = (V_{\eta+\omega})^{M_{\alpha^*}}$$

where $\eta = \text{SPT}(E_\alpha; \gamma)$. We shall discuss this more later.

(6) One could require in (3b) that if $\alpha^* < \alpha$ then

$$\text{SPT}(E_\alpha) + 1 \leq \min\left\{\rho^*(E_\beta) \mid \alpha^* \leq \beta < \alpha\right\},$$

where $\rho^*(E) = \sup\left\{\eta \mid \mathcal{P}(\eta) \subset M_E\right\}$. □

We come to the definition of a $(+\theta)$-iteration tree where $\theta \in \text{Ord}$. The definition is a direct generalization of the corresponding definition in [1] and we shall mostly be concerned in the case where $\theta = 2$.

Definition 2.8: Suppose that (M, δ) is a premouse and that $\mathcal{T}$ is an iteration tree on (M, δ) with associated sequence,

$$\langle M_\alpha, E_\beta, j_{\gamma,\alpha} : \alpha < \eta, \beta + 1 < \eta, \gamma <_\mathcal{T} \alpha\rangle$$

Suppose that $\theta \in \text{Ord}$. Then the iteration tree, $\mathcal{T}$, is a $(+\theta)$-*iteration tree* if for all $\alpha + 1 < \eta$,

$$\sup\left\{\text{SPT}(E_\beta) \mid \alpha + 1 \leq \beta \text{ and } \beta^* \leq \alpha\right\} + \theta \leq \rho(E_\alpha)$$

where for each $\beta + 1 < \eta$, β^* is the $\mathcal{T}$ predecessor of $\beta + 1$. □

Remark 2.9: Suppose that (M, δ) is a premouse and that $\mathcal{T}$ is an iteration tree on (M, δ) with associated sequence,

$$\langle M_\alpha, E_\beta, j_{\gamma,\alpha} : \alpha < \eta, \beta + 1 < \eta, \gamma <_{\mathcal{T}} \alpha \rangle$$

(1) It is easily verified that for all $\alpha + 1 < \eta$,

$$\sup \left\{ \mathrm{SPT}(E_\beta) + 1 \mid \alpha + 1 \leq \beta \text{ and } \beta^* \leq \alpha \right\} \leq \rho(E_\alpha).$$

So every iteration tree is a $(+0)$-iteration tree and every iteration tree of finite length is necessarily a $(+1)$-iteration tree.

(2) If $\mathcal{T}$ is a $(+1)$-iteration tree and if $\rho(E_\beta)$ is a limit ordinal for all $\beta + 1 < \eta$, then for all $n < \omega$, $\mathcal{T}$ is a $(+n)$-iteration tree. $\qquad\square$

Remark 2.10: Suppose that $\mathcal{T}$ is an iteration tree on (M, δ) of length η where condition (3b) of Definition 2.6 is replaced by the weaker condition:

If $\alpha^* < \alpha$ then for all $\gamma < \mathrm{LTH}(E_\alpha)$,

$$\mathrm{SPT}(E_\alpha; \gamma) + 1 \leq \min \left\{ \rho(E_\beta) \mid \alpha^* \leq \beta < \alpha \right\}.$$

Then for $\mathcal{T}$ the natural definition that $\mathcal{T}$ be a $(+i)$-iteration tree would be that $\mathcal{T}$ satisfies the requirement that for all for all $\alpha + 1 < \eta$,

$$\sup \left\{ \mathrm{SPT}(E_\beta; \gamma) \mid \gamma < \mathrm{LTH}(E_\beta), \alpha + 1 \leq \beta \text{ and } \beta^* \leq \alpha \right\} + i \leq \rho(E_\alpha)$$

where for each $\beta + 1 < \eta$, β^* is the $\mathcal{T}$ predecessor of $\beta + 1$.

Thus, since

$$\mathrm{SPT}(E) = \sup \left\{ \mathrm{SPT}(E; \gamma) \mid \gamma < \mathrm{LTH}(E) \right\},$$

if $\mathcal{T}$ is a $(+i)$-iteration tree in this sense and $i > 0$ then $\mathcal{T}$ is an $(+i)$-iteration tree in the sense of Definition 2.6, and so by restricting to $(+i)$-iteration trees for $i \geq 1$, the two definitions of iteration trees coincide. Since we shall almost always be dealing with iteration trees which are $(+\theta)$-iteration trees for $\theta \geq 1$, we could have used the more general definition. $\qquad\square$

Suppose that $\mathcal{T}$ is an iteration tree of length η on (M, δ) where η is a (nonzero) limit ordinal. A subset, $b \subseteq \eta$, defines a *maximal branch* of $\mathcal{T}$ if b totally ordered under the tree relation, $<_{\mathcal{T}}$, and b is maximal. The maximal branch, b, is *wellfounded* if the limit of the models M_α over $\alpha \in b$ is wellfounded; the maximal branch, b, is *proper* if b has limit length.

Notice that if $0 < \eta' < \eta$ and η' is a limit ordinal then $\mathcal{T}|\eta'$ is an iteration tree of length η' and the set

$$\{\alpha \mid \alpha <_{\mathcal{T}} \eta'\}$$

is a cofinal wellfounded maximal branch of $\mathcal{T}|\eta'$.

Our main result on iterability is that the fundamental theorem of [1] generalizes to the case of long extenders, for this definition of an iteration tree. This extends earlier results of Steel [4] which deal with generalized iteration trees of length ω and for ultimately a different purpose (wellfoundedness of the Mitchell order), not for the purpose of identifying a notion of iteration that might be relevant to the inner model theory of coherent sequences of (long) extenders.

The proof of Theorem 2.12 easily adapts to the iteration trees defined using the parameters, $\mathrm{sp}(E)$ and $\rho^*(E)$, as indicated in Remark 2.7(4). For iteration trees of length $\leq \omega$, this is precisely the main theorem of [4].

Definition 2.11: Suppose that (M, δ) is a premouse,

$$\pi : M \to V_\Theta$$

is an elementary embedding, T is an iteration tree on (M, δ), and that b is a maximal branch of T. Let M_b be the direct limit given by b and let

$$j_b : M \to M_b$$

be the associated elementary embedding.

The branch, b, is π-*realizable* if there exists an elementary embedding,

$$\pi_b : M_b \to V_\Theta$$

such that $\pi = \pi_b \circ j_b$. $\qquad\qquad\qquad\qquad\qquad\qquad\qquad\Box$

Theorem 2.12: *Suppose that (M, δ) is a countable premouse,*

$$\pi : M \to V_\Theta$$

is an elementary embedding,

$$T = \langle M_\alpha, E_\beta, j_{\gamma,\alpha} : \alpha < \eta, \beta + 1 < \eta, \gamma <_T \alpha \rangle$$

is a countable $(+2)$-iteration tree on (M, δ) and that T has no proper maximal π-realizable branch.

Then $\eta = \gamma + 1$ and for all extenders $E \in M_\gamma \cap V_{j_{0,\gamma}(\delta)}$, for all $\gamma^* \leq \gamma$, if $\gamma^* = \gamma$ or if $\gamma^* < \gamma$ and

$$\mathrm{SPT}(E) + 2 \leq \min \left\{ \rho(E_\alpha) \mid \gamma^* \leq \alpha < \gamma \right\},$$

then $\mathrm{Ult}(M_{\gamma^*}, E)$ is wellfounded and moreover the corresponding maximal branch of the induced iteration tree of length $\gamma + 2$ is π-realizable. $\qquad\Box$

Note that the requirement of Theorem 2.12 on the extender E is simply the requirement that if one extends the iteration tree, $\mathcal{T}$, to an iteration tree of length $\gamma+2$ by setting $E_\gamma = E$, then the extended iteration tree is a $(+2)$-iteration tree (modulo wellfoundedness of the ultrapower, $\mathrm{Ult}(M_{\gamma^*}, E)$). This is the induced iteration tree of length $\gamma + 2$.

So this generalization of iteration trees to the case of long extenders seems reasonable. For example any $(+1)$-iteration tree on V of length ω has an infinite branch. This implies that the Mitchell order on extenders is wellfounded which is the main theorem of Steel, [4], see [1] for some special cases of this. In contrast, Neeman has shown that with more general rules for the construction of an iteration tree (allowing extenders from one model to be applied to another whenever the ultrapower is defined), if there is a cardinal δ which is $\beth_\omega(\delta)$-supercompact then there is an iteration tree on V of length ω and *no* branch of length 4, c.f. Remark 2.7(5). For an explanation see Remark 2.10.

The central open question on iterability concerns the extent to which Theorem 2.12 can be strengthened.

2.3. *Branch conjectures*

Martin and Steel, [1], proposed two hypotheses with regard to iteration trees on V.

(UBH) *The Unique Branch Hypothesis*:

Suppose $\mathcal{T}$ is an iteration tree on a premouse (V_Θ, δ). Then $\mathcal{T}$ does not have two distinct cofinal wellfounded branches. □

(CBH) *The Cofinal Branch Hypothesis*:

Suppose $\mathcal{T}$ is an iteration tree on a premouse (V_Θ, δ). Then:

(1) If $\mathcal{T}$ has limit length then $\mathcal{T}$ has a cofinal wellfounded branch;
(2) If $\mathcal{T}$ has successor length, $\eta + 1$, then $\mathcal{T}$ can be freely extended to an iteration tree of length $\eta + 2$. □

Unfortunately if there is a supercompact cardinal then these hypotheses are each false in essentially the simplest cases. Define an iteration tree on V to be *short* if no extender occurring in the iteration tree is a long extender. Both UBH and CBH refer only to iteration trees which are short.

An iteration tree, $\mathcal{T}$, is *non-overlapping* if for all $\alpha + 1 <_{\mathcal{T}} \beta + 1$, $\mathrm{LTH}(E_\alpha) \leq \mathrm{CRT}(E_\beta)$. The iteration tree, $\mathcal{T}$, is *totally non-overlapping* if for all $\alpha + 1 <_{\mathcal{T}} \beta + 1$, $j_{E_\alpha}(\mathrm{SPT}(E_\alpha)) < \mathrm{CRT}(E_\beta)$.

Theorem 2.13: *Suppose that there is a supercompact cardinal.*

(1) *There is a short, totally non-overlapping, (+2)-iteration tree on V of length ω with only two cofinal branches and each is wellfounded.*

(2) *There is a short, totally non-overlapping, (+2)-iteration tree on V of length $\omega \cdot \omega$ with only one cofinal branch and this branch is illfounded.*$\Box$

If δ is supercompact then the counterexamples of Theorem 2.13 can be constructed such that for a given set $\mathcal{E} \subset V_\delta$ of extenders which witnesses that δ is a Woodin cardinal and which is closed under initial segments, each extender, E, of the iteration tree, except for the first extender E_0, has the following properties in the model from which E is selected.

(1) $E \in \mathcal{E}^*$;

(2) $\text{LTH}(E) = \rho(E)$ and $\text{LTH}(E)$ is strongly inaccessible;

(3) $\text{CRT}(E) > \text{LTH}(E_0)$;

where $\mathcal{E}^*$ is the image of $\mathcal{E}$ in that model. By (3) the iteration trees are necessarily given by iteration trees on $\text{Ult}(V, E_0)$. Further E_0 can be chosen to be very "short":

(1) $\text{LTH}(E_0) < \left(2^{2^\kappa}\right)^{\text{Ult}(V,E_0)}$ where $\kappa = \text{CRT}(E_0)$.

Let $\mathcal{F}_\mathcal{E}$ be the set of all short extenders $F \in V_\delta$ such that F satisfies (2) and such that if $\gamma = \rho(F)$, then

$$j_F(\mathcal{E}) \cap V_\gamma = \mathcal{E} \cap V_\gamma$$

and $(V_\gamma, \mathcal{E} \cap V_\gamma) \prec (V_\delta, \mathcal{E})$.

In the case of the counterexample to UBH still more can be required. If $\mathcal{T}$ is the iteration tree on $\text{Ult}(V, E_0)$ with exactly two cofinal branches, b and c, each of which are wellfounded, then there exists an extender $F_0 \in \mathcal{F}_\mathcal{E}$, and elementary embeddings,

$$\pi_b : \text{Ult}(V, E_0) \to \text{Ult}(V, F_0)$$

and

$$\pi_c : \text{Ult}(V, E_0) \to \text{Ult}(V, F_0)$$

each determined by their restrictions to $\text{LTH}(E_0)$ such that

(1) $\mathcal{T}$ copies by π_b to an iteration tree on $\text{Ult}(V, F_0)$ for which c copies to an illfounded branch,

(2) $\mathcal{T}$ copies by π_c to an iteration tree on $\text{Ult}(V, F_0)$ for which b copies to an illfounded branch,

(3) T copied by π_b yields an iteration tree on V (with first extender given by F_0) which is totally nonoverlapping,

(4) T copied by π_c yields an iteration tree on V (with first extender given by F_0) which is totally nonoverlapping.

The point here is that in the counterexample to UBH, the two wellfounded branches are wellfounded in essentially the strongest possible sense (their wellfoundedness is *certified* by the additional extender F_0 together with the embeddings, π_b and π_c). In particular if (V_Θ, δ) is a premouse, these branches copy to wellfounded branches on any premouse (M, δ_M) under any Σ_0-embedding,

$$\pi : V_\delta \to M_{\delta_M}$$

(chosen in any extension of V). This is inherited by all elementary substructures of (V_Θ, δ) which contain (E_0, T, F_0).

In contrast to this stronger version of Theorem 2.13, Steel [3] has proved that in any iterable Mitchell-Steel extender model, if δ is strongly inaccessible and $\mathcal{E} \subset V_\delta$ is the set of the extenders in V_δ which are the initial segment of some extender on the sequence, then UBH *holds* for all countable non-overlapping $(+2)$-iteration trees based on $(V_\Theta, \delta, \mathcal{E})$ where $\Theta > \delta$ and (V_Θ, δ) is a premouse. Because of the shortness of E_0, Steel's proof easily adapts to rule out the kind of strong counterexample to UBH indicated above.

We propose revised versions of these iteration hypotheses which are suggested by the specifics of the construction of the counterexamples of Theorem 2.13. We formulate these hypotheses for a restricted class of iteration trees. The utility of the iteration hypotheses is arguably not affected by this restriction.

Definition 2.14: An iteration tree, T, is *strongly closed* if:

(1) T is a $(+1)$-iteration tree; and
(2) each extender, E, occurring in T is LTH(E)-strong in the model from which it is selected and LTH(E) is strongly inaccessible in that model.$\square$

Lemma 2.15: *Suppose that (M, δ) is a premouse and that*

$$T = \langle M_\alpha, E_\beta, j_{\gamma,\alpha} : \alpha < \eta, \beta + 1 < \eta, \gamma <_T \alpha \rangle$$

is a strongly closed iteration tree on (M, δ). Let

$$\theta = \min \left\{ \mathrm{CRT}(E_\beta) \mid \beta + 1 < \eta \right\}.$$

Then T is a $(+\theta)$-iteration tree. $\square$

It is not known if UBH can fail for strongly closed iteration trees of length ω. There is a very indirect argument that shows UBH does fail for countable strongly closed iteration trees if there is a supercompact cardinal with a Woodin cardinal above. The counterexample, as is the case of the counterexamples given by Theorem 2.13, is an iteration tree with no long extenders which is non-overlapping. It is this counterexample which motivates the revised branch conjectures which we formulate below.

Suppose that (V_Θ, κ) is a premouse and that $X \prec (V_\Theta, \kappa)$ is a countable elementary substructure with transitive collapse (M, κ_M). Let

$$\pi : M \to V_\Theta$$

invert the transitive collapse.

Suppose $\delta_M < \kappa_M$. Then $\mathcal{G}(\pi, M, \delta_M, \kappa_M)$ is the set of all $(\mathcal{T}, b)$ such that

(1) $\mathcal{T}$ is a strongly closed iteration tree on (M, κ_M) of countable length and with all critical points above δ_M,
(2) $\mathcal{T}$ can be copied by π to an iteration tree, $\mathcal{T}_\pi$, on (V_Θ, κ),
(3) b is a cofinal branch of $\mathcal{T}$ such that the induced cofinal branch b_π of $\mathcal{T}_\pi$ is wellfounded — this is trivially implied by (2) if $\mathcal{T}$ has successor length.

Lemma 2.16: *Suppose that δ is supercompact, (V_Θ, κ) is a premouse with $\kappa > \delta$, and that $X \prec (V_\Theta, \kappa, \delta)$ is a countable elementary substructure with transitive collapse (M, κ_M, δ_M). Let*

$$\pi : M \to V_\Theta$$

invert the transitive collapse and let A be the set of $x \in \mathbb{R}$ such that x codes an element of $\mathcal{G}(\pi, M, \kappa_M, \delta_M)$. Then A is universally Baire. □

The following are the revised branch conjectures.

Revised Cofinal Branch Hypothesis:

Suppose that δ is supercompact and that $\mathcal{T}$ is a countable iteration tree on V which is strongly closed and with critical points above δ. Then:

(1) If $\mathcal{T}$ has limit length then $\mathcal{T}$ has a cofinal wellfounded branch;
(2) If $\mathcal{T}$ has successor length, $\eta + 1$, then $\mathcal{T}$ can be freely extended to an iteration tree of length $\eta + 2$ subject only to the constraints of being strongly closed. □

Revised Unique Branch Hypothesis:

Suppose that δ is supercompact, (V_Θ, κ) is a premouse with $\kappa > \delta$, and that $X \prec (V_\Theta, \kappa, \delta)$ is a countable elementary substructure with transitive collapse (M, κ_M, δ_M). Let $\pi : M \to V_\Theta$ invert the transitive collapse.

Suppose that $(\mathcal{T}, b) \in \mathcal{G}(\pi, M, \delta_M, \kappa_M)$ and let $(\mathcal{T}_\pi, b_\pi)$ be the result of copying $(\mathcal{T}, b)$ by π.

Let $j_b : M \to M_b$ be embedding given by the cofinal branch b and let

$$\pi_b : M_b \to (V_\Theta)_{b_\pi}$$

be the embedding given by π from copying. Let γ_b be the lim-inf of the critical points along b;

$$\gamma_b = \sup \left\{ \min \left\{ \mathrm{CRT}(E_\alpha^{\mathcal{T}}) \,\middle|\, \beta + 1 <_{\mathcal{T}} \alpha + 1 \in b \right\} \,\middle|\, \beta + 1 \in b \right\},$$

and let $(\kappa_b, \delta_b) \triangleq j_b(\kappa_M, \delta_M)$.

Let $\mathcal{H}_b$ be the set of $(\mathcal{T}^*, b^*) \in \mathcal{G}(\pi, M, \delta_M, \kappa_M)$ such that

(i) $\mathcal{T}^*|\eta = \mathcal{T}$, and b is the branch $\{\alpha < \eta \mid \alpha <_{\mathcal{T}^*} \eta\}$,
(ii) $\mathcal{T}^*|[\eta, \eta^*)$ is an iteration tree on M_b, where η^* the length of $\mathcal{T}^*$.

Let $\mathcal{H}_b|\gamma_b$ be the set of $(\mathcal{T}^*, b^*) \in \mathcal{H}_b$ such that $\mathcal{T}^*|[\eta, \eta^*)$ is an iteration tree based on $M_b \cap V_{\gamma_b}$; i.e., for all $\eta \le \alpha + 1 < \eta^*$, $E_\alpha \in j_{\eta,\alpha}^{\mathcal{T}^*}(M_b \cap V_\xi)$ for some $\xi < \gamma_b$.

Then:

(1) If $\mathcal{T}$ has limit length then the cofinal branch, b, is uniquely specified (in the obvious sense) by $\mathcal{H}_b|\gamma_b$ among all cofinal branches c of $\mathcal{T}$ such that $(\mathcal{T}, c) \in \mathcal{G}(\pi, M, \delta_M, \kappa_M)$.
(2) There is an elementary embedding

$$\pi^b : M_b \to V_\Theta$$

such that $\pi^b \circ j_b = \pi$ and such that $\mathcal{H}_b$ corresponds exactly with the set $\mathcal{G}(\pi^b, M_b, \kappa_b, \delta_b)$. $\square$

Remark 2.17: The natural conjecture is that if $\mathcal{T}$ is an iteration tree which is a counterexample to the *Revised Cofinal Branch Hypothesis*, then a proper initial segment of $\mathcal{T}$ yields (by taking a countable elementary substructure of a suitable V_η) a counterexample to the *Revised Unique Branch Hypothesis*. This would in essence be a strong form of Theorem 2.12 and a proof of this conjecture would probably yield proofs of both the *Revised Cofinal Branch Hypothesis* and the *Revised Unique Branch Hypothesis*.

The *Revised Cofinal Branch Hypothesis* is simply the *Cofinal Branch Hypothesis* of [1] restricted to countable iteration trees which are strongly closed, with all critical points above a supercompact cardinal, but generalized to allow long extenders. By a theorem of Steel, [4], the *Revised Cofinal Branch Hypothesis* holds for strongly closed iteration trees of length ω.

The *Revised Unique Branch Hypothesis* is admittedly a bit technical. But the underlying idea is a simple one. Note that this hypothesis is really two hypotheses which we state more informally for iteration trees of limit length–part (1) of the *Revised Unique Branch Hypothesis* is trivial for iteration trees of successor length and if κ is a limit of measurable cardinals then part (2) reduces to the limit cases.

First: At every limit stage of a good iteration tree, every good cofinal branch b, this is a cofinal branch which copies to a wellfounded branch on V, is uniquely specified (in the set of all good cofinal branches) by the set of all possible good continuations of the iteration tree prolonged by the branch and which are based just on the rank initial segment of the limit model up to the lim-inf, γ_b, of the critical points along the branch. We note if b_1 and b_2 are each cofinal wellfounded branches of T with limit models, M_{b_1} and M_{b_2} respectively, then $\gamma_{b_1} = \gamma_{b_2}$ and

$$V_{\gamma_{b_1}} \cap M_{b_1} = V_{\gamma_{b_2}} \cap M_{b_2}.$$

Therefore there is an obvious sense in which to compare the sets of good continuations corresponding to (T, b) and to (T, c) (where c is another good cofinal branch of T) which are based on just the common part,

$$V_{\gamma_b} \cap M_b = V_{\gamma_c} \cap M_c,$$

and this is the intended meaning of part (1).

Second: The limit model given by a good branch looks very much like the initial model with respect to the structure of the good iteration trees which exist on that limit model (now with no further restriction). □

The stronger version of Theorem 2.13 indicated just after the statement of Theorem 2.13 yields counterexamples to UBH and to CBH which fail to be a strongly closed iteration trees *only* in the choice of the first extender.

In fact the construction gives a counterexample (T, b) to part (2) of the *Revised Unique Branch Hypothesis*, naturally interpreted without the requirement of strong closure on T, where the iteration tree T has (formally) length 2; i.e, with only one ultrapower and this is internal.

3. Generalized Martin-Steel extender sequences

3.1. *Martin-Steel extender sequences with long extenders*

An important precursor to the fine structural models of Mitchell-Steel of [2] are the Martin-Steel inner models of [1], these extender models are of the form $L[\tilde{E}]$ where

$$\tilde{E} \subset (\mathrm{Ord} \times \mathrm{Ord}) \times V$$

is a predicate defining a sequence of (total) extenders. More precisely if $(\alpha, \beta) \in \mathrm{dom}(\tilde{E})$ then the set

$$\left\{ a \in V \mid ((\alpha, \beta), a) \in \tilde{E} \right\}$$

is an extender which we denote by E_β^α. In the case of the Martin-Steel inner models, the extender E_β^α is the (κ, α)-extender derived from an elementary embedding

$$j : V \to M$$

such that $\mathcal{P}^\omega(\alpha) \subset M$ and such that $\alpha < j(\kappa)$.

Definition 3.1: An extender sequence,

$$\tilde{E} = \langle E_\beta^\alpha : (\alpha, \beta) \in \mathrm{dom}(\tilde{E}) \rangle$$

is a *generalized Martin-Steel extender sequence* if for each pair $(\alpha, \beta) \in \mathrm{dom}(\tilde{E})$:

(1) (Coherence) There exists an extender F such that

    (a) $\alpha + \omega \leq \rho(F)$,
    (b) $E_\beta^\alpha = F|\alpha$
    (c) $j_F(\tilde{E})|(\alpha + 1, 0) = \tilde{E}|(\alpha, \beta)$.

(2) (Novelty) For all $\beta^* < \beta$, $(\alpha, \beta^*) \in \mathrm{dom}(\tilde{E})$ and

$$E_{\beta^*}^\alpha \cap L[\tilde{E}|(\alpha, \beta)] \neq E_\beta^\alpha \cap L[\tilde{E}|(\alpha, \beta)]$$

(3) (Initial Segment Condition) Suppose that

$$\kappa < \alpha^* < \alpha$$

where κ is the critical point associated to E_β^α.

Then there exists β^* such that $(\alpha^*, \beta^*) \in \mathrm{dom}(\tilde{E})$ and such that

$$E_{\beta^*}^{\alpha^*} \cap L[\tilde{E}|(\alpha^* + 1, 0)] = (E_\beta^\alpha|\alpha^*) \cap L[\tilde{E}|(\alpha^* + 1, 0)]. \qquad \square$$

If one requires in condition 1(a) that $\rho(F) < j_F(\mathrm{CRT}(F))$ then one obtains a Martin-Steel extender sequence. So the generalization, which is an obvious one, is simply to allow long extenders on the sequence.

A *generalized Martin-Steel premouse* is a structure, $\langle M, \tilde{E}, \delta \rangle$, such that:

(1) (M, δ) is a premouse;
(2) $\tilde{E} \in M_\delta$ and $M_\delta \vDash$ "$\tilde{E}$ *is a generalized Martin-Steel extender sequence*".

Suppose that $\langle M_0, \tilde{E}_0, \delta_0 \rangle$ and $\langle M_1, \tilde{E}_1, \delta_1 \rangle$ are generalized Martin-Steel premice. Then

$$\tilde{E}_0 \trianglelefteq \tilde{E}_1$$

if either

$$\tilde{E}_1 \cap (L[\tilde{E}_0])^{M_0} = \tilde{E}_0 \cap (L[\tilde{E}_0])^{M_0},$$

or for some $(\alpha, \beta) \in \mathrm{dom}(\tilde{E}_1)$,

$$(\tilde{E}_1 | (\alpha, \beta)) \cap (L[\tilde{E}_0])^{M_0} = \tilde{E}_0 \cap (L[\tilde{E}_0])^{M_0}.$$

These two possibilities are not in general mutually exclusive, though if $\mathrm{dom}(\tilde{E}_1) \subset M_0$ then they are.

The attempts to generalize the theorems on comparison even granting iterability have failed to date because of several obstacles. Suppose that $\langle M_0, \tilde{E}_0, \delta_0 \rangle$ and $\langle M_1, \tilde{E}_1, \delta_1 \rangle$ are countable generalized Martin-Steel premice such that there exist iteration strategies of order $\omega_1 + 1$ for both (M_0, δ_0) and (M_1, δ_1).

Generally, to prove *comparison* one attempts to construct iteration maps

$$\pi_0 : (M_0, \delta_0) \to (M_0^*, \delta_0^*)$$

and

$$\pi_1 : (M_1, \delta_1) \to (M_1^*, \delta_1^*)$$

such that either

$$\pi_0(\tilde{E}_0) \trianglelefteq \pi_1(\tilde{E}_1)$$

or

$$\pi_1(\tilde{E}_1) \trianglelefteq \pi_0(\tilde{E}_0).$$

Further the iteration is usually constructed not using extenders on the image of the extender sequence but the corresponding "background" extenders, these are extenders (in the image model) which witness the coherence condition for some extender on the image sequence (usually with minimum possible length). However one also achieves a much stronger version of comparison, producing iterations,

$$\pi_0 : (M_0, \delta_0) \to (M_0^*, \delta_0^*)$$

and

$$\pi_1 : (M_1, \delta_1) \to (M_1^*, \delta_1^*),$$

with the property that (in addition to achieving comparison as defined above) for all $(\alpha, \beta) \in \pi_0(\text{dom}(\tilde{E}_0)) \cap \pi_1(\text{dom}(\tilde{E}_1))$,

$$\left(\pi_0(\text{dom}(\tilde{E}_0))\right)(\alpha, \beta) \cap M_0^* \cap M_1^* = \left(\pi_1(\text{dom}(\tilde{E}_1))\right)(\alpha, \beta) \cap M_0^* \cap M_1^*.$$

This we will use to motivate an abstract version of comparison in Section 4.3.

Martin and Steel proved this is possible for Martin-Steel premice. The iteration trees are constructed by choosing the extenders of *least disagreement* at every stage and choosing the models to which the corresponding background extenders are to be applied based on the requirement of not *moving the generators* of previously selected extenders. The doctrine of not moving generators is central to the theory of the comparison of extender models.

The attempts to generalize the Martin-Steel construction to the context of extender sequences with long extenders have failed because of two obstacles each of which arises because the constraint of not moving generators.

(Steel) The moving spaces problem:

> In the case of generalized Martin-Steel premice, the requirement of not moving generators is not compatible with the rules for constructing iteration trees; even allowing the most general rules possible for constructing iteration trees.

The difficulty, for example, is that if

$$T = \langle M_\alpha, E_\beta, j_{\gamma, \alpha} : \alpha < \eta, \beta + 1 < \eta, \gamma <_T \alpha \rangle$$

is an iteration tree on (M, δ) then there could exist (a limit ordinal) $\alpha < \eta$ such that for all $\beta < \alpha$,

(1) $M_\beta \cap \text{Ord} < \text{SPT}(E_\alpha)$
(2) $\text{CRT}(E_\alpha) \leq \text{CRT}(E_\beta)$.

Not moving generators would demand that $\alpha^* = 0$, but the rules for iteration trees require $\alpha^* = \alpha$. In fact in this situation, $\mathrm{Ult}(M_\beta, E_\alpha)$ simply cannot be defined for any $\beta < \alpha$.

There is a restricted class of generalized Martin-Steel premice for which the requirement of not moving generators is compatible with the rules for forming an iteration tree in a comparison iteration *if* one allows a more general notion of an iteration tree. However this approach is also doomed.

(Neeman) Iteration trees cannot be generalized:

Allowing the most general possible rules for constructing iterations trees leads to a total failure of iterability.

3.2. *The failure of comparison*

There is a simple reason for the problem of moving spaces, comparison is (consistently) false. Suppose $\tilde{E}$ is a generalized Martin-Steel extender sequence. Let $T_{\tilde{E}}$ denote the Σ_1-theory of the structure,

$$(L[\tilde{E}], \tilde{E} \cap L[\tilde{E}]).$$

Definition 3.2: *Comparison* holds for generalized Martin-Steel extender sequences if for all generalized Martin-Steel extender sequences $\tilde{E}$ and $\tilde{F}$, either $T_{\tilde{E}} \subseteq T_{\tilde{F}}$ or $T_{\tilde{F}} \subseteq T_{\tilde{E}}$. □

We fix some notation.

Definition 3.3: Suppose that $\tilde{E}$ is an extender sequence. For each δ let $o_{\mathrm{LONG}}^{\tilde{E}}(\delta)$ be the supremum of the ordinals α such that there exists β with:

(1) $\delta < \alpha$,
(2) $(\alpha, \beta) \in \mathrm{dom}(\tilde{E})$,
(3) $j_E(\kappa_E) = \delta$,

where $E = E_\beta^\alpha$ and κ_E is the critical point of j_E. □

For generalized Martin-Steel extender sequences, the moving spaces problem is unavoidable when the following occurs.

There exists $(\alpha, \beta) \in \mathrm{dom}(\tilde{E})$ such that if κ is the critical point associated to E_β^α then

$$o_{\mathrm{LONG}}^{\tilde{E}}(\delta) \geq \kappa$$

for some $\delta < \kappa$.

This is also where comparison can fail. Notice that if

$$o^{\tilde{E}}_{\text{LONG}}(\delta) \geq \kappa$$

then (since κ is a critical point of an extender on the sequence) by the initial segment condition, $o^{\tilde{E}}_{\text{LONG}}(\delta) \geq \kappa + 1$.

Definition 3.4: let $\mathcal{E}^{(\infty)}$ be the set of all generalized Martin-Steel extender sequences, $\tilde{E}$, such that there exists $(\kappa + 1, \eta) \in \text{dom}(E)$ such that

(1) $(\kappa + 1, \eta)$ is the maximum element of $\text{dom}(\tilde{E})$,
(2) $\text{CRT}(E^{\kappa+1}_\eta) = \kappa$,
(3) for all $(\alpha, \beta) \in \text{dom}(\tilde{E}|(\kappa + 1, \eta))$, for all $\gamma < \text{CRT}(E^\alpha_\beta)$,

$$o^{\tilde{E}}_{\text{LONG}}(\gamma) < \text{CRT}(E^\alpha_\beta) < \kappa,$$

(4) there exists $\delta < \kappa$ such that $o^{\tilde{E}}_{\text{LONG}}(\delta) = \kappa + 1$.   □

Theorem 3.5: *Suppose that there exist a supercompact cardinal with a strong cardinal above it. Then there exists a generalized Martin-Steel extender sequence, $\tilde{E}$, such that in the associated inner model, $L[\tilde{E}]$, comparison fails for generalized Martin-Steel extender sequences in $\mathcal{E}^{(\infty)}$.*   □

One might hope that the failure of comparison for sequences in $\mathcal{E}^{(\infty)}$ is the result of the necessity of having to refine the definition of generalized Martin-Steel extender sequences by adding additional requirements beyond the requirements of novelty and the initial segment condition. Unfortunately this is not the case, even increasing the strength of the extenders witnessing the coherence condition cannot help because this is exactly the source of the problem.

Definition 3.6:

(1) A generalized Martin-Steel extender sequence, $\tilde{E}$, is *strongly backgrounded* if for each $(\alpha, \beta) \in \text{dom}(\tilde{E})$ and for each $\eta \in \text{Ord}$ there is an extender F which witnesses the coherence condition for E^α_β such that $\rho(F) > \eta$.
(2) $\mathcal{E}^+_{(\infty)}$ denotes the set of $\tilde{E} \in \mathcal{E}^{(\infty)}$ such that $\tilde{E}$ is strongly backgrounded.   □

A cardinal, δ, is an *extendible* cardinal if for each $\alpha > \delta$ there exists an elementary embedding,

$$j : V_{\alpha+1} \to V_{j(\alpha)+1},$$

such that $\text{CRT}(j) = \delta$ and such that $j(\delta) > \alpha$.

Lemma 3.7: *Suppose that there is an extendible cardinal. Then* $\mathcal{E}^+_{(\infty)} \neq \emptyset$. □

Restricting to strongly backgrounded extender sequences, the following variation of Theorem 3.5 arguably rules out any possible inner model theory at the level of extender sequences in $\mathcal{E}^{(\infty)}$, based on a refinement of the definition of generalized Martin-Steel sequences, for which there is a comparison theorem based on iterability (or *any* other Σ_2 property) of the associated premice of small cardinality (cardinality below κ where κ is Σ_2-definable suffices). For each $\tilde{E} \in \mathcal{E}^+_{(\infty)}$ let $T_{\tilde{E}}$ denote the Σ_1-theory of the structure,

$$(L[\tilde{E}], \tilde{E} \cap L[\tilde{E}]).$$

Notice that if M is an uncountable transitive set such that

$$M \vDash \text{ZFC} + \text{``There is an extendible cardinal''}$$

and if $\tilde{F} \in \left(\mathcal{E}^+_{(\infty)}\right)^M$, then $(T_{\tilde{F}})^M$ is the Σ_1-theory of the structure,

$$(L[\tilde{F}], \tilde{F} \cap L[\tilde{F}])$$

and so $T_{\tilde{F}}$ is the same computed in M or in V. This is because

$$M \vDash \text{``For all } a \subset \text{Ord, } a^\# \text{ exists''}$$

and so since M is uncountable and $M \vDash \text{ZFC}$, for all $a \subset \text{Ord}$, if $a \in M$ then $a^\# \in M$. Therefore we can unambiguously write $T_{\tilde{F}}$ for $(T_{\tilde{F}})^M$.

Theorem 3.8: *Suppose that there is a proper class of huge cardinals. Then for each* $\gamma \in \text{Ord}$ *such that* γ *is* Σ_2-*definable, there exists a transitive set* M *such that*

(1) $M \vDash \text{ZFC}$ *and* $V_\gamma \in M$,
(2) $M \vDash$ *"There is an extendible cardinal"*,
(3) *for all* $\tilde{E} \in \mathcal{E}^+_{(\infty)}$, *for all* $\tilde{F} \in \left(\mathcal{E}^+_{(\infty)}\right)^M$, $T_{\tilde{E}} \nsubseteq T_{\tilde{F}}$ *and* $T_{\tilde{F}} \nsubseteq T_{\tilde{E}}$. □

4. Suitable extender sequences

The moving spaces problem must arise for generalized Martin-Steel extender sequences, $\tilde{E}$, if there exists $(\alpha, \beta) \in \text{dom}(\tilde{E})$ such that for some $\delta < \text{CRT}(E^\alpha_\beta)$

$$\text{CRT}(E^\alpha_\beta) \leq o^{\tilde{E}}_{\text{LONG}}(\delta).$$

Further Theorem 3.8 shows that comparison though iteration must fail at this level no matter how one tries to refine the definition of such sequences.

This is our approach, we shall restrict the length of the long extenders on the sequence in the obvious fashion in an attempt to avoid the moving spaces problem. However we shall also reorganize the sequences so that long extenders have priority, in particular at every stage α, we essentially avoid putting short extenders of length α on the sequence whenever possible. The result will define a class of "suitable extender sequences". One could work with generalized Martin-Steel extender sequences with the obvious restriction, and this was our original approach ("promising extender sequences"), but these changes simplify a number of details and allow a reasonable definition of sequences, $\tilde{E}$, of length Ord with the property that $o_{\mathrm{LONG}}^{\tilde{E}}(\delta) = \infty$ for some δ (which is necessarily uniquely specified by $\tilde{E}$). We also increase the strength requirement on the extenders witnessing the coherence condition. Here again we treat short extenders differently, for short extenders we require slightly more strength on the extenders witnessing the coherence condition. This is necessary in order to conform with the definition of an iteration tree (in particular requirement (3b) of Definition 2.6 which defines the cancellation rule). Increasing the strength of the extenders witnessing coherence fundamentally alters the nature of the associated iteration trees (again because of the precise formulation of cancellation rules for iteration trees). This does not seem to have a parallel for extender sequences without long extenders.

Definition 4.1: An extender sequence,

$$\tilde{E} = \langle E_\beta^\alpha : (\alpha, \beta) \in \mathrm{dom}(\tilde{E}) \rangle$$

is a *suitable extender sequence* if for each pair $(\alpha, \beta) \in \mathrm{dom}(\tilde{E})$:

Case I: E_β^α is not a long extender.

(1) (Coherence) There exists an extender F such that

(a) $\mathrm{CRT}(F) < \alpha \leq j_F(\mathrm{CRT}(F))$ and $\lambda < \rho(F)$ where λ is least such that $\alpha < \lambda = |V_\lambda|$,

(b) $o_{\mathrm{LONG}}^{\tilde{E}|(\alpha,\beta)}(\delta) < \mathrm{CRT}(F)$ for each $\delta < \mathrm{CRT}(F)$,

(c) $E_\beta^\alpha = F|\alpha$

(d) $j_F(\tilde{E})|(\alpha + 1, 0) = \tilde{E}|(\alpha, \beta)$.

(2) (Novelty) For all $\beta^* < \beta$, $(\alpha, \beta^*) \in \mathrm{dom}(\tilde{E})$ and

$$E_{\beta^*}^\alpha \cap L[\tilde{E}|(\alpha, \beta)] \neq E_\beta^\alpha \cap L[\tilde{E}|(\alpha, \beta)]$$

(3) (Long Extender Priority) α is a limit ordinal and for all $\delta < \alpha$, $o_{\text{LONG}}^{\tilde{E}|(\alpha,\beta)}(\delta) < \alpha$.

(4) (Initial Segment Condition) Suppose that

$$\text{CRT}(E_\beta^\alpha) < \alpha^* < \alpha,$$

α^* is a limit ordinal, and for all $\delta < \alpha^*$, $o_{\text{LONG}}^{\tilde{E}|(\alpha,\beta)}(\delta) < \alpha^*$.
Then there exists β^* such that $(\alpha^*, \beta^*) \in \text{dom}(\tilde{E})$ and such that

$$E_{\beta^*}^{\alpha^*} \cap L[\tilde{E}|(\alpha^* + 1, 0)] = (E_\beta^\alpha | \alpha^*) \cap L[\tilde{E}|(\alpha^* + 1, 0)].$$

Case II: E_β^α is a long extender.

(1) (Coherence) There exists an extender, F, such that

    (a) $j_F(\text{CRT}(F)) < \alpha$ and $\lambda \leq \rho(F)$ where λ is least such that $\alpha < \lambda = |V_\lambda|$,

    (b) $o_{\text{LONG}}^{\tilde{E}|(\alpha,\beta)}(\delta) < \text{CRT}(F)$ for each $\delta < \text{CRT}(F)$,

    (c) $E_\beta^\alpha = F|\alpha$,

    (d) $j_F(\tilde{E})|(\alpha + 1, 0) = \tilde{E}|(\alpha, \beta)$.

(2) (Novelty) For all $\beta^* < \beta$, $(\alpha, \beta^*) \in \text{dom}(\tilde{E})$ and

$$E_{\beta^*}^\alpha \cap L[\tilde{E}|(\alpha, \beta)] \neq E_\beta^\alpha \cap L[\tilde{E}|(\alpha, \beta)]$$

(3) (Initial Segment Condition) Suppose that

$$\text{CRT}(E_\beta^\alpha) < \alpha^* < \alpha.$$

Let $j = j_{E_\beta^\alpha}$ and suppose that either,

    (a) $j(\text{CRT}(E_\beta^\alpha)) < \alpha^*$, or

    (b) $\alpha^* \leq j(\text{CRT}(E_\beta^\alpha))$, α^* is a limit ordinal and for all $\delta < \alpha^*$,

$$o_{\text{LONG}}^{\tilde{E}|(\alpha,\beta)}(\delta) < \alpha^*.$$

Then there exists β^* such that $(\alpha^*, \beta^*) \in \text{dom}(\tilde{E})$ and such that

$$E_{\beta^*}^{\alpha^*} \cap L[\tilde{E}|(\alpha^* + 1, 0)] = (E_\beta^\alpha | \alpha^*) \cap L[\tilde{E}|(\alpha^* + 1, 0)]. \qquad \square$$

The next lemma answers a natural question concerning the coherence condition in the case that E_β^α is a short extender. One corollary is that if F witnesses the coherence condition for E_β^α, E_β^α is a short extender, and if $F|\eta$ fails to witness the coherence condition for E_β^α for all $\eta < \text{LTH}(F)$, then F is also a short extender.

Lemma 4.2: *Suppose that $\tilde{E}$ is a suitable extender sequence, $(\alpha, \beta) \in \text{dom}(\tilde{E})$, and that E_β^α is not a long extender.*

Suppose that F is an extender which witnesses the coherence condition for E_β^α. Then $\alpha < j_F(\mathrm{CRT}(F))$.   □

Thus in Case 1 of Definition 4.1, one could strengthen the coherence condition and require that $\alpha < j_F(\mathrm{CRT}(F))$.

Remark 4.3: The definition of a suitable extender sequence provides an example where there is a genuine distinction between short extenders and degenerate long extenders. Recall from page 199 that a long extender E is a degenerate long extender if

$$j_E = j_F$$

where $F = E|j_E(\mathrm{CRT}(E))$.

Suppose $\tilde{E}$ is a suitable extender sequence, $(\alpha, \beta) \in \mathrm{dom}(\tilde{E})$ and that

$$\alpha = j_E(\mathrm{CRT}(E)) + 1$$

where $E = E_\beta^\alpha$. Then the extender, E_β^α, is a degenerate long extender. However if F is an extender which witnesses the coherence condition for E_β^α then necessarily F is a long extender which is not a degenerate long extender.   □

An extender sequence, $\tilde{E}$, of class length is a suitable extender sequece if for all ordinals α, $\tilde{E}|(\alpha, 0)$ is a suitable extender sequence. The definition of a suitable premouse is formulated to cover the case of extender sequences of class length.

Definition 4.4: A *suitable premouse* is a structure, $\langle M, \tilde{E}, \delta \rangle$, such that:

(1) (M, δ) is a premouse,
(2) $\tilde{E} \in M$, $\tilde{E} \subset M_\delta$ and for all $\alpha < \delta$,

$$M_\delta \vDash \text{``}\tilde{E}|(\alpha, 0) \text{ is a suitable extender sequence.''}$$ □

It is reasonable expectation that however comparison is proved, the methodology will involve not moving the short generators of previously selected extenders. Recall that for an extender, E, the short generators of E are the generators of $E|j_E(\mathrm{CRT}(E))$, see Definition 2.1 on page 198.

For generalized Martin-Steel extender sequences even this weaker requirement can be in conflict with the rules for constructing iteration tree, for example if there is an extender E on the sequence such that

$$\mathrm{SPT}(E) \geq \mathrm{CRT}(F) > \mathrm{CRT}(E)$$

for some other extender, F, on the sequence then it is easy construct iteration trees where this conflict arises.

As we have indicated in the discussion before Definition 4.1, priority is given to long extenders in order to avoid a restricted version moving spaces problem. But this alone does not suffice, and this is the reason for requiring more strength in the coherence condition in Case 1 of Definition 4.1 versus Case 2 of Definition 4.1.

The conditions on the iteration tree based on a suitable premouse, $(M, \tilde{E}, \delta)$, which are specified in the hypothesis of Lemma 4.5, abstractly characterize the iteration trees based on (M, δ) which arise in an attempt to compare $(M, \tilde{E}, \delta)$ with another suitable premouse, by the usual construction based on *least disagreement*.

Lemma 4.5: *Suppose that $(M, \tilde{E}, \delta)$ is a suitable premouse and*

$$\mathcal{T} = \langle M_\alpha, F_\beta, j_{\gamma,\alpha} : \alpha \leq \eta, \beta < \eta, \gamma <_T \alpha \rangle$$

is an iteration tree on (M, δ) of length $\eta + 1$ and there exists an increasing sequence,

$$\langle (\xi_\beta, \theta_\beta) : \beta \leq \eta \rangle$$

such that for all $\beta \leq \eta$:

(i) $(\xi_\beta, \theta_\beta) \in \mathrm{dom}(j_{0,\beta}(\tilde{E}))$
(ii) *if $\beta < \eta$ then F_β witnesses the coherence condition for*

$$\left(j_{0,\beta}(\tilde{E}) \right) (\xi_\beta, \theta_\beta)$$

with $\mathrm{LTH}(F_\beta)$ as short as possible.

Suppose that $F \in M_\eta$ witnesses the coherence condition for $\left(j_{0,\eta}(\tilde{E}) \right)$ (ξ_η, θ_η) with $\mathrm{LTH}(F)$ as short as possible. Let $\eta^* \leq \eta$ be least such that either $\eta^* = \eta$ or $\eta^* < \eta$ and

$$\mathrm{SPT}(F) + 1 \leq \min \left\{ \rho(F_\beta) \mid \eta^* \leq \beta < \eta \right\}.$$

Then for all $\beta < \eta^*$,

$$\mathrm{CRT}(F) \geq \min \left\{ \xi_\beta, j_{F_\beta}(\mathrm{CRT}(F_\beta)) \right\}. \qquad \square$$

Thus for iterations of suitable premice which arise in a comparison by least disagreement, the requirement of not moving the short generators of previously selected extenders is compatible with the rules for forming the iteration tree.

This will be relevant in the sequel to [7]. Proving comparison for iterable suitable premice through iterations based on least disagreement is a slightly subtle problem, see Lemma 5.8 on page 241 and the discussion which precedes it.

The next lemma is the basic lemma on the existence of suitable extender sequences.

Lemma 4.6: *Suppose that $\eta > \delta$ and that δ is V_λ-supercompact for some $\lambda > \eta$ such that $|V_\lambda| = \lambda$. Then there exists a suitable extender sequence $\tilde{E}$ such that $o^{\tilde{E}}_{\text{LONG}}(\delta) = \eta$.* ☐

It could be that one needs to require that the extenders which witness the coherence condition be even stronger. Define a suitable extender sequence, $\tilde{E}$, to be *strongly backgrounded* if for each $(\alpha, \beta) \in \text{dom}(\tilde{E})$ and for each λ there exists an extender F which witnesses the Coherence Condition for E^α_β such that $\rho(F) > \lambda$. The following variation of Lemma 4.6 suggests this is not unreasonable.

Lemma 4.7: *Suppose that $\eta > \delta$ and that δ is extendible. Then there exists a suitable extender sequence $\tilde{E}$ such that $o^{\tilde{E}}_{\text{LONG}}(\delta) = \eta$ and such that $\tilde{E}$ is strongly backgrounded.* ☐

By Lemma 4.6, if $\delta < \kappa$, κ is strongly inaccessible and δ is supercompact then there exists a suitable extender sequence,

$$\tilde{E} \subset V_\kappa,$$

such that:

(1) $o^{\tilde{E}}_{\text{LONG}}(\delta) = \kappa$;
(2) For each $(\alpha, \beta) \in \text{dom}(\tilde{E})$, $\tilde{E}|(\alpha, \beta)$ is a suitable extender sequence in V_κ.

We now consider the situation that $\tilde{E}$ is an extender sequence, $\tilde{E}$ is a proper class, every initial segment of $\tilde{E}$ is a suitable extender sequence, and that there exists δ such that $o^{\tilde{E}}_{\text{LONG}}(\delta) = \infty$. The next lemma records some simple properties of such sequences.

Theorem 4.8: *Suppose that $\tilde{E}$ is an extender sequence, $\tilde{E}$ is a proper class, every initial segment of $\tilde{E}$ is a suitable extender sequence, and $o^{\tilde{E}}_{\text{LONG}}(\delta) = \infty$.*

Suppose $\alpha > \delta$. Then:

(1) *Suppose $(\alpha, \beta) \in \text{dom}(\tilde{E})$. Then $\text{CRT}(E^\alpha_\beta) < \delta$.*
(2) $\mathcal{P}(\alpha) \cap L[\tilde{E}] \subset L[\tilde{E}|(\alpha + 1, 0)]$.
(3) $|\mathcal{P}(\alpha) \cap L[\tilde{E}]|^{L[\tilde{E}]} = |\text{dom}(\tilde{E}|(\alpha + 1, 0))|^{L[\tilde{E}]}$. ☐

4.1. Closure of $L[\tilde{E}]$ under extenders and the Ω-logic of $L[\tilde{E}]$

We survey some of the typical results on the closure of $L[\tilde{E}]$ under extenders and begin with a fairly general version.

Theorem 4.9: *Suppose $\tilde{E}$ is a suitable extender sequence, $o^{\tilde{E}}_{\text{LONG}}(\delta) = \infty$, and that $\gamma > \delta$. Suppose that $\gamma > \delta$ and γ is a cardinal of $L[\tilde{E}]$.*

Suppose that M is a transitive set, and that

$$j : \left(H(\gamma^+)\right)^{L[\tilde{E}]} \to M$$

is an elementary embedding with $\text{CRT}(j) \geq \delta$. Suppose that $\lambda \leq j(\gamma)$ and $\mathcal{P}(\lambda) \cap M \subset L[\tilde{E}]$.

Let F be the $L[\tilde{E}]$-extender of length λ given by j. Then $F \in L[\tilde{E}]$. □

Theorem 4.9 yields several variations as immediate corollaries.

Theorem 4.10: *Suppose $\tilde{E}$ is a suitable extender sequence, $o^{\tilde{E}}_{\text{LONG}}(\delta) = \infty$, and that $\gamma > \delta$. Suppose that $\gamma > \delta$ and γ is a cardinal of $L[\tilde{E}]$.*

Suppose that $M \in L[\tilde{E}]$, M is a transitive set and that

$$j : \left(H(\gamma^+)\right)^{L[\tilde{E}]} \to M$$

is an elementary embedding with critical point $\kappa \geq \delta$. Then $j \in L[\tilde{E}]$. □

One application of Theorem 4.10 is the following theorem which shows that the proof relation, $\vdash_\Omega$, is absolute to $L[\tilde{E}]$. The theorem is also true with real parameters. The proof requires developing a representation theorem for universally Baire sets in terms of elementary embeddings $j : H(\kappa^+) \to H(j(\kappa)^+)$.

Theorem 4.11: *Suppose $\tilde{E}$ is a suitable extender sequence, $o^{\tilde{E}}_{\text{LONG}}(\delta) = \infty$, and there exists an $(\omega + 2)$-extendible cardinal above δ. Suppose that ϕ is a sentence and that T is a theory with $T \in L[\tilde{E}]$. Then the following are equivalent.*

(1) $T \vdash_\Omega \phi$.
(2) $L[\tilde{E}] \vDash$ "$T \vdash_\Omega \phi$". □

The second variation of Theorem 4.10 shows that $L[\tilde{E}]$ is closed under extenders which in general because of the smallness requirement on the extenders of $\tilde{E}$, *cannot* be on the sequence $\tilde{E}$. This version of Theorem 4.10 is best suited for transferring the strongest hypotheses from V to $L[\tilde{E}]$ is the following.

Theorem 4.12: *Suppose $\tilde{E}$ is a suitable extender sequence, $o^{\tilde{E}}_{\mathrm{LONG}}(\delta) = \infty$, and that $\gamma > \delta$. Suppose that*

$$j : L[\tilde{E}] \cap V_{\gamma+1} \to L[\tilde{E}] \cap V_{j(\gamma)+1}$$

is an elementary embedding with $\mathrm{CRT}(j) \geq \delta$. Then $j \in L[\tilde{E}]$. $\hspace{2em}\square$

Using Theorem 4.12 one can quite easily transfer many large cardinals from V to $L[\tilde{E}]$. For some large cardinal hypotheses it may be necessary to assume a slightly stronger hypothesis holds in V but for the analysis of which large cardinal hypotheses can hold in $L[\tilde{E}]$ this is an irrelevant issue.

For example, it is a straightforward corollary of Theorem 4.12 that for each $n < \omega$, it there exists an $(n + 1)$-huge cardinal in V above δ (where $o^{\tilde{E}}_{\mathrm{LONG}}(\delta) = \infty$) then in $L[\tilde{E}]$, there is an n-huge cardinal above δ. For transferring the existence of an ω-huge cardinal from V to $L[\tilde{E}]$ a similar argument works. If there exists an elementary embedding,

$$j : V_{\lambda+1} \to V_{\lambda+1},$$

with $\lambda > \delta$ then there exists an elementary embedding,

$$j' : V_{\lambda'} \to V_{\lambda'},$$

such that $\lambda' > \delta$ and such that

$$j'(\tilde{E}|(\lambda', 0)) = \tilde{E}|(\lambda', 0),$$

and so by Theorem 4.12 and absoluteness, in $L[\tilde{E}]$, there exists $\lambda'' > \delta$ and an elementary embedding,

$$j'' : V_{\lambda''} \to V_{\lambda''}.$$

However, Theorem 4.12 also suggests difficulties arise in trying to transfer the strongest largest cardinal hypotheses from V to $L[\tilde{E}]$. The point is that if $\tilde{E}$ is a suitable extender sequence, $o^{\tilde{E}}_{\mathrm{LONG}}(\delta) = \infty$, and if

$$j : L[\tilde{E}] \cap V_{\lambda+1} \to L[\tilde{E}] \cap V_{j(\lambda)+1}$$

is an elementary embedding with critical point $\kappa \geq \delta$, then $j \in L[\tilde{E}]$ (since the induced $L[\tilde{E}]$-extender is in $L[\tilde{E}]$).

In particular, Theorem 4.12 essentially rules out directly transferring hypotheses such as the following which are among the strongest currently known (and not known to refute the *Axiom of Choice*).

(1) There exists a nontrivial elementary embedding,

$$j : V_{\lambda+1} \to V_{\lambda+1}.$$

(2) There exists a nontrivial elementary embedding,

$$j : L(V_{\lambda+1}) \to L(V_{\lambda+1})$$

with critical point below λ.

The problem is that the corresponding requirement;

$$j(L[\tilde{E}] \cap V_{\lambda+1}) = L[\tilde{E}] \cap V_{\lambda+1},$$

is not a reasonable requirement, indeed if

$$j : V_{\lambda+1} \to V_{\lambda+1}$$

is an elementary embedding with critical point above δ such that $\mathrm{cof}(\lambda)^{L[\tilde{E}]} > \omega$ then by Theorem 4.12, the requirement *cannot* hold.

Nevertheless the large cardinal hypotheses listed above do transfer to $L[\tilde{E}]$ under reasonable conditions. To show this we must establish a stronger version of Theorem 4.12 and the method is to exploit the following closure condition which follows from an analysis of $L[\tilde{E}]$ using the *Extender Algebra*.

Theorem 4.13: *Suppose $\tilde{E}$ is a suitable extender sequence and that $o_{\mathrm{LONG}}^{\tilde{E}}(\delta) = \infty$.*

Then there exists $\mathcal{F} \subset \delta$ such that:

(1) $V_\delta \in L[\mathcal{F}]$;
(2) $L[\tilde{E}][\mathcal{F}]$ *is a generic extension of $L[\tilde{E}]$ for a partial order*

$$\mathbb{P} = (\delta, <_{\mathbb{P}})$$

such that $\mathbb{P} \in L[\tilde{E}]$ and such that $\mathbb{P}$ is δ-cc in $L[\tilde{E}]$;
(3) $\left(L[\tilde{E}][\mathcal{F}]\right)^{<\delta} \subset L[\tilde{E}][\mathcal{F}]$. $\qquad\qquad\square$

Theorem 4.13 can be used to transfer the existence of large cardinals from V to $L[\tilde{E}]$ which are essentially as strong as possible; well beyond ω-huge cardinals. The method uses a generalization of a theorem of Steel.

Definition 4.14: An extender, E, of length η is λ-*complete* if

$$\eta^\lambda \subset M$$

where $M = \mathrm{Ult}(V, E)$. $\qquad\qquad\square$

Suppose that E is a (κ, η)-extender, $\mathbb{P} \in V_\kappa$, and $G \subset \mathbb{P}$ is V-generic. Then E naturally defines a (κ, η)-extender in $V[G]$ and

$$(j_E)^{V[G]} | V = (j_E)^V.$$

The special case of Lemma 4.15 where the extender E has length 1 (i.e., for measures) is due to Steel.

Lemma 4.15: *Suppose that $\delta < \kappa$, E is an extender which is δ-complete with critical point κ, and that*

$$j : V \to M \subset V[G]$$

is a generic elementary embedding such that

(i) $M - \{j(f)(\alpha) \mid \alpha < \delta\}$,

(ii) *G is V-generic for some partial order $\mathbb{P} \in V$ such that $|\mathbb{P}| \leq \delta$ in V.*

Then $(j_E)^{V[G]}|M = (j_F)^M$ where $F = j(E)$. □

Using Lemma 4.15 and the generic elementary embeddings given by the *stationary tower* one obtains the next theorem.

Theorem 4.16: *Suppose $\tilde{E}$ is a suitable extender sequence, $o^{\tilde{E}}_{\text{LONG}}(\delta) = \infty$, $\lambda_0 > \delta$ and*

$$j_0 : L(V_{\lambda_0+1}) \to L(V_{\lambda_0+1})$$

is an elementary embedding with critical point below λ_0. Then:

(1) *In $L[\tilde{E}]$, for some $\lambda > \delta$ there exists a (nontrivial) elementary embedding*

$$j : V_{\lambda+1} \to V_{\lambda+1};$$

(2) *If in addition, $j_0(\tilde{E}|(\lambda_0, 0)) = \tilde{E}|(\lambda_0, 0)$, then in $L[\tilde{E}]$, for some $\lambda > \delta$ there exists an elementary embedding*

$$j : L(V_{\lambda+1}) \to L(V_{\lambda+1})$$

with critical point below λ. □

The hypothesis,

$$j_0(\tilde{E}|(\lambda_0, 0)) = \tilde{E}|(\lambda_0, 0),$$

for part (2) of Theorem 4.16 is not that restrictive and is easily arranged. One can also simply eliminate this assumption by strengthening the large cardinal hypothesis, as is illustrated by the next theorem.

Theorem 4.17: *Suppose $\tilde{E}$ is a suitable extender sequence such that $o^{\tilde{E}}_{\text{LONG}}(\delta) = \infty$.*

Suppose $\kappa > \delta$, κ is strongly inaccessible and for each $A \subset V_\kappa$ there exist $\lambda < \kappa$ and

$$j : L(V_{\lambda+1}) \to L(V_{\lambda+1})$$

such that j is an elementary embedding with critical point below λ,

$$j(A \cap V_\lambda) - A \cap V_\lambda,$$

and such that $(V_\lambda, A \cap V_\lambda) \prec (V_\kappa, A)$.

Then in $L[\tilde{E}]$, for each $A \subset V_\kappa$ there exist $\lambda < \kappa$ and

$$j : L(V_{\lambda+1}) \to L(V_{\lambda+1})$$

such that j is an elementary embedding with critical point below λ,

$$j(A \cap V_\lambda) = A \cap V_\lambda,$$

and such that $(V_\lambda, A \cap V_\lambda) \prec (V_\kappa, A)$. □

There are a number of ways one might strengthen the axiom that there is an elementary embedding

$$j : L(V_{\lambda+1}) \to L(V_{\lambda+1})$$

with critical point below λ. This and the associated transference problems are explored in some detail in [7].

The theorems above collectively show that $L[\tilde{E}]$ is strongly closed within V both with respect to Ω-logic and with respect to large cardinals.

4.2. *Iteration hypotheses and comparison conjectures*

A finer calculation of the reals of $L[\tilde{E}]$ is possible under additional hypotheses of iterability and comparison.

We give the definition of $(\omega_1 + 1)$-iterable premice adapted to iteration trees with long extenders. In this paper we will have limited use for this since we will not prove comparison for the premice we shall define assuming iterability, we shall deal with this problem in the sequel to this paper.

Definition 4.18:

Suppose that (M, δ) is a premouse. An *iteration strategy of order γ* for (M, δ) is a function $\mathcal{I}$ such that the following hold:

(1) If $\mathcal{T}$ is a $(+1)$-iteration tree on (M, δ) of length η for some limit ordinal $\eta < \gamma$. Then $\mathcal{T} \in \mathrm{dom}(\mathcal{I})$ and $\mathcal{I}(\mathcal{T})$ is a maximal wellfounded branch of $\mathcal{T}$ which is either cofinal in η or cofinal in η' for some limit ordinal $\eta' < \eta$.

(2) Suppose that $\mathcal{T}$ is a $(+1)$-iteration tree on (M, δ) of length $\eta < \gamma$, η is a limit ordinal, and that for all limit ordinals $\eta' < \eta$, $\mathcal{I}(\mathcal{T}|\eta')$ is the maximal cofinal branch,

$$\{\alpha < \eta' \mid \alpha <_{\mathcal{T}} \eta'\}.$$

Then $\mathcal{I}(\mathcal{T})$ is a cofinal maximal branch of $\mathcal{T}$. □

For the most part we shall only be concerned with iteration strategies of order γ where γ is at least $\omega_1 + 1$.

Definition 4.19: Suppose that (M, δ) is a premouse. The premouse, (M, δ), is $(\omega_1 + 1)$-*iterable* if there exists an iteration strategy for (M, δ) of order γ where $\gamma = \omega_1 + 1$.   □

The natural, and most ambitious, conjecture is that if (V_Θ, δ) is a premouse and $X \prec (V_\Theta, \delta)$ is a countable elementary substructure with transitive collapse, (M, δ_M), then the countable premouse, (M, δ_M), is $(\omega_1 + 1)$-iterable.

Definition 4.20: $((\omega_1 + 1)$-*Iteration Hypothesis*) Suppose that (M, δ) is a countable premouse and that

$$\pi : M \to V_\Theta$$

is an elementary embedding. Then (M, δ) has an iteration strategy of order $\omega_1 + 1$.   □

There is a weaker version of the $(\omega_1 + 1)$-Iteration Hypothesis which one can naturally consider.

Definition 4.21: (*Weak $(\omega_1 + 1)$-Iteration Hypothesis*) Suppose there is a proper class of Woodin cardinals, (M, δ) is a countable premouse, and that

$$\pi : M \to V_\Theta$$

is an elementary embedding. Then it is Ω-consistent that (M, δ) has an iteration strategy of order $\omega_1 + 1$.   □

At this time there is no substantial evidence that the *Weak $(\omega_1 + 1)$-Iteration Hypothesis* is any more plausible than the $(\omega_1 + 1)$-*Iteration Hypothesis*. Below we describe a scenario for proving the latter which might in the details distinguish between these two iteration hypotheses but even this is far from clear and this is all based on the revised branch conjectures. Nevertheless the *Weak $(\omega_1 + 1)$-Iteration Hypothesis* does suffice for most of the following analysis of $L[\tilde{E}]$ and so because of this it seems worthwhile to isolate the weaker iteration hypothesis.

We conjecture the following which concerns the revised branch conjectures of Section 2.3 on page 209. The point of this conjecture is to suggest that in the end, the essence of establishing $(\omega_1 + 1)$-iterability (in either sense above) might lie in simply understanding the *countable* iteration trees on V.

Suppose that δ_0 is a supercompact cardinal, and the revised branch conjectures hold. Then the Weak (ω_1+1)-Iteration Hypothesis *holds for all pairs $\langle (M,\delta), \pi \rangle$ such that $\pi(\delta) > \delta_0$ and restricting to iteration trees where all critical points have image under π above δ_0.*

The hypothesis is not vacuous (assuming supercompact cardinals are consistent), for example if there is no cardinal $\gamma > \delta_0$ which is a Woodin cardinal in $L(V_\gamma)$ and δ_0 is the only supercompact cardinal then the revised branch conjectures hold (but this is not a very interesting case).

The scenario for proving this conjecture is to use the determinacy of long games (clopen games of length ω_1 for which the set of winning conditions is universally Baire in the codes). The point is that the *Revised Unique Branch Conjecture* identifies auxiliary information which naturally could be played as the tree is constructed and from which in a canonical fashion the branches at limit stages can be computed. Obviously if one could prove such a conjecture then the iterability problem would be reduced to the revised branch conjectures. Further the proof could well give important insight as to the difference (if any) between the two iteration hypotheses. It is entirely possible that one, or both, of the iteration hypotheses stated above is provable only in the special case where one restricts the extenders, for example one might need to require that the extenders used are each on (the images of) some fixed suitable extender sequence. But if such a restriction is relevant, this too should be revealed in the proof of this conjecture.

Remark 4.22: Suppose that there is a proper class of measurable Woodin cardinals and that the Ω Conjecture holds. Then for each countable premouse, (M, δ), there is an iteration strategy for (M, δ) of order ω_1 if and only if it is Ω-consistent that there be an iteration strategy for (M, δ) of order ω_1. This equivalence is in general false for the case of iteration strategies of order $\omega_1 + 1$. $\qquad\square$

We formulate the natural conjecture for the comparison of iterable suitable premice.

Definition 4.23: *Comparison Conjecture* Suppose that

$$\langle M_0, \tilde{E}_0, \delta_0 \rangle, \langle M_1, \tilde{E}_1, \delta_1 \rangle$$

are countable suitable premice and:

(1) (M_0, δ_0) and (M_1, δ_1) each have $(\omega_1 + 1)$-iteration strategies;

(2) $A \subset \eta$, A is generic over M_0 and M_1 for partial orders of cardinality η in each model, and $\tilde{E}_0|(\eta+1,0) = \emptyset = \tilde{E}_1|(\eta+1,0)$;

Let $\mathcal{T}_0$ be the set of pairs, (a,T), such that $a \in \mathcal{P}(\eta) \cap L[A][\tilde{E}_0]$ and T is the Σ_1-theory of the structure

$$(L[A][\tilde{F}_0], \{a\}, A, \tilde{F}_0 \cap L[A][\tilde{F}_0])$$

with parameters from η and let $\mathcal{T}_1$ be the set of pairs, (a,T), such that $a \in \mathcal{P}(\eta) \cap L[A][\tilde{E}_1]$ and T is the Σ_1-theory of the structure

$$(L[A][\tilde{E}_1], \{a\}, A, \tilde{E}_1 \cap L[A][\tilde{E}_1])$$

with parameters from η. Then either $\mathcal{T}_0 \subseteq \mathcal{T}_1$ or $\mathcal{T}_1 \subseteq \mathcal{T}_1$. $\qquad\square$

It is possible that comparison is sensitive to additional requirements on the extenders, for example one may reasonably restrict to the case of suitable premice, $(M, \delta, \tilde{E})$, such that $\tilde{E}$ is strongly backgrounded in M_δ, or for which $\tilde{E}$ satisfies some other first order property in M_δ. Of course, to avoid trivialities, the existence of such premice should be *provable* from some large cardinal hypothesis (the existence of an extendible cardinal suffices in the case that the property is strong backgrounding).

The following theorem is immediate from the results of [1], and we shall in effect prove a strong form of this theorem in Section 5.1, see Theorem 5.7 on page 241.

Theorem 4.24: (Martin, Steel]) *The* Comparison Conjecture *holds for Martin-Steel premice.* $\qquad\square$

The following theorems are immediate from the definitions.

Theorem 4.25: *Suppose that the* $(\omega_1 + 1)$*-Iteration Hypothesis and the* Comparison Conjecture *both hold. Then strong comparison holds for all pairs of suitable extender sequences.* $\qquad\square$

Theorem 4.26: *Suppose that there is a proper class of Woodin cardinals, the* Weak $(\omega_1 + 1)$*-Iteration Hypothesis holds and that the* Comparison Conjecture *is provable. Then strong comparison holds for all pairs of suitable extender sequences.* $\qquad\square$

Assuming the $(\omega_1 + 1)$-*Iteration Hypothesis* and the *Comparison Conjecture* each provable then one obtains a finer calculation of the complexity of $\mathbb{R} \cap L[\tilde{E}]$.

Definition 4.27: Suppose that $\tilde{E}$ and $\tilde{F}$ are suitable extender sequences. Then $\tilde{F} < \tilde{E}$ if there exists an increasing function

$$\pi : \text{dom}(\tilde{F}) \to \text{dom}(\tilde{E})$$

such that the following hold.

(1) For all α, $(\tilde{F}|(\alpha,0)) \cap L[\tilde{E}] \in L[\tilde{E}]$ and

$$L[\tilde{E}] \vDash \text{``}(\tilde{F}|(\alpha,0)) \cap L[\tilde{E}] \text{ is a suitable extender sequence''}.$$

(2) $\pi \in L[\tilde{F}]$ and for all $(\alpha,\beta) \in \text{dom}(\tilde{F})$, the ordertype of $\text{dom}(\tilde{F}|(\alpha,\beta))$ is less than α' where $(\alpha',\beta') = \pi(\alpha,\beta)$.

(3) For all $(\alpha,\beta) \in \text{dom}(\tilde{F})$, $j_E(\pi)|(\alpha+1,0) = \pi|(\alpha,\beta)$, where $E = \tilde{E}(\pi(\alpha,\beta)))$.

(4) For all $(\alpha,\beta) \in \text{dom}(\tilde{F})$ $F_\beta^\alpha = E|\alpha$ where $E = \tilde{E}(\pi(\alpha,\beta)))$.   □

The definition of $\tilde{F} < \tilde{E}$ makes sense even if $\tilde{E}$ is a proper class. If both $\tilde{E}$ and $\tilde{F}$ are proper classes, we define $\tilde{F} < \tilde{E}$ in the obvious fashion.

Definition 4.28: Suppose that $\tilde{E}$ and $\tilde{F}$ are suitable extender sequences (possibly each of class length). Then $\tilde{F} < \tilde{E}$ if for all $(\alpha,\beta) \in \text{dom}(\tilde{F})$, $\tilde{F}|(\alpha,\beta) < \tilde{E}$.   □

Suppose that $\tilde{F}$ is a suitable extender sequence such that $o_{\text{LONG}}^{\tilde{F}}(\delta) = \infty$, $\tilde{E}$ is a suitable extender sequence, and that $\tilde{F} < \tilde{E}$. Then necessarily, $o_{\text{LONG}}^{\tilde{E}}(\delta) = \infty$.

Lemma 4.29: *Suppose that $\tilde{E}_0 < \tilde{E}_1 < \tilde{E}_2$. Then $\tilde{E}_0 < \tilde{E}_2$.*

Lemma 4.30: *Suppose that $\tilde{E}$ is a suitable extender sequence such that $o_{\text{LONG}}^{\tilde{E}}(\delta) = \infty$. Then for each $\eta < \delta$, there exists a suitable extender sequence, $\tilde{F}$, such that*

$$\tilde{F} < \tilde{E},$$

$o_{\text{LONG}}^{\tilde{F}}(\delta) = \infty$, *and such that* $\text{dom}(\tilde{F}) \cap V_\eta = \emptyset$.

Definition 4.31: Suppose that $\tilde{E}$ is a suitable extender sequence such that $o_{\text{LONG}}^{\tilde{E}}(\delta) = \infty$. Then $\tilde{E}$ is *stable* if for all suitable extender sequences $\tilde{F} < \tilde{E}$,

$$\mathbb{R} \cap L[\tilde{E}] = \mathbb{R} \cap L[\tilde{F}].$$ □

Lemma 4.32: *Suppose that $\tilde{E}$ is a suitable extender sequence such that $o^{\tilde{E}}_{\text{LONG}}(\delta) = \infty$. Then there exists a suitable extender sequence $\tilde{F}$ such that $\tilde{F} < \tilde{E}$, $o^{\tilde{F}}_{\text{LONG}}(\delta) = \infty$, and such that $\tilde{F}$ is stable.*

Theorem 4.33: *Suppose there is a proper class of cardinals κ which are $(\omega + 2)$-extendible, $\tilde{E}$ is a suitable extender sequences such that*

$$o^{\tilde{E}}_{\text{LONG}}(\delta) = \infty,$$

$\tilde{E}$ is stable, and that the $(\omega_1 + 1)$-Iteration Hypothesis and the Comparison Conjecture both hold in all generic extensions of $L[\tilde{E}]$.

Then there is a wellordering of $\mathbb{R} \cap L[\tilde{E}]$ which is Δ^2_3-definable in $L[\tilde{E}]$ and for each $x \in \mathbb{R}$ the following are equivalent.

(1) *$x \in L[\tilde{E}]$.*
(2) *There exists $\alpha < \omega_1$ such that $\text{ZFC} \vdash_\Omega$ "x is Δ^2_3-definable from α".*
(3) *There exists $\alpha < (\omega_1)^{L[\tilde{E}]}$ such that*

$$L[\tilde{F}] \models \ \text{ZFC} \vdash_\Omega \text{ "}x \text{ is } \Delta^2_3\text{-definable from } \alpha\text{".}$$

Finally there is the version of Theorem 4.33 for the *Weak $(\omega_1 + 1)$-Iteration Hypothesis.*

Theorem 4.34: *Suppose there is a proper class of cardinals κ which are $(\omega + 2)$-extendible, $\tilde{E}$ is a suitable extender sequences such that*

$$o^{\tilde{E}}_{\text{LONG}}(\delta) = \infty,$$

$\tilde{E}$ is stable, and that the Weak $(\omega_1 + 1)$-Iteration Hypothesis and the Comparison Conjecture both hold in all generic extensions of $L[\tilde{E}]$.

Then for each $x \in \mathbb{R}$ the following are equivalent.

(1) *$x \in L[\tilde{E}]$.*
(2) *$\text{ZFC} \vdash_\Omega$ "$x \in \text{HOD}$".*
(3) *$x \in L[\tilde{E}]$ and $L[\tilde{E}] \models \ \text{ZFC} \vdash_\Omega$ "$x \in \text{HOD}$".*

Remark 4.35: There is in fact no distinction between conclusions of Theorem 4.33 and Theorem 4.34. A very involved analysis shows the set

$$\{\phi \mid \emptyset \vdash_\Omega \phi\}$$

is either $\Sigma^2_2(\mathcal{I}_{\text{NS}})$ or $\Pi^2_2(\mathcal{I}_{\text{NS}})$; i.e., it is either Σ_2 definable or Π_2 definable in the structure

$$\langle H(c^+), \mathcal{I}_{\text{NS}} \rangle$$

where $\mathcal{I}_{\text{NS}}$ is the nonstationary ideal on ω_1. Similarly the proof relation for Ω-logic, this the set

$$\{(T,\phi) \mid T \vdash_\Omega \phi\},$$

is either $\Sigma^2_2(\mathcal{I}_{\text{NS}})$ or $\Pi^2_2(\mathcal{I}_{\text{NS}})$.

Using these results, Theorem 4.33 can be rephrased with Δ^2_3 replaced by $\Delta^2_2(\mathcal{I}_{\text{NS}})$ and Theorem 4.34 can be similarly restated.

A natural question is whether one can actually replace Δ^2_3 by Δ^2_2. By analogy to the case of Martin-Steel extender sequences one might expect this to be the case. But for suitable extender sequences, this is related to the question of whether the calculation of the complexity of proof relation for Ω-logic can be improved to show that this relation is necessarily either Σ^2_2 or Π^2_2 (because of the absoluteness of Ω-logic to $L[\tilde{E}]$).

A related question concerns the complexity of the set of all Σ^2_2-sentences which are Ω-consistent with CH; specifically is this set Ω-recursive? For this question one defines a set $X \subset \omega$ to be Ω-*recursive* if there exists a universally Baire set $A \subset \mathbb{R}$ such that

$$L(A, \mathbb{R}) \vDash \text{AD}^+$$

and such that X is Σ_1 definable in $L(A, \mathbb{R})$ from $\{\mathbb{R}\}$ (equivalently, if X is a Δ^2_1 set in $L(A, \mathbb{R})$). This question is a variation of the question of whether conditional Σ^2_2-generic absoluteness is possible, see [6].

If this set of sentences is not Ω-recursive and if the $(\omega_1 + 1)$-*Iteration Hypothesis* and the *Comparison Conjecture* are both provable, then in Theorem 4.33, Δ^2_3 cannot be replaced by Δ^2_2 and so the improvement with Δ^2_3 replaced by $\Delta^2_2(\mathcal{I}_{\text{NS}})$ would be best possible. □

4.3. $L[\tilde{E}]$ *in the context of comparison hypotheses*

We continue our analysis of the inner model, $L[\tilde{E}]$, assuming $\tilde{E}$ is a proper class, every initial segment of $\tilde{E}$ is a suitable extender sequence, and that there exists δ such that $o^{\tilde{E}}_{\text{LONG}}(\delta) = \infty$. Now we shall assume in addition that various forms of comparison hold for $\tilde{E}$.

We define the abstract version of comparison which we shall use to provide a more detailed analysis of $L[\tilde{E}]$ for those suitable extender sequences which satisfy this comparison condition. The two main consequences will be that

$$L[\tilde{E}] \vDash \text{The } \Omega \text{ Conjecture}$$

and an analysis of $\mathbb{R} \cap L[\tilde{E}]$, though for the latter the analysis in general will only apply to $\mathbb{R} \cap L[\tilde{F}]$ where $\tilde{F}$ is a suitable extender sequence obtained from $\tilde{E}$ (but still with the property that $o^{\tilde{F}}_{\text{LONG}}(\delta) = \infty$ for some δ).

Suppose that $\tilde{E}$ and $\tilde{F}$ are each suitable extender sequences. Then *strong comparison* holds for the pair $(\tilde{E}, \tilde{F})$ if there exists a function

$$H : V_\kappa \to V_\kappa$$

such that the following holds where κ is least such that $\kappa = |V_\kappa|$ and such that $(\tilde{E}, \tilde{F}) \in V_\kappa$.

Suppose $X, Y \prec V_\kappa$ are countable elementary substructures, each closed under H, with $\tilde{E} \in X$ and $\tilde{F} \in Y$. Let $\tilde{E}_X, \tilde{F}_Y$ be the images of $\tilde{E}, \tilde{F}$ under the respective transitive collapses.

Suppose that $\gamma \in M_X \cap M_Y \cap \text{Ord}$, $A \subset \gamma$, A is generic over M_X and M_Y for partial orders of cardinality γ in each model, and that

$$\tilde{E}_X|(\gamma + 1, 0) = \emptyset = \tilde{F}_Y|(\gamma + 1, 0),$$

where M_X is the transitive collapse of X and M_Y is the transitive collapse of Y.

Let $\eta_X = M_X \cap \text{Ord}$ and let T^A_X be the set of pairs, (a, T), such that

(1) $a \in \mathcal{P}(\gamma) \cap L_{\eta_X}[A][\tilde{E}_X]$,
(2) T is the Σ_1-theory of the structure

$$(L_{\eta_X}[A][\tilde{E}_X], \{a\}, \tilde{E}_X \cap L_{\eta_X}[A][\tilde{E}_X])$$

with parameters from γ;

let $\eta_Y = M_Y \cap \text{Ord}$ and let T^A_Y be the set of pairs, (b, T), such that

(3) $b \in \mathcal{P}(\gamma) \cap L_{\eta_Y}[A][\tilde{F}_Y]$,
(4) T is the Σ_1-theory of the structure

$$(L_{\eta_Y}[A][\tilde{F}_Y], \{b\}, \tilde{F}_Y \cap L_{\eta_Y}[A][\tilde{F}_Y])$$

with parameters from γ.

Then either $T^A_X \subseteq T^A_Y$ or $T^A_Y \subseteq T^A_X$.

The point of the definition of strong comparison is to isolate an abstract consequence of comparison through iterations based on disagreement, see page 214. Assuming the $(\omega_1 + 1)$-*Iteration Hypothesis* (which is defined on page 228), strong comparison holds for all pairs of Martin-Steel extender sequences.

Theorem 4.36: *Suppose that the* $(\omega_1 + 1)$-Iteration Hypothesis *and the* Comparison Conjecture *both hold. Then strong comparison holds for all pairs of suitable extender sequences.* □

Theorem 4.37: *Suppose that there is a proper class of Woodin cardinals, the* Weak $(\omega_1 + 1)$-Iteration Hypothesis *holds and that the* Comparison Conjecture *is provable. Then strong comparison holds for all pairs of suitable extender sequences.* □

Notice that, in contrast to the case for comparison, the assertion that strong comparison holds for $(\tilde{E}, \tilde{F})$ is in general nontrivial even if $\tilde{E} = \tilde{F}$. In fact this is really the only case we shall be interested in except of course we would like to require that for some δ, $o^{\tilde{E}}_{\text{LONG}}(\delta) = \infty$. Before giving the relevant definition we state a preliminary lemma which motivates the definition and which also confirms a consequence of (strong) comparison which one would expect to hold.

Lemma 4.38: *Suppose that* $\tilde{E}$ *and* $\tilde{F}$ *are each suitable extender sequences and that strong comparison holds for the pair* $(\tilde{E}, \tilde{F})$.
Then for all $(\alpha, \beta) \in \text{dom}(\tilde{E})$ *and for all* $(\eta, \gamma) \in \text{dom}(\tilde{F})$, *strong comparison holds for the pair* $(\tilde{E}|(\alpha, \beta), \tilde{F}|(\eta, \gamma))$. □

Definition 4.39: Suppose that $\tilde{E}$ is a suitable extender sequence such that $o^{\tilde{E}}_{\text{LONG}}(\delta) = \infty$. Then *strong comparison* holds for the pair $(\tilde{E}, \tilde{E})$ if for all $\gamma \in \text{Ord}$, strong comparison holds for the pair $(\tilde{E}|(\gamma, 0), \tilde{E}|(\gamma, 0))$. □

Theorem 4.40: *Suppose that* δ_0 *is a supercompact cardinal,* $\tilde{E}$ *is a suitable extender sequence,* $o^{\tilde{E}}_{\text{LONG}}(\delta_1) = \infty$, $\delta_0 < \delta_1$, *and that strong comparison holds for the pair* $(\tilde{E}, \tilde{E})$. *Suppose there is an* $(\omega + 2)$-extendible cardinal above δ_1 and that $\tilde{E}|(\delta_0, 0) = \emptyset$. *Then*

$$L[\tilde{E}] \vDash \text{``The } \Omega \text{ Conjecture''}.$$ □

Thus to show that the Ω Conjecture is not refuted by any known large cardinal hypothesis it suffices to prove that if δ is a supercompact cardinal then there exists *one* suitable extender sequence, $\tilde{E}$, such that $o^{\tilde{E}}_{\text{LONG}}(\delta) = \infty$ and such that strong comparison holds for the pair $(\tilde{E}, \tilde{E})$.

There is a curious consequence of strong comparison. Suppose that strong comparison holds for all pairs $(\tilde{E}, \tilde{E})$ such that $\tilde{E}$ is a suitable extender sequence, then for all suitable extender sequences, $\tilde{E}$, if $\tilde{E}$ is strongly backgrounded and if

$$o^{\tilde{E}}_{\text{LONG}}(\delta) = \infty$$

for some δ, then $L[\tilde{E}]$ is a generic extension. In fact we shall show that

$$L[\tilde{E}] = L[\tilde{F}][g]$$

for some suitable extender sequence $\tilde{F}$ and for some g which is set generic over $L[\tilde{F}]$ (with $g \notin L[\tilde{F}]$). Given $\tilde{E}$, $\tilde{F}$ is an explicit suitable extender sequence derived from $\tilde{E}$ and with no assumptions regarding strong comparison, $L[\tilde{E}] = L[\tilde{F}]$ or $L[\tilde{E}]$ is generic extension of $L[\tilde{F}]$. Assuming strong comparison holds for the pair $(\tilde{F}, \tilde{F})$ it follows that

$$L[\tilde{E}] \neq L[\tilde{F}]$$

and therefore it follows that $L[\tilde{E}]$ is a (nontrivial) generic extension. For this last step a much weaker property than that of being strongly backgrounded is all that is required.

Remark 4.41: Suppose M is an iterable Mitchell-Steel model and that

$$M \vDash \text{ZFC} + \text{"There is a Woodin cardinal"}.$$

Then M is generic extension of an iterable Mitchell-Steel model, $N \subsetneq M$.

So this curious consequence of strong comparison is not evidence for a failure of comparison for suitable extender sequences.   □

Suppose $\tilde{E}$ is a suitable extender sequence. Then for each $\lambda \in \text{Ord}$, one can obtain from $\tilde{E}$ a suitable extender sequence $(\tilde{E})_\lambda$ with the property that every critical point of an extender on the sequence $(\tilde{E})_\lambda$ is above λ.

Definition 4.42: Suppose that $\tilde{E}$ is a suitable extender sequence and that $\lambda \in \text{Ord}$. Then $(\tilde{E})_\lambda$ is the suitable extender obtained from $\tilde{E}$ as follows. We define

$$(\tilde{E}|(\alpha, \beta))_\lambda$$

by induction on $(\alpha, \beta) \in \text{dom}(\tilde{E}) \cup \{(\infty, \infty)\}$.

Suppose $(\tilde{E}|(\alpha, \beta))_\lambda$ is defined. Let (α^*, β^*) be the least element of $\text{dom}(\tilde{E})$ above (α, β) or be (∞, ∞) if (α, β) is the maximum element of $\text{dom}(\tilde{E})$. Suppose that

(1) $\text{CRT}(E_\beta^\alpha) > \lambda$,

(2) for all $(\alpha, \eta) \in \text{dom}\left((\tilde{E}|(\alpha, \beta))_\lambda\right)$,

$$E_\beta^\alpha \cap L[\tilde{E}^*] \neq E^* \cap L[\tilde{E}^*]$$

where $\tilde{E}^* = (\tilde{E}|(\alpha, \beta))_\lambda|(\alpha, \eta)$ and $E^* = (\tilde{E}|(\alpha, \beta))_\lambda(\alpha, \eta)$.

Let

$$\xi = \sup\left\{\eta \mid (\alpha, \beta^*) \in \operatorname{dom}\left((\tilde{E}|(\alpha, \beta))_\lambda\right)\right\}.$$

Then $(\alpha, \xi) \in \operatorname{dom}\left((\tilde{E}|(\alpha^*, \beta^*))_\lambda\right)$ and $(\tilde{E}|(\alpha^*, \beta^*))_\lambda(\alpha, \xi) = \tilde{E}(\alpha, \beta)$. Otherwise $(\tilde{E}|(\alpha^*, \beta^*))_\lambda = (\tilde{E}|(\alpha, \beta))_\lambda$. □

The suitable extender sequence, $(\tilde{E})_\lambda$, is the *maximal* suitable extender sequence obtained from $\tilde{E}$ by deleting all extenders with critical point below $\lambda + 1$. Note that if $o_{\mathrm{LONG}}^{\tilde{E}}(\delta) = \infty$ and if $\lambda < \delta$ then

$$o_{\mathrm{LONG}}^{(\tilde{E})_\lambda}(\delta) = \infty.$$

The next theorem shows that under fairly general conditions, if $\tilde{E}$ is a suitable extender sequence such that $o_{\mathrm{LONG}}^{\tilde{E}}(\delta) = \infty$ for some δ, then $L[\tilde{E}]$ is a generic extension of an inner model which is also of the same form.

Theorem 4.43: *Suppose that $\tilde{E}$ is a suitable extender sequence and that $o_{\mathrm{LONG}}^{\tilde{E}}(\delta) = \infty$. Let λ be such that for all $(\alpha, \beta) \in \operatorname{dom}(\tilde{E})$, if $\mathrm{CRT}(E_\beta^\alpha) \le \lambda$ then $\alpha < \lambda$.*
Then

(1) $o_{\mathrm{LONG}}^{(\tilde{E})_\lambda}(\delta) = \infty$,
(2) *Either $L[\tilde{E}] = L[(\tilde{E})_\lambda]$ or $L[\tilde{E}]$ is a generic extension of $L[(\tilde{E})_\lambda]$.* □

Strong comparison guarantees that the requirement, $L[\tilde{E}] \ne L[(\tilde{E})_\lambda]$, can be satisfied.

Theorem 4.44: *Suppose that $\tilde{E}$ is a suitable extender sequence and $o_{\mathrm{LONG}}^{\tilde{E}}(\delta) = \infty$,*
Let λ be the least inaccessible cardinal such that for cofinally many $\gamma < \lambda$, the set

$$\left\{E_\beta^\alpha \cap L[\tilde{E}] \mid (\alpha, \beta) \in \operatorname{dom}(\tilde{E}) \cap V_\lambda\right\}$$

witnesses in $L[\tilde{E}]$ that γ is a Woodin cardinal.
Suppose that λ is a limit of strong cardinals and that strong comparison holds for the pair $(\tilde{F}, \tilde{F})$ where $\tilde{F} = (\tilde{E})_\lambda$.
Then

(1) $o_{\mathrm{LONG}}^{(\tilde{E})_\lambda}(\delta) = \infty$ *and $L[(\tilde{E})_\lambda] \ne L[\tilde{E}]$,*
(2) *$L[\tilde{E}]$ is a generic extension of $L[(\tilde{E})_\lambda]$.* □

5. HOD and supercompact cardinals

5.1. *Suitable Doddages and the Doddage Conjecture*

The Martin-Steel extender models are actually defined in [1] as $L[P]$ where P is a predicate defined from a sequence of sets of extenders. Such sequences are called Doddages and the approach of constructing extender models from Doddages has the advantage that the resulting inner model can be ordinal definable. Whether or not this can be generalized to the situation of long extenders emerges as the key question and is the subject of the *Doddage Conjecture* which we shall define shortly in Definition 5.12.

Definition 5.1: A *Doddage* is a sequence $\tilde{\mathcal{E}}$ such that

$$\mathrm{dom}(\tilde{\mathcal{E}}) \subset \mathrm{Ord} \times \mathrm{Ord}$$

and such that for all $(\alpha, \beta) \in \mathrm{dom}(\tilde{\mathcal{E}})$, $\tilde{\mathcal{E}}(\alpha, \beta)$ is a set of extenders of length α.

Definition 5.2: Suppose that $\tilde{\mathcal{E}}$ is a Doddage. Then $L[\tilde{\mathcal{E}}] = L[P_{\tilde{\mathcal{E}}}]$ where $P_{\tilde{\mathcal{E}}}$ is the set of all (α, β, s, a) such that

(1) $(\alpha, \beta) \in \mathrm{dom}(\tilde{\mathcal{E}})$,
(2) $s \in [\alpha]^{<\omega}$,
(3) $a \in E(s)$ for all $E \in \tilde{\mathcal{E}}(\alpha, \beta)$. $\qquad\qquad\square$

Suppose $\tilde{\mathcal{E}}$ is a Doddage. For each $(\alpha, \beta) \in \mathrm{dom}(\tilde{\mathcal{E}})$ we denote $\tilde{\mathcal{E}}(\alpha, \beta)$ by $\mathcal{E}_\beta^\alpha$. For each ordinal, δ, $o_{\mathrm{LONG}}^{\tilde{\mathcal{E}}}(\delta)$ is the supremum of the ordinals, $\mathrm{LTH}(E)$, where

(1) $j_E(\mathrm{CRT}(E)) = \delta$
(2) $\delta < \mathrm{LTH}(E)$,
(3) $E \in \mathcal{E}_\beta^\alpha$ for some $(\alpha, \beta) \in \mathrm{dom}(\tilde{\mathcal{E}})$.

Definition 5.3: A Doddage,

$$\tilde{\mathcal{E}} = \langle \mathcal{E}_\beta^\alpha : (\alpha, \beta) \in \mathrm{dom}(\tilde{\mathcal{E}}) \rangle$$

is a *suitable Doddage* if for each pair $(\alpha, \beta) \in \mathrm{dom}(\tilde{\mathcal{E}})$ and for each extender $E \in \mathcal{E}_\beta^\alpha$,

Case I: E is not a long extender.

(1) (Coherence) There exists an extender F such that

    (a) $\mathrm{CRT}(F) < \alpha \leq j_F(\mathrm{CRT}(F))$ and $\lambda < \rho(F)$ where λ is least such that $\alpha < \lambda = |V_\lambda|$,

(b) $o_{\mathrm{LONG}}^{\tilde{\mathcal{E}}|(\alpha,\beta)}(\delta) < \mathrm{CRT}(F)$ for each $\delta < \mathrm{CRT}(F)$,

(c) $E = F|\alpha$

(d) $j_F(\tilde{\mathcal{E}})|(\alpha+1,0) = \tilde{\mathcal{E}}|(\alpha,\beta)$.

(2) (Novelty) For all $\beta^* < \beta$, $(\alpha,\beta^*) \in \mathrm{dom}(\mathcal{E})$ and for all $E^* \in \mathcal{E}_{\beta^*}^{\alpha}$,

$$E^* \cap L[\tilde{\mathcal{E}}|(\alpha,\beta)] \neq E \cap L[\tilde{\mathcal{E}}|(\alpha,\beta)]$$

(3) (Long Extender Priority) α is a limit ordinal and for all $\delta < \alpha$, $o_{\mathrm{LONG}}^{\tilde{\mathcal{E}}|(\alpha,\beta)}(\delta) < \alpha$.

(4) (Initial Segment Condition) Suppose that

$$\mathrm{CRT}(E) < \alpha^* < \alpha,$$

α^* is a limit ordinal, and for all $\delta < \alpha^*$, $o_{\mathrm{LONG}}^{\tilde{\mathcal{E}}|(\alpha,\beta)}(\delta) < \alpha^*$.
Then there exists $(\alpha^*,\beta^*) \in \mathrm{dom}(\tilde{\mathcal{E}})$ and there exists $E^* \in \mathcal{E}_{\beta^*}^{\alpha^*}$ such that

$$E^* \cap L[\tilde{\mathcal{E}}|(\alpha^*+1,0)] = (E|\alpha^*) \cap L[\tilde{\mathcal{E}}|(\alpha^*+1,0)].$$

Case II: E is a long extender.

(1) (Coherence) There exists an extender, F, such that

(a) $j_F(\mathrm{CRT}(F)) < \alpha$ and $\lambda \le \rho(F)$ where λ is least such that $\alpha < \lambda = |V_\lambda|$,

(b) $o_{\mathrm{LONG}}^{\tilde{\mathcal{E}}|(\alpha,\beta)}(\delta) < \mathrm{CRT}(F)$ for each $\delta < \mathrm{CRT}(F)$,

(c) $E = F|\alpha$,

(d) $j_F(\tilde{\mathcal{E}})|(\alpha+1,0) = \tilde{\mathcal{E}}|(\alpha,\beta)$.

(2) (Novelty) For all $\beta^* < \beta$, $(\alpha,\beta^*) \in \mathrm{dom}(\mathcal{E})$ and for all $E^* \in \mathcal{E}_{\beta^*}^{\alpha}$,

$$E \cap L[\tilde{\mathcal{E}}|(\alpha,\beta)] \neq E \cap L[\tilde{\mathcal{E}}|(\alpha,\beta)]$$

(3) (Initial Segment Condition) Suppose that

$$\mathrm{CRT}(E) < \alpha^* < \alpha.$$

Let $j = j_E$ and suppose that either,

(a) $j(\mathrm{CRT}(E)) < \alpha^*$, or

(b) $\alpha^* \le j(\mathrm{CRT}(E))$, α^* is a limit ordinal and for all $\delta < \alpha^*$,

$$o_{\mathrm{LONG}}^{\tilde{\mathcal{E}}|(\alpha,\beta)}(\delta) < \alpha^*.$$

Then there exists $(\alpha^*,\beta^*) \in \mathrm{dom}(\tilde{\mathcal{E}})$ and there exists $E^* \in \mathcal{E}_{\beta^*}^{\alpha^*}$ such that

$$E^* \cap L[\tilde{\mathcal{E}}|(\alpha^*+1,0)] = (E|\alpha^*) \cap L[\tilde{\mathcal{E}}|(\alpha^*+1,0)]. \qquad \square$$

Definition 5.4: Suppose $\tilde{\mathcal{E}}$ is a suitable Doddage. Then $\tilde{\mathcal{E}}$ is *good* if for all $(\alpha, \beta) \in \mathrm{dom}(\tilde{\mathcal{E}})$, for all $E_0, E_1 \in \mathcal{E}_\beta^\alpha$,

$$E_0 \cap L[\tilde{\mathcal{E}}] = E_1 \cap L[\tilde{\mathcal{E}}]. \qquad \Box$$

The initial segment and novelty conditions for suitable Doddages really only make sense when the relevant initial segments are good.

The entire analysis of $L[\tilde{E}]$ carries over to $L[\tilde{\mathcal{E}}]$. The only exception is that the proof of Theorem 4.13 does not work for $L[\tilde{\mathcal{E}}]$, indeed we do not know if the conclusion of Theorem 4.13 holds for $L[\mathcal{E}]$; i.e., if there must exist $\mathcal{F} \subset \delta$ such that $V_\delta \subset L[\mathcal{F}]$ and such that $\mathcal{F}$ is $L[\tilde{\mathcal{E}}]$-generic for a partial order $\mathbb{P} = (\delta, <_{\mathbb{P}})$ such that $\mathbb{P} \in L[\tilde{\mathcal{E}}]$ where $o_{\mathrm{LONG}}^{\tilde{\mathcal{E}}}(\delta) = \infty$.

However, $L\tilde{[}\mathcal{E}](V_\delta)$ is a symmetric extension of $L[\tilde{\mathcal{E}}]$ for a boolean algebra $\mathbb{B} \in L[\tilde{\mathcal{E}}] \cap V_{\delta+1}$. This suffices in place of Theorem 4.13 for the analysis of $L[\tilde{\mathcal{E}}]$.

Theorem 5.5: *Suppose $\tilde{\mathcal{E}}$ is a good suitable Doddage and that $o_{\mathrm{LONG}}^{\tilde{\mathcal{E}}}(\delta) = \infty$. Then:*

(1) *$L[\tilde{\mathcal{E}}](V_\delta)$ is a symmetric generic extension of $L[\tilde{\mathcal{E}}]$ for a Boolean algebra $\mathbb{B} \in L[\tilde{\mathcal{E}}]$ such that $\mathbb{B} \subset L_\delta[\tilde{\mathcal{E}}]$ and such that $\mathbb{B}$ is δ-cc in $L[\tilde{\mathcal{E}}]$;*

(2) *$\left(L[\tilde{\mathcal{E}}](V_\delta) \right)^{<\delta} \subset L[\tilde{\mathcal{E}}](V_\delta).$*

(3) *Suppose that*

$$j : L[\tilde{\mathcal{E}}] \cap V_{\gamma+1} \to L[\tilde{\mathcal{E}}] \cap V_{j(\gamma)+1}$$

is an elementary embedding with $\mathrm{CRT}(j) \geq \delta$. Then $j \in L[\tilde{\mathcal{E}}]$. $\qquad \Box$

In summary, the entire analysis we have given for suitable extender sequences carries over to good suitable Doddages.

A suitable Doddage, $\tilde{\mathcal{E}}$, is a *Martin-Steel Doddage* if for all $(\alpha, \beta) \in \mathrm{dom}(\tilde{\mathcal{E}})$, for all $E \in \mathcal{E}_\beta^\alpha$, E is not a long extender.

The following theorem is an immediate corollary of the results of [1]. The requirement that $\tilde{\mathcal{E}}_0 \in V_\delta$ for some strongly inaccessible cardinal δ is necessary only because of the precise formulation of the $(\omega_1 + 1)$-*Iteration Hypothesis*.

Theorem 5.6: *(Martin-Steel) Suppose that the $(\omega_1 + 1)$-Iteration Hypothesis holds, $\tilde{\mathcal{E}}$ is a Suitable Doddage such that $\tilde{\mathcal{E}} \in V_\delta$ for some strongly inaccessible cardinal δ, and that $\tilde{\mathcal{E}}$ contains no long extenders. Then $\tilde{\mathcal{E}}$ is good.* $\qquad \Box$

We prove a stronger version of this theorem. Theorem 5.7 is the generalization to suitable Doddages (with no long extenders) of the uniqueness of $L[\mu]$.

Theorem 5.7: *Suppose that $\tilde{\mathcal{E}}_0$ and $\tilde{\mathcal{E}}_1$ are suitable Doddages such that*

$$\mathrm{dom}(\tilde{\mathcal{E}}_0) = \mathrm{dom}(\tilde{\mathcal{E}}_1).$$

Suppose that (V_Θ, δ) is a premouse such that $\tilde{\mathcal{E}}_0 \in V_\delta$ and such that there exists a countable elementary substructure,

$$X \prec (V_\Theta, \delta)$$

such that (M, δ_M) has an $(\omega_1 + 1)$-iteration strategy where (M, δ_M) is the transitive collapse of X.

Suppose that $\tilde{\mathcal{E}}_0$ and $\tilde{\mathcal{E}}_1$ each contain only short extenders. Then

$$L[\tilde{\mathcal{E}}_0] = L[\tilde{\mathcal{E}}_1],$$

and moreover for all $(\alpha, \beta) \in \mathrm{dom}(\tilde{\mathcal{E}}_0)$, for all $E_0 \in \tilde{\mathcal{E}}_0(\alpha, \beta)$, for all $E_1 \in \tilde{\mathcal{E}}_1(\alpha, \beta)$,

$$E_0 \cap L[\tilde{\mathcal{E}}_0] = E_1 \cap L[\tilde{\mathcal{E}}_1].$$

The second theorem fails if either of the Doddages contains a long extender, in fact it fails at the first possible stage (one long extender); this arguably explains why proving the *Comparison Conjecture* is subtle. This does not necessarily argue against the *Comparison Conjecture*, it just shows that certain aspects of the comparison argument must change. Even if more severe obstacles arise, comparison through iteration could still be possible if $L[\tilde{E}]$ can be enlarged to a more complicated fine-structural inner model, say of the form $L[\mathbb{E}]$, where $\mathbb{E}$ is a sequence of *partial* extenders, such that $L[\tilde{E}] \subset L[\mathbb{E}]$ and such that $\tilde{E} \cap L[\mathbb{E}]$ is definable in the structure $(L[\mathbb{E}], \mathbb{E})$. Such a restriction could still yield a reasonable definable class of suitable extender sequences.

Lemma 5.8: *Suppose that δ is a supercompact cardinal. Then there exists suitable extender sequences, $\tilde{E}$ and $\tilde{F}$, such that:*

(1) $(\delta + 1; 0)$ *is the maximum element of $\mathrm{dom}(\tilde{E})$,*
(2) *for all $(\alpha, \beta) \in \mathrm{dom}(\tilde{E}|(\delta + 1, 0))$, E_β^α is not a long extender,*
(3) $\tilde{F}|(\delta + 1, 0) = \tilde{E}|(\delta + 1, 0)$ *and $\mathrm{dom}(\tilde{F}) = \mathrm{dom}(\tilde{E})$,*
(4) $\mathrm{CRT}(F_0^{\delta+1}) \neq \mathrm{CRT}(E_0^{\delta+1})$. $\qquad\qquad\qquad\qquad\square$

As an immediate corollary we obtain the following theorem.

Theorem 5.9: *Suppose that there is a supercompact cardinal. Then there is a suitable Doddage which is not good.*

Recall that if E is an extender then the three ordinals, $\mathrm{CRT}(E)$, $\mathrm{SPT}(E)$, and $\mathrm{LTH}(E)$, are the same as computed in V or in any inner model M such that $E \cap M \in M$.

Define a Doddage, $\tilde{\mathcal{E}}$ to be *weakly good* if for all $(\alpha, \beta) \in \mathrm{dom}(\tilde{\mathcal{E}})$, for all $E_0, E_1 \in \tilde{\mathcal{E}}(\alpha, \beta)$,

(1) $\mathrm{LTH}(E_0) = \mathrm{LTH}(E_1) = \alpha$,
(2) $\mathrm{CRT}(E_0) = \mathrm{CRT}(E_1)$,
(3) $\mathrm{SPT}(E_0) = \mathrm{SPT}(E_1)$.

One might hope that Theorem 5.9 could be avoided by restricting to suitable Doddages which are weakly good. Unfortunately such a restriction does not help.

Lemma 5.10: *Suppose that δ is a supercompact cardinal, $\lambda > \delta$, $|V_\lambda| = \lambda$ and $\mathrm{cof}(\lambda) = \omega$.*

Then there exists suitable extender sequences, $\tilde{E}$ and $\tilde{F}$, such that:

(1) $(\lambda, 0)$ *is the maximum element of* $\mathrm{dom}(\tilde{E})$ *and* $o^{\tilde{E}}_{\mathrm{LONG}}(\delta) = \lambda + 1$,
(2) $\tilde{F}|(\lambda, 0) = \tilde{E}|(\lambda, 0)$ *and* $\mathrm{dom}(\tilde{F}) = \mathrm{dom}(\tilde{E})$,
(3) $\mathrm{CRT}(F_0^\lambda) = \mathrm{CRT}(F_0^\lambda)$ *and* $\mathrm{SPT}(F_0^\lambda) = \mathrm{SPT}(F_0^\lambda)$,
(4) $j_{E_0^\lambda}|\mathrm{SPT}(E_0^\lambda) \neq j_{F_0^\lambda}|\mathrm{SPT}(F_0^\lambda)$.

Theorem 5.11: *Suppose that there is a supercompact cardinal. Then there is a weakly good suitable Doddage which is not good.*

Theorem 5.9 does not rule out the following conjecture.

Definition 5.12: *(Doddage Conjecture)* Suppose that there is a supercompact cardinal. Then for all λ such that $V_\lambda \prec_{\Sigma_3} V$, there exists a good suitable Doddage, $\tilde{\mathcal{E}} \subset V_\lambda$, such that

(1) $o^{\tilde{\mathcal{E}}}_{\mathrm{LONG}}(\delta) = \lambda$ for some $\delta < \lambda$,
(2) $\tilde{\mathcal{E}}$ is definable in V_λ. $\square$

Suppose that δ is a supercompact cardinal. Then the *Doddage Conjecture* holds at δ if for all $\lambda > \delta$ such that $V_\lambda \prec \Sigma_3$, there exists a good suitable Doddage, $\tilde{\mathcal{E}} \subset V_\lambda$, such that $o^{\tilde{\mathcal{E}}}_{\mathrm{LONG}}(\delta) = \lambda$, and such that $\tilde{\mathcal{E}}$ is OD in V_λ. The *Doddage Conjecture* is equivalent to the assertion that for some δ, the *Doddage Conjecture* holds at δ.

It is a corollary of Theorem 5.28, which we state below, that if the *Doddage Conjecture* holds at δ for some δ then the *Doddage Conjecture* holds at δ for *all* δ such that δ is HOD-supercompact. This suggests that the *Doddage Conjecture* is equivalent to the possibly stronger conjecture that if δ is supercompact then the *Doddage Conjecture* holds at δ.

It is the following theorem which motivates the *Doddage Conjecture* and it is this theorem, combined with the natural expectations for a definable inner model theory for supercompact cardinals, which suggests that the *Doddage Conjecture* might actually be true.

Theorem 5.13: *Suppose that δ is an extendible cardinal. Then the following are equivalent.*

(1) *The* Doddage Conjecture *holds at δ.*
(2) *There exists an inner model, $N \subseteq \mathrm{HOD}$, such that*

$$N \vDash \mathrm{ZFC} + \text{"δ is a supercompact cardinal"}$$

and this is witnessed by the class of all $E \cap N$ such that E is an extender such that $\rho(E) = \mathrm{LTH}(E)$ and such that $E \cap N \in N$. $\square$

5.2. *Ramifications for the strongest hypotheses*

Whether or not the *Doddage Conjecture* holds is a key question. If the *Doddage Conjecture* is provable then the corollaries are striking. We list three the last two of which are theorems of ZF. The statements refer to extendible cardinals. Working in just ZF, δ is an extendible cardinal if for all α there is an elementary embedding,

$$j : V_{\kappa+\alpha} \to V_{j(\kappa)+j(\alpha)}$$

such that $\kappa = \mathrm{CRT}(j)$ and such that $j(\kappa) > \alpha$. Assuming the *Axiom of Choice* this is equivalent to the assertion that for all α there is an elementary embedding,

$$j : V_{\kappa+\alpha} \to V_{j(\kappa)+j(\alpha)}$$

such that $\kappa = \mathrm{CRT}(j)$.

Suppose that the *Doddage Conjecture* is provable. Then:

(I) Suppose that δ is an extendible cardinal. Then:

(a) For each $a \in [\mathrm{Ord}]^{<\delta}$, then there exists $b \in [\mathrm{Ord}]^{<\delta} \cap \mathrm{HOD}$ such that $a \subset b$.

(b) If $\lambda > \delta$ and $\mathrm{cof}(\lambda) < \delta$ then λ^+ is correctly computed by HOD.

(II) (ZF) Suppose that $\delta_0 < \delta_1$ are extendible cardinals. Then for each $a \in [\mathrm{Ord}]^{<\delta_0} \cap V_{\delta_1}$, then there exists $b \in [\mathrm{Ord}]^{<\delta_0} \cap \mathrm{HOD}$ such that $a \subset b$.

(III) (ZF) Suppose that λ is a limit of extendible cardinals, $\mathrm{cof}(\lambda) < \lambda$, and there is a extendible cardinal above λ. Then λ^+ is correctly computed by HOD.

A consequence of (III) is that if there exist a proper class of supercompact cardinals, then there exists no nontrivial elementary embedding $j : V \to V$ (as a theorem of ZF).

It will be convenient to define the notion that κ is a supercompact cardinal in just ZF. If the *Doddage Conjecture* is provable then very likely it is also provable that the *Doddage Conjecture* holds at every supercompact cardinal. In this case, (I)–(III), hold with "extendible" replaced by "supercompact".

Definition 5.14: (ZF) A cardinal, κ, is supercompact if for each $\alpha > \kappa$ there exist $\beta > \alpha$ and an elementary embedding,

$$j : V_\beta \to N$$

such that

(1) N is a transitive set and $N^{V_\alpha} \subset N$,

(2) j has critical point κ,

(3) $\alpha < j(\kappa)$.   □

Remark 5.15:

(1) It is clear that assuming the *Axiom of Choice* this definition of a supercompact cardinal coincides with the usual definition.

(2) (ZF) Suppose that

$$j : V \to M$$

is an elementary embedding with $\mathrm{CRT}(j) = \kappa$, $M^{V_\alpha} \subset M$, and with $\alpha < j(\kappa)$. Let $\beta = j(\kappa)$, let $N = M \cap V_{j(\beta)}$ and let $k = j|V_\beta$. Then $N^{V_\alpha} \subset N$ and

$$k : V_\beta \to N$$

is an elementary embedding.

(3) (ZF) If κ is an extendible cardinal then κ is supercompact. □

Theorem 5.16 shows that (II) and (III) are corollaries of (I).

Theorem 5.16: (ZF) *Suppose that δ_0 is a supercompact cardinal. Then there is a homogeneous partial order $\mathbb{Q}$ such that $\mathbb{Q}$ is Σ_3-definable in V_{δ_0} and such that if $G \subset \mathbb{Q}$ is V-generic then:*

(1) $V[G]_{\delta_0} \vDash ZFC$;
(2) *If δ is a supercompact cardinal in V and $\delta < \delta_0$ then δ is a supercompact cardinal in $V[G]_{\delta_0}$;*
(3) *Suppose $V_\lambda \prec V_{\delta_0}$ and*

$$j : V_{\lambda+1} \to V_{\lambda+1}$$

is an elementary embedding with $\lambda = \kappa_\omega(j)$. Then $V[G]_\lambda \prec V[G]_{\delta_0}$ and in $V[G]$, j lifts to an elementary embedding,

$$j_G : V[G]_{\lambda+1} \to V[G]_{\lambda+1}.$$

Assuming the *Doddage Conjecture* is provable then one can in a natural hierarchy of large cardinal axioms give a threshold for inconsistency in just ZF which more closely parallels the Kunen inconsistency in ZFC.

Suppose

$$j : V_\gamma \to V_\gamma$$

is a nontrivial elementary embedding. Then $\kappa_\omega(j)$ is the ω-th element of the critical sequence of j; i.e.,

$$\kappa_\omega(j) = \sup\{\kappa_n(j) \mid n < \omega\}$$

where $\kappa_0(j) = \text{CRT}(j)$ and for all $n \geq 0$, $\kappa_{n+1}(j) = j(\kappa_n(j))$. Assuming the *Axiom of Choice*, Kunen's Theorem shows that if

$$j : V_\gamma \to V_\gamma$$

is a nontrivial elementary embedding then $\gamma = \kappa_\omega(j)$ or $\gamma = \kappa_\omega(j) + 1$.

The proof of Kunen's Theorem does not obviously adapt to rule out any of the following axioms (for $n < \omega$) which are therefore, in the classical view of the large cardinal hierarchy, among the strongest large cardinal axioms believed to be consistent with the *Axiom of Choice*:

There exists $\lambda \in \text{Ord}$ such that

$$V_\lambda \prec_{\Sigma_n} V$$

and such that there is a nontrivial elementary embedding

$$j : V_{\lambda+1} \to V_{\lambda+1}.$$

By Kunen's Theorem, necessarily $\lambda = \kappa_\omega(j)$. Notice that the requirement,

$$V_\lambda \prec_{\Sigma_n} V,$$

is equivalent to the requirements:

(1) $V_{\mathrm{CRT}(j)} \prec_{\Sigma_n} V$;
(2) for all $\alpha < \lambda$, if $V_\alpha \prec_{\Sigma_n} V$ then $V_{j(\alpha)} \prec_{\Sigma_n} V$.

If the Doddage Conjecture is provable then these axioms cannot be significantly strengthened even without the *Axiom of Choice*.

A sentence ϕ is Ω-valid from ZFC if for all complete Boolean algebras, $\mathbb{B}$, and for all $\alpha \in \mathrm{Ord}$ if

$$V_\alpha^{\mathbb{B}} \vDash \mathrm{ZFC}$$

then $V_\alpha^{\mathbb{B}} \vDash \phi$. This definition is in the context of ZF.

Theorem 5.17: (ZF) *Suppose the* Doddage Conjecture *is Ω-valid from* ZFC *and that*

$$V_\lambda \prec_{\Sigma_4} V.$$

Then there is no nontrivial elementary embedding

$$j : V_{\lambda+2} \to V_{\lambda+2}$$

such that $\lambda = \kappa_\omega(j)$. $\qquad\qquad\square$

5.3. *Closure properties of* HOD

We begin with a definition.

Definition 5.18: A cardinal δ is HOD-supercompact if if for all $\lambda > \delta$ there exists an elementary embedding,

$$j : V \to M$$

such that

(1) $\mathrm{CRT}(j) = \delta$ and $\lambda < j(\delta)$,
(2) $M^{V_\lambda} \subset M$ and $j(\mathrm{HOD} \cap V_\delta) \cap V_\lambda = \mathrm{HOD} \cap V_\lambda$. $\qquad\square$

We shall examine the situation that there exists a cardinal δ such that δ is HOD-supercompact. If $V =$ HOD then trivially every supercompact cardinal is HOD-supercompact and more generally the existence of HOD-supercompact cardinals follows from the existence of an extendible cardinal.

Lemma 5.19: *Suppose that δ is an extendible cardinal. Then the cardinal δ is HOD-supercompact.* □

Suppose that κ is an uncountable regular cardinal which is not measurable in HOD and $S \subset \kappa$ is a stationary set such that $S \in$ HOD. Then for each $\gamma < \kappa$ such that $(2^\gamma)^{\text{HOD}} < \kappa$, there exists a sequence

$$\langle S_\alpha : \alpha < \gamma \rangle \in \text{HOD}$$

of pairwise disjoint subsets of S such that for each $\alpha < \gamma$, S_α is stationary in S. This motivates the next definition.

Definition 5.20: Suppose that κ is an uncountable regular cardinal. Then κ is *ω-strongly measurable in* HOD if there exists $\gamma < \kappa$ such that:

(1) $(2^\gamma)^{\text{HOD}} < \kappa$;
(2) There does not exist a sequence

$$\langle S_\alpha : \alpha < \gamma \rangle \in \text{HOD}$$

of pairwise disjoint subsets of κ such that for each $\alpha < \gamma$, S_α is stationary in $\{\eta < \kappa \mid \text{cof}(\eta) = \omega\}$. □

The following conjecture is a strong version of a conjecture from [5].

Conjecture (ZFC) *There is a proper class of uncountable regular cardinals κ which are not ω-strongly measurable in* HOD.

Remark 5.21:

(1) If one can prove the weaker conjecture that there is at least one cardinal $\kappa > \omega_1$ that is not ω-strongly measurable in HOD then one can prove this conjecture.
(2) It is unknown whether there can even exist *one* singular strong limit cardinal κ of uncountable cofinality such that κ^+ is ω-strongly measurable in HOD.
(3) If there is a supercompact cardinal then this conjecture is implied by the *Doddage Conjecture*. □

Theorem 5.22:

Suppose that δ is a supercompact cardinal and that there exists a proper class of regular cardinals which are not ω-strongly measurable in HOD. Then:

(1) *For each $a \in [\mathrm{Ord}]^{<\delta}$, then there exists $b \in [\mathrm{Ord}]^{<\delta} \cap \mathrm{HOD}$ such that $a \subseteq b$;*

(2) *Suppose that $\lambda > \delta$, λ is a limit cardinal such that $\mathrm{cof}(\lambda) < \delta$, and that δ is HOD-supercompact. Then λ^+ is correctly computed by HOD.*

(3) *Suppose that $\gamma \in \mathrm{Ord}$ and that*

$$j : \mathrm{HOD} \cap V_{\gamma+1} \to \mathrm{HOD} \cap V_{j(\gamma)+1}$$

is an elementary embedding with critical point $\kappa \geq \delta$. Then $j \in \mathrm{HOD}$.□

An immediate corollary of Theorem 5.22 is that if δ is HOD-supercompact and there exists a proper class of regular cardinals which are not ω-strongly measurable in HOD, then there is no elementary embedding,

$$j : \mathrm{HOD} \to \mathrm{HOD}$$

with critical point above δ. We can strengthen this a bit and obtain (from the hypotheses above) that there exists an ordinal λ such that there is no nontrivial elementary embedding

$$j : \mathrm{HOD} \to \mathrm{HOD}$$

with the property that $j(\lambda) = \lambda$.

Theorem 5.23: *Suppose that there is a HOD-supercompact cardinal and there exists a proper class of regular cardinals which are not ω-strongly measurable in HOD.*

Then there is an ordinal λ such that for all $\gamma > \lambda$, if

$$j : \mathrm{HOD} \cap V_{\gamma+1} \to \mathrm{HOD} \cap V_{j(\gamma)+1}$$

is an elementary embedding with $\lambda < \gamma$ and $j(\lambda) = \lambda$, then $j \in \mathrm{HOD}$.   □

There are two corollaries of Theorem 5.23.

Corollary 5.24: *Suppose that there is a HOD-supercompact cardinal and there exists a proper class of regular cardinals which are not ω-strongly measurable in HOD.*

Then there is no nontrivial sequence $\langle j_k : k < \omega \rangle$ of elementary embeddings,

$$j_k : \mathrm{HOD} \to \mathrm{HOD},$$

with wellfounded limit.   □

Corollary 5.25: *Suppose that there is a HOD-supercompact cardinal and there exists a proper class of regular cardinals which are not ω-strongly measurable in* HOD.

Then there is no nontrivial elementary embedding

$$j : (\mathrm{HOD}, T) \to (\mathrm{HOD}, T)$$

where T is the Σ_2-theory of V in ordinal parameters. □

We conjecture that it is impossible for there to exist a nontrivial elementary embedding

$$j : \mathrm{HOD} \to \mathrm{HOD}.$$

A weaker conjecture is simply that if there is a proper class of cardinals which are not ω-strongly measurable in HOD then there is no nontrivial elementary embedding

$$j : \mathrm{HOD} \to \mathrm{HOD}.$$

Two more immediate corollaries of Theorem 5.22 block natural attempts to find (strong) variations of the axiom that there exists a nontrivial elementary embedding

$$j : V_{\lambda+1} \to V_{\lambda+1}.$$

Theorem 5.26: *Suppose that λ is a cardinal, there is a HOD-supercompact cardinal below λ, and that there exists a proper class of regular cardinals which are not ω-strongly measurable in* HOD.

Then there is no nontrivial elementary embedding,

$$j : \mathrm{HOD}_{V_{\lambda+1}} \cap V_{\lambda+2} \to \mathrm{HOD}_{V_{\lambda+1}} \cap V_{\lambda+2}.$$ □

Theorem 5.27: *Suppose that λ is a cardinal, there is a HOD-supercompact cardinal below λ, and that there exists a proper class of regular cardinals which are not ω-strongly measurable in* HOD.

Then there is no nontrivial elementary embedding,

$$j : V_{\lambda+1} \to V_{\lambda+1},$$

such that for all Σ_2-formulas $\phi(x)$, for all $a \in V_{\lambda+1}$,

$$V \vDash \phi[a]$$

if and only if $V \vDash \phi[j(a)]$. □

Note that if it is consistent for there to exist an elementary embedding

$$j : L(V_{\lambda+1}) \to L(V_{\lambda+1})$$

with critical point below λ then it is consistent for there to exist a nontrivial elementary embedding,

$$j : V_{\lambda+1} \to V_{\lambda+1},$$

such that for all Σ_2-formulas $\phi(x)$, for all $a \in V_{\lambda+1}$,

$$V \vDash \phi[a]$$

if and only if $V \vDash \phi[j(a)]$, for if $G \subset \mathrm{Coll}(\lambda^+, \lambda^+)$ is $L(V_{\lambda+1})$-generic then this latter axiom holds in $L(V_{\lambda+1})[G]$ and

$$L(V_{\lambda+1})[G] \vDash \mathrm{ZFC}.$$

Further the cardinals of $(\mathrm{HOD})^{L(V_{\lambda+1})[G]}$ coincide exactly with the cardinals of $L(V_{\lambda+1})[G]$ above λ^+. Thus in Theorem 5.27, the assumption that there is a supercompact cardinal below λ is necessary.

A natural conjecture is that for all λ, if there is a supercompact cardinal below λ then there is no nontrivial elementary embedding

$$j : V_{\lambda+1} \to V_{\lambda+1}$$

such that for all formulas $\phi(x)$, for all $a \in V_{\lambda+1}$,

$$V_{\lambda+2} \vDash \phi[a]$$

if and only if $V_{\lambda+2} \vDash \phi[j(a)]$.

We now continue the analysis of HOD under the assumptions that there exists a HOD-supercompact cardinal and there exists a proper class of regular cardinals which are not ω-strongly measurable in HOD.

Suppose δ is HOD-supercompact and there exists a proper class of regular cardinals which are not ω-strongly measurable in HOD. Then HOD has closure properties which are very similar to those which $L[\tilde{E}]$ has, where $\tilde{E}$ is a suitable extender sequence such that $o^{\tilde{E}}_{\mathrm{LONG}}(\delta) = \infty$. The next theorem explains why the covering properties hold.

Theorem 5.28: *Suppose that δ is* HOD-*supercompact. Then the following are equivalent.*

(1) *The* Doddage Conjecture *holds at δ.*
(2) *There is a good suitable Doddage, $\tilde{\mathcal{E}}$, such that $o^{\tilde{\mathcal{E}}}_{\mathrm{LONG}}(\delta) = \infty$ and such that $\tilde{\mathcal{E}}$ is Σ_3-definable from δ.*

(3) *There exists a proper class of regular cardinals which are not measurable in* HOD.

(4) *There exists a proper class of regular cardinals which are not ω-strongly measurable in* HOD. □

A more elegant version of Theorem 5.28 is the following which is an immediate corollary.

Theorem 5.29: *Suppose that δ is an extendible cardinal. Then the following are equivalent.*

(1) *The* Doddage Conjecture *holds at δ.*

(2) *There is a good suitable Doddage, $\tilde{\mathcal{E}}$, such that $o^{\tilde{\mathcal{E}}}_{\text{LONG}}(\delta) = \infty$ and such that $\tilde{\mathcal{E}}$ is Σ_3-definable from δ.*

(3) *There exists a regular cardinal $\kappa > \delta$ such that κ is not measurable in* HOD.

(4) *There exists a regular cardinal $\kappa > \delta$ such that κ is not ω-strongly measurable in* HOD. □

The next theorem is a corollary of Theorem 5.29.

Theorem 5.30: *Suppose there is a proper class of extendible cardinals. Then*

$$V \vDash \text{``The Doddage Conjecture''},$$

if and only if for all complete Boolean algebras, $\mathbb{B}$,

$$V^{\mathbb{B}} \vDash \text{``The Doddage Conjecture.''}$$

If δ is HOD-supercompact and there exists a proper class of regular cardinals which are not ω-strongly measurable in HOD then large cardinal axioms are downward absolute from V to HOD. If in addition one can arrange that every OD set $A \subset \mathbb{R}$ is universally Baire then one obtains that these large cardinal axioms cannot refute the Ω Conjecture.

Theorem 5.31:

Suppose that there is a proper class of Woodin cardinals and that for every set $A \subseteq \mathbb{R}$, if A is OD then A is universally Baire.

Then HOD $\vDash$ *"Ω Conjecture."*. □

Thus if some large cardinal hypothesis, such as there exists a nontrivial elementary embedding,

$$j : V_{\lambda+1} \to V_{\lambda+1},$$

refutes the Ω Conjecture then assuming sufficient cardinal hypotheses hold in V either:

(1) There is an OD set $A \subset \mathbb{R}$ such that A is not universally Baire; or
(2) there exists α such that every regular cardinal $\delta > \alpha$ is ω strongly measurable in HOD

This would be a rather pathological dichotomy from the viewpoint of inner model theory. This is the kind of phenomenon which is evidence for the claim that a large cardinal hypothesis which refutes the Ω Conjecture cannot have a reasonable inner model theory.

6. Conclusions

The prominent open problems are the *Comparison Conjecture* and the *Weak* $(\omega_1 + 1)$-*Iteration Hypothesis*. Of course there is also the $(\omega_1 + 1)$-*Iteration Hypothesis*, but assuming the *Comparison Conjecture* is provable, most of the metamathematical implications of the $(\omega_1 + 1)$-*Iteration Hypothesis* are already implications of the *Weak* $(\omega_1 + 1)$-*Iteration Hypothesis*. The only potential exception concerns the definability of the the reals of the inner model but this would depend on the details of the proof of the *Comparison Conjecture*.

A clearly related problem is to generalize the fine-structural models of Mitchell-Steel to long extenders and construct, assuming the appropriate iteration hypothesis and that δ is supercompact, a backgrounded fine-structural inner model $L[\tilde{E}]$ such that $o_{\text{LONG}}^{\tilde{E}}(\delta) = \infty$ (where $\tilde{E}$ is now a sequence of *partial* extenders). As remarked in the introduction, such an inner model would be a reasonable analog of L in the context of $0^\#$ does not exist, but with no limiting assumption on the large cardinals of V.

In conclusion, the *Doddage Conjecture* and more importantly the fine-structural version of the *Doddage Conjecture* are the key questions for the answers have truly profound consequences for both V and the large cardinal hierarchy. These questions are the topic of the sequel to [7].

Acknowledgments

I would like to thank both the IMS and *The National University of Singapore* for the support provided, and I would like to thank the participants of the *Computational Prospects of Infinity* program for their patience, comments, and suggestions.

References

1. D. A. Martin and J. Steel. Iteration trees. *J. Amer. Math. Soc.*, 7:1–74, 1994.
2. William J. Mitchell and John R. Steel. *Fine structure and iteration trees*. Springer-Verlag, Berlin, 1994.
3. John R. Steel. Local K^c Constructions. *Notes, November 2003*, pages 1–20.
4. John R. Steel. Wellfoundedness of the Mitchell Order. *J. Symbolic Logic*.
5. W. Hugh Woodin. *The Axiom of Determinacy, forcing axioms, and the non-stationary ideal*. Walter de Gruyter & Co., Berlin, 1999.
6. W. Hugh Woodin. Beyond Σ_1^2 Absoluteness. In *Proceedings of the International Congress of Mathematicians, Beijing, 2002*, volume I, pages 515–524. Higher Education Press, 2003.
7. W. Hugh Woodin. Suitable extender sequences. *Preprint*, pages 1–542, 2008.